Praise for *Designing Circuit Boards with EAGLE*

"Matt Scarpino has succeeded where scores of others have failed—he's managed to make the formidable EAGLE software understandable and, more importantly, useable. His presentation is not only approachable and logical, but it's complete. When you've finished his book, you'll be able to do something meaningful with EAGLE. This book belongs on every engineer's bookshelf or tablet."

—**Bryan Bergeron**, Editor, *Nuts & Volts Magazine*

"Matt Scarpino's *Designing Circuit Boards with EAGLE* is a great resource for electronics enthusiasts who are ready to get serious and produce their own circuit boards. Matt's sensible instructions take readers through the steps to design simple and not-so-simple circuit boards, and you can really tell that he's been using EAGLE for 10 years and loves it. I'm recommending this book to all my maker friends."

—**John Baichtal**, Author of *Arduino for Beginners: Essential Skills Every Maker Needs*

"With the rising popularity of open source hardware projects, the EAGLE circuit board software has become a vital tool for both hobbyists and professional engineers alike. *Designing Circuit Boards with EAGLE* provides all the information you'll need to get up to speed with the EAGLE software, and to start creating your own circuit board designs. Matt Scarpino has provided a great tool for the hobbyist starting out in the circuit board design world, demonstrating all of the features you'll need to know to create your own circuit board projects. However, the experienced engineer will also benefit from the book, as it also serves as a complete reference guide to all the EAGLE software configuration settings and features. His insightful guidance helps simplify difficult tasks in the EAGLE software, and his handy tips will help save you hours of trial-and-error experimenting in your circuit board designs."

—**Rich Blum**, Author of *Sam's Teach Yourself Arduino Programming in 24 Hours* and *Sams Teach Yourself Python Programming for Raspberry Pi in 24 Hours*

Designing Circuit Boards with EAGLE

Designing Circuit Boards with EAGLE

Make High-Quality PCBs at Low Cost

Matthew Scarpino

Upper Saddle River, NJ • Boston • Indianapolis • San Francisco
New York • Toronto • Montreal • London • Munich • Paris • Madrid
Cape Town • Sydney • Tokyo • Singapore • Mexico City

For information about buying this title in bulk quantities, or for special sales opportunities (which may include electronic versions; custom cover designs; and content particular to your business, training goals, marketing focus, or branding interests), please contact our corporate sales department at corpsales@pearsoned.com or (800) 382-3419.

For government sales inquiries, please contact governmentsales@pearsoned.com.

For questions about sales outside the U.S., please contact international@pearsoned.com.

Visit us on the Web: informit.com/ph

Library of Congress Control Number: 2013957510

ISBN-13: 978-0-13-381999-1
ISBN-10: 0-13-381999-X

Text printed in the United States on recycled paper at Edward Brothers Malloy in Ann Arbor, Michigan.
Second printing: April 2014

Executive Editor
Bernard Goodwin

Managing Editor
Kristy Hart

Senior Project Editor
Betsy Gratner

Copy Editor
Apostrophe Editing Services

Indexer
Tim Wright

Proofreader
Kathy Ruiz

Technical Reviewers
John Baichtal
Bryan Bergeron
Rich Blum

Editorial Assistant
Michelle Housley

Interior Designer
Matthew Scarpino

Cover Designer
Chuti Prasertsith

Compositor
Matthew Scarpino

Senior Compositor
Gloria Schurick

Contents

Part IV: Automating EAGLE

Part V: The BeagleBone Black

Preface

As I write this in late 2013, the Maker Movement has flourished from a tiny group of tinkerers into a passionate community of millions. Hobbyists have become entrepreneurs and entrepreneurs have become large-scale manufacturers. 3-D printers have fallen into the price range of the average consumer, and the printers' capabilities have risen to such an extent that they're being used to fabricate high-precision aircraft parts and medical equipment. With good reason, many economists and journalists have likened the rise of the Maker Movement to a second Industrial Revolution.

Nothing better illustrates the movement's success than the popularity of the Arduino platform. The first Arduino board design, the Arduino USB, was released in 2005, giving students and hobbyists a low-cost means of programming Atmel microcontrollers. Since then, hundreds of thousands of Arduino boards have been sold, and the Arduino family has expanded to include a vast array of boards, shields, kits, and accessories. Arduino boards have found their way into robots, musical instruments, game platforms, and even unmanned aerial vehicles. The boards have become so popular that many hobbyists-turned-entrepreneurs use them to build prototypes of new inventions.

But Makers still demand more: more capability, more affordability, and more customization. This means designing new circuit boards, a task that requires specialized knowledge and software. Most professional design tools are beyond the price range of the average Maker, but not EAGLE. Since its release in 1988, EAGLE has grown steadily in features and stability while maintaining a price that even cash-strapped students can afford. EAGLE wins legions of admirers with every new version, and the analogy couldn't be clearer: What Arduino is to hardware, EAGLE is to software. It's no wonder that all open-source Arduino designs are released in EAGLE's format.

In writing this book, my mission is to show Makers how to take full advantage of EAGLE's capabilities. This requires a basic understanding of circuit theory, including Ohm's law and Kirchoff's laws, but nothing beyond that. You won't find any transistor analysis or differential equations here. Instead, my goal is to provide a practical, hands-on exploration of EAGLE so that readers can design practical circuit boards, thereby bringing exciting new gadgets to the marketplace and continuing the extraordinary momentum of the Maker Movement.

Matthew Scarpino

Structure of This Book

This book presents EAGLE by walking through a series of circuit design projects, starting with a simple inverting amplifier and proceeding to a six-layer, single-board computer. As the circuits grow in complexity, I'll explain more advanced features of EAGLE and show how to automate repetitive tasks. This book also includes a great deal of material to help readers understand the fundamentals of circuit boards and the theory behind the example circuits.

Chapters 1, 2, and 3 introduce the topics of EAGLE and circuit board design. Their primary purpose is to familiarize you with EAGLE's capabilities and present the terminology used throughout the book. Chapter 3 breezes through the complete design of a trivially simple circuit.

Chapters 4 through 7 present the design of a practical circuit board: the Arduino Femtoduino. These chapters take a hands-on approach to explaining the four fundamental steps of circuit board design: drawing a schematic, laying out components, routing connections, and generating Gerber/Excellon files.

Chapters 8 through 13 discuss an assortment of topics related to EAGLE circuit design. These include circuit simulation, the process of creating custom components, and the all-important subject of design automation. Design automation is one of the most powerful aspects of EAGLE, but it's also one of the most overlooked. For this reason, I highly recommend becoming familiar with editor commands and User Language programs.

Chapters 14 and 15 present the book's advanced example design: the BeagleBone Black. The name may sound silly but there's nothing silly about the circuit. It has six board layers, hundreds of components, and thousands upon thousands of routed connections. As I present the design, I'll discuss EAGLE's advanced capabilities and ways to take advantage of design automation.

Example File Archive

To supplement the text, all the circuit designs, programs, and support files in this book are provided in a zip file called eagle-book.zip. This can be freely downloaded from http://eagle-book.com. As you follow the discussion, I recommend that you compare the theoretical discussion to the real-world EAGLE designs. In addition, the color figures for this book can be accessed at www.informit.com/title/9780133819991.

Acknowledgments

First and foremost, I'd like to thank Bernard Goodwin of Pearson North America for his support and sage wisdom during the creation of this book. Thanks to his deft handling, the development process was as frictionless as could be asked.

I'd like to express my deep appreciation to San Dee Phillips of Apostrophe Editing Services, who caught so many of my formatting, spelling, and grammar errors. I'd also like to thank Betsy Gratner for her cheerful yet firm masterminding of this book's production, Gloria Schurick for her painstaking efforts in compositing, Kathy Ruiz for her eagle-eyed proofreading, and Laura Robbins for managing this book's images.

Last but not least, I'd like to extend my gratitude to Bryan Bergeron of *Nuts & Volts Magazine*, Richard Blum, author of *Sams Teach Yourself Arduino Programming in 24 Hours*, and John Baichtal, author of *Arduino for Beginners*. These reviewers bravely made their way through the book in its rough state and provided many useful suggestions. Their comments have improved the book's accessibility to newcomers and expanded the number of topics.

About the Author

Matthew Scarpino is an engineer with more than 12 years of experience designing hardware and software. He has a Master's degree in electrical engineering and is an Advanced Certified Interconnect Designer (CID+). He currently resides in Massachusetts where he develops software for embedded systems. In his spare time, he uses EAGLE to design accessories for his Android smartphone and the Google Glass.

Chapter 1

Introducing EAGLE

Circuit design applications can be divided into two categories: those intended for large design firms and those intended for everyone else. Applications in the first category provide high reliability, a wide range of features, and responsive technical support. But these advantages come with a hefty price tag. A perpetual license for Altium Designer costs more than $7,000 and Cadence's OrCad suite costs nearly $10,000.

Applications in the second category are less expensive, and this makes them accessible to students, individuals, and small-to-medium businesses. Unfortunately, they tend to be unreliable and plagued with bugs. Without technical support, there may be no way to work around these difficulties. What's worse, the companies that release these tools tend to be as flaky as their software and may disappear before their support contracts expire.

But not CadSoft's EAGLE. The Easily Applicable Graphical Layout Editor provides the best of both worlds: the quality of a first-tier design application for the price of a second-tier application. EAGLE has been around since 1988, and with each year, it has improved in capability and reliability. It provides a complete set of features for designing circuit boards, and despite thousands of hours of use, it has never crashed on me. If problems arise, users can visit multiple online forums or read through the many online articles.

EAGLE has one major drawback: its user interface. If you're a frequent Windows user, you're accustomed to applications behaving in a certain manner. You're used to a common set of toolbar items and mouse gestures. But EAGLE has its own unique behavior, and it's impossible to simply start the application and figure out how everything works. It takes time to understand the many editors, dialogs, menus, and commands. And because circuit design is such a complex task to begin with, many newcomers to EAGLE give up.

The goal of this book is to ease the process of learning EAGLE. In these chapters, I'm going to walk through the process of designing circuits, starting with a simple circuit (a noninverting amplifier), proceeding to an intermediate circuit (the Arduino Femtoduino), and finally reaching an advanced circuit (the BeagleBone Black). During the course of this presentation, I'll describe both the EAGLE interface and the general process of desiging circuit boards.

In addition to point-and-click design, a significant portion of this book is devoted to automation. EAGLE has a rich command language that can be accessed through scripts and User Language programs, or ULPs. When you have a solid grasp of how to create circuit designs in code, you can perform long, repetitive tasks with a single command. With this automation, your errors will decrease and your productivity will skyrocket.

1.1 A Whirlwind Tour of EAGLE

EAGLE is a software application that makes it possible to design circuit boards. Boiled down to its essentials, EAGLE consists of six features:

- **Component library**—The set of devices that can be inserted into a design
- **Schematic editor**—An editor that makes it possible to draw the circuit's preliminary design
- **Board editor**—An editor that defines the circuit board's physical layout and routing
- **Device editors**—Editors used to design new components
- **Autorouter**—A tool that automatically determines how circuit elements can be connected
- **CAM (Computer Aided Manufacturing) processor**—A tool that reads in a board design and produces files for the board's fabrication

This section briefly describes each of these features and how they relate to the overall process of circuit design.

1.1.1 The Component Library

One of the most important features of any circuit design tool is the set of available parts. This set of components is called a *library*, and the larger the library, the less time the designer needs to spend defining new devices.

Thanks to its longevity, EAGLE's set of libraries has expanded to thousands and thousands of components, from vacuum tubes to field programmable gate arrays. No matter how complex the design, the odds are that EAGLE will have most of the required parts. If it doesn't, the site http://www.cadsoftusa.com/downloads/libraries provides more libraries for free download. If a part still can't be found, Chapter 8, "Creating Libraries and Components," explains how to design custom parts.

One new feature of EAGLE 6 is the format used by the library files. Each library is defined within a *.lbr file, and the format for this file is the eXtensible Markup Language (XML), which is popular throughout the world of computing. Appendix A, "EAGLE Library Files," describes the XML schema that defines the structure of EAGLE's library files.

1.1.2 The Schematic Editor

After you verified that your circuit's components are available, you can select and connect them inside a schematic design, as shown in Figure 1.1.

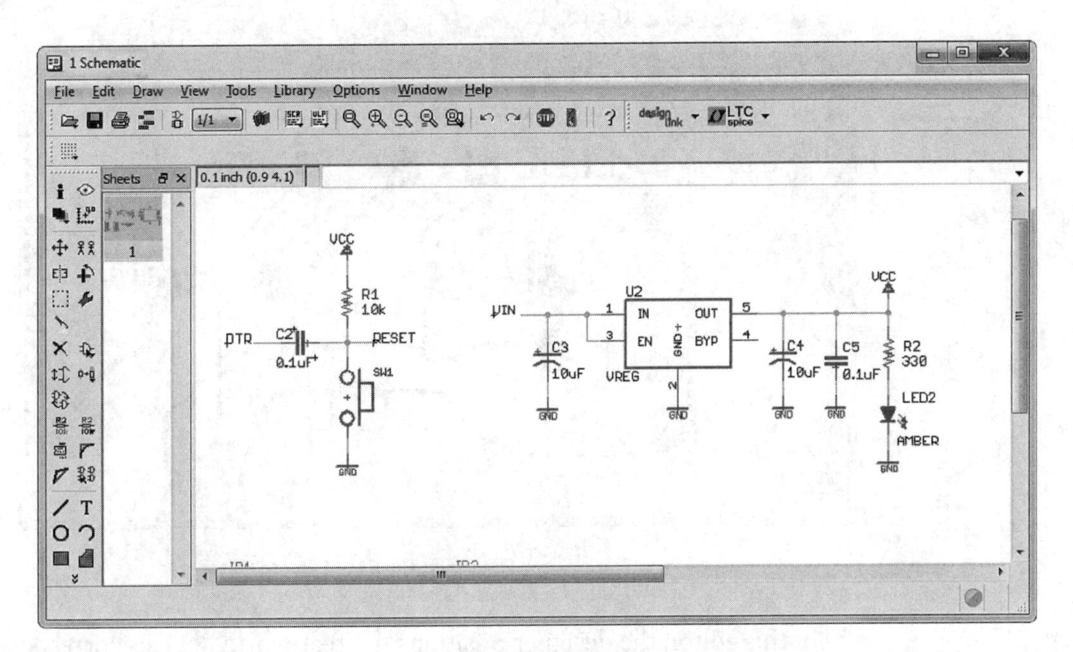

Figure 1.1: The EAGLE Schematic Editor

As with most schematic editors, this keeps track of four important pieces of information:

- Which components are present in the design
- Connections between the components' pins
- Names and values associated with the components
- Properties of the components' connections

EAGLE's schematic editor makes it easy to design a preliminary circuit. Just select a part from the library, move it to a position, and draw connections between it and other components. Afterward, you may assign names and values to the component, such as a resistor's resistance in ohms. Chapter 3, "Designing a Simple Circuit," and Chapter 4, "Designing the Femtoduino Schematic," discuss the schematic editor in detail.

1.1.3 The Board Editor

After a schematic design is complete, EAGLE can generate a board file (*.brd) that defines the layout of the actual circuit board. Board files are modified in EAGLE's board editor, as shown in Figure 1.2.

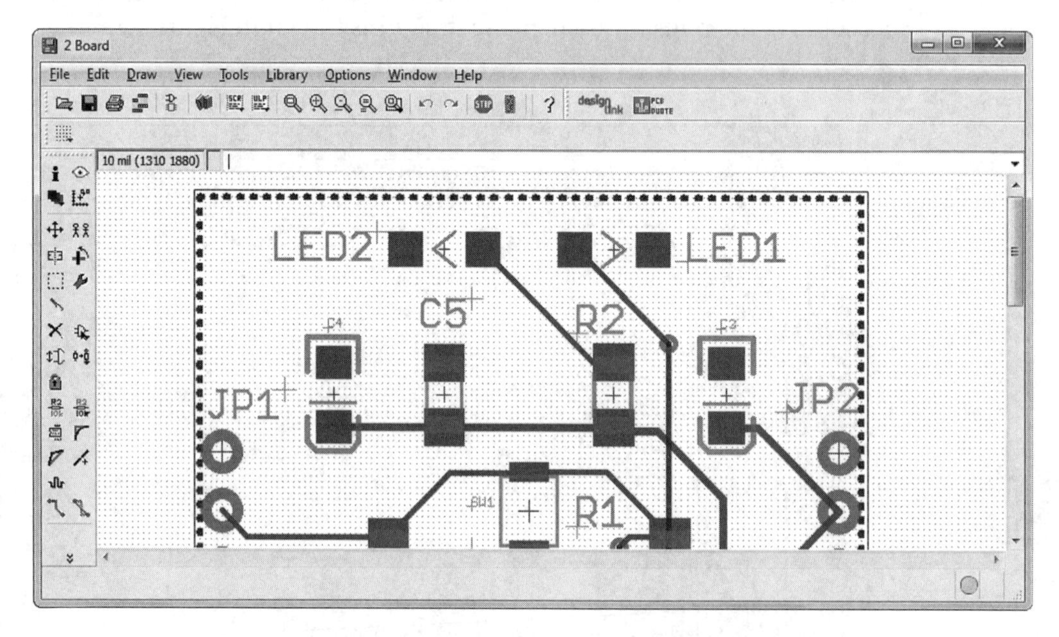

Figure 1.2: The EAGLE Board Editor

In this editor, the designer positions the real-world devices corresponding to the components in the schematic. This position includes not only x and y coordinates, but also whether the components are on the top or bottom layer.

1.1.4 The Device Editors

If the EAGLE library doesn't contain a crucial part, the device editors make it possible to design a new one. This process has three steps:

1. Create a design for the schematic editor. This is called a *symbol.*
2. Create a design for the board editor. This is called a *package.*
3. Create an association between the symbol and its package. This is called a *device.*

EAGLE provides editors for laying out a component's symbol and package. These are collectively called the device editors, as shown in Figure 1.3.

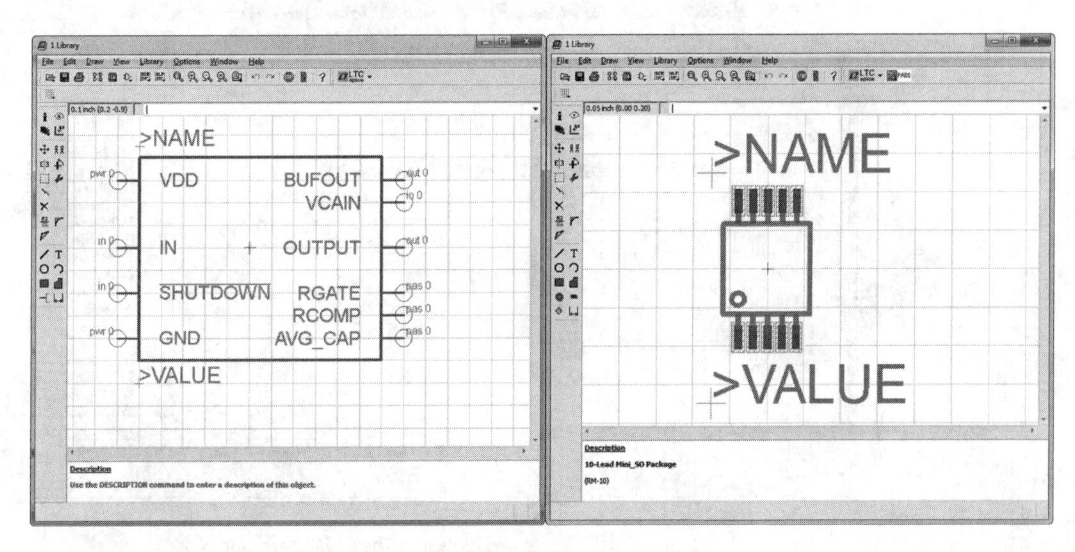

Figure 1.3: The EAGLE Device Editors

The left side of the figure displays the symbol for Analog Devices' SSM2167 component. The right side displays the component's package, which can be used in the board design.

Don't be concerned about terms like symbol, package, and device just yet. These topics will be explored throughout this book, and Chapter 8 presents the entire process of designing new components. Appendix A explains the file format used by EAGLE to store these designs.

1.1.5 The Autorouter

After the boards' devices are in place, the next step is to create the connections between them. This is called *routing*, and even with high-end design tools, this process can be complex and time-consuming.

EAGLE's autorouter simplifies the routing process and provides insight into how circuit components can be connected. But for large-scale circuits, it generally isn't capable of completely routing a board on its own. However, if a designer manually creates initial routes, it will help the autorouter do its job. Chapter 6, "Routing," explains all the different routing methods supported by EAGLE.

1.1.6 The CAM Processor

Most fabrication facilities don't accept EAGLE design files, so EAGLE's CAM (Computer Aided Manufacturing) Processor converts EAGLE designs into different formats. Figure 1.4 shows the processor's dialog.

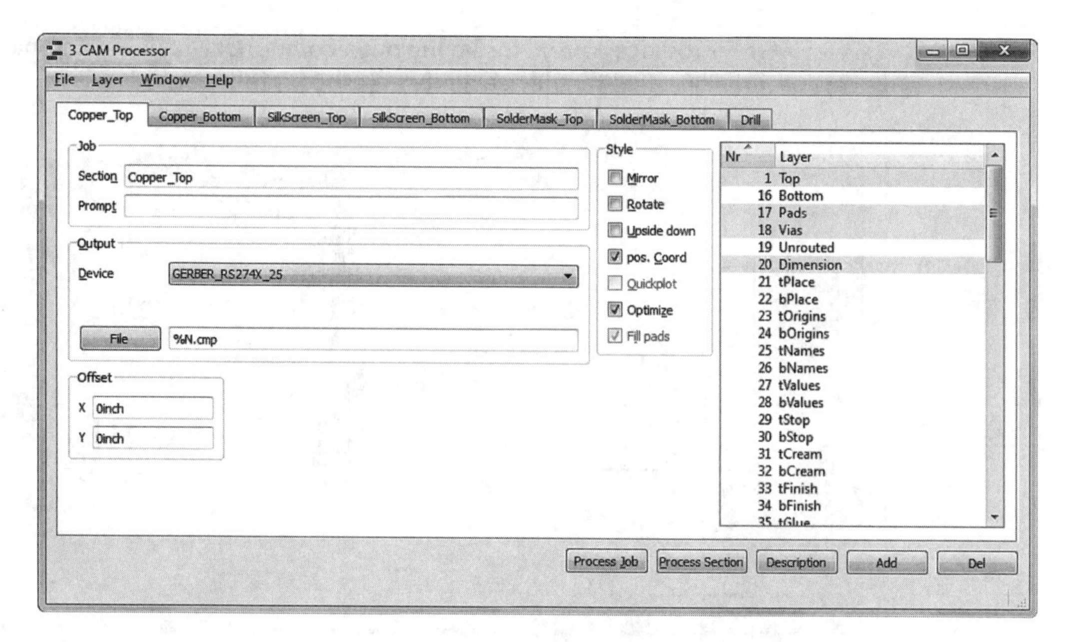

Figure 1.4: The CAM Processor

When the Process Job button is pressed, the processor executes a sequence of tasks called a *job*. A designer can load a job from a file (*.cam) or create a new job from scratch. As a job executes, each of its tasks reads a portion of the board design and creates a file of the selected type.

To fabricate a circuit board, most manufacturers require two types of files. To define a circuit's geometry and connections, the accepted file format is RS-274X, also called the Gerber format. To specify drill diameters and drill locations, the accepted format is the Excellon format. The CAM Processor generates files of both types.

1.2 Obtaining EAGLE

CadSoft makes it easy to start with EAGLE. After you download the executable, you can try it out without registering or paying anything. If you're interested in more features, you can make decisions regarding purchasing, licensing, and registration.

If you run a supported operating system and connect to the Internet, you need to download only a single file. Currently, EAGLE can run on any of the following operating systems:

- Windows 8, Windows 7, Windows Vista, or Windows XP
- Mac OS 10.6, 10.7 on Intel-based processors
- Linux (kernel 2.6, Intel processors, 32-bit runtime environment)

CadSoft's primary web site is http://www.cadsoft.de but the company can also be accessed at locale-specific sites such as http://www.cadsoftusa.com. At these sites, you can download EAGLE by finding the Downloads link in the upper menu and selecting Download EAGLE. This takes you to a page with download links, and you can choose between the Windows, Linux, and Mac OS offerings.

1.3 Licensing

When you first launch EAGLE, a dialog appears and gives you the option of providing a license key or running the tool as freeware. The Freeware option enables you to run EAGLE in a special configuration called the Freeware version of EAGLE Light. This enables you to access EAGLE's editing and routing for free, but with a limited set of features. In addition, this version can be used only for evaluation or nonprofit purposes. If you intend to make money through your PCB design, CadSoft asks that you purchase a license.

EAGLE provides four types of licenses that appeal to different segments of the PCB design community. Each has a different price and set of features (the higher the price, the more features). Specifically, the license type determines the maximum number of schematic sheets, the maximum number of board layers, and the maximum routing area. Table 1.1 lists the characteristics of the different licenses.

Table 1.1

EAGLE License Features

License	Number of Sheets	Number of Layers	Routing Area in mm²
Light	1	2	100 x 80
Hobbyist	99	6	160 x 100
Standard	99	6	160 x 100
Professional	999	16	4000 x 4000

Looking at this table, you may wonder what the difference is between the Hobbyist and Standard licenses. The Hobbyist license is much, much less expensive but carries the requirement that EAGLE can be used only for noncommerical purposes. CadSoft requires a signed statement to this effect.

Table 1.1 doesn't list the prices for these licenses for three reasons:

1. EAGLE's prices change over time and any listed price will prove inaccurate in the near future.
2. For the Standard and Professional licenses, CadSoft doesn't sell EAGLE as an integrated application. Instead, it splits EAGLE into three parts (schematic editor, autorouter, and board editor), and sells them separately.
3. For the Standard and Professional licenses, pricing depends on how many users can use the tool at once.

For a full presentation of the EAGLE pricing structure, visit the CadSoft web site. For prices in American dollars, the link is http://www.cadsoftusa.com/shop/pricing/?language=en.

In writing this book, I have made no assumptions regarding which license readers have purchased or if any license has been purchased at all. But this book covers every aspect of EAGLE, so if one or more features are unavailable in your installation, skip over the corresponding material. The first and second example circuits can be designed with any of the licenses, but the final design (discussed in Chapters 14, "Schematic Design for the BeagleBone Black," and 15, "Board Design for the BeagleBone Black") requires more advanced capabilities.

1.4 Organization of This Book

This book is structured so that the material proceeds from the simple to the complex and from the fundamentally important to the esoteric. More specifically, the chapters in this book can be divided into five parts, each of which focuses on a different task or aspect of EAGLE.

Part I: Preliminary Introduction

The first part of this book provides essential information for readers new to circuit board design and EAGLE. Chapter 2, "An Overview of Circuit Boards and EAGLE Design," explains what circuit boards are and how they're manufactured, thereby establishing the vocabulary that will be used throughout this book. It also explains the overall circuit board design process with EAGLE.

Chapter 3, "Designing a Simple Circuit," expands on this introduction and walks through the schematic design and board design for a simple amplifier circuit. This circuit isn't intended to be manufactured, but the design process will be helpful to inexperienced readers.

Part II: Designing the Arduino Femtoduino

The second and largest part of this book centers on designing an Arduino Femtoduino. The Arduino family of circuit boards enjoys a great deal of popularity among amateurs and professionals, and Chapters 4 through 7 explain how to design one for yourself. Chapter 4, "Designing the Femtoduino Schematic," explains how to create the schematic and Chapter 5, "Layout and Design Rules," explains how to position the packages in the board editor.

Chapter 6, "Routing," discusses the process of design rule checking and shows how to route the connections on the Arduino Femtoduino. Lastly, Chapter 7, "Generating and Submitting Output Files," presents the Computer Aided Manufacturing (CAM) processor and explains how to generate the final artwork files for the Femtoduino.

It also presents five different fabrication services that accept these files and deliver finished circuit boards.

Part III: Advanced Capabilities

The next part of the book covers two topics that go beyond regular schematic/board design. Chapter 8, "Creating Libraries and Components," explains how to create custom components for EAGLE and walks through two designs. The first creates a symbol and package for a through-hole component and the second creates a symbol and package for a surface-mount component.

Chapter 9, "Simulating Circuits with LTspice," delves into one of EAGLE's newest and most interesting features: circuit simulation with LTspice. LTspice is a freely downloadable simulation tool that makes it possible to draw circuits, assign inputs, and simulate the circuit's operation. By combining EAGLE and LTspice, designers can test a design before sending it out for fabrication.

Part IV: Automating EAGLE

The fourth part of this book focuses on automating EAGLE using scripts and program files. Chapter 10, "Editor Commands," presents the EAGLE command language, which executes design operations in text. For example, the `add` command adds a new component to a schematic or board design, and the `auto` command launches the autorouter.

Chapters 11 through 13 explain how to write User Language programs (ULPs), which make it possible to examine circuit designs automatically. These chapters provide many useful examples that can simplify EAGLE usage and reduce time associated with the design process.

Part V: The BeagleBone Black

The last two chapters of this book focus on designing the BeagleBone Black. This advanced circuit board has six layers and hundreds of components, some of which have high-density ball grid array (BGA) pins. Though difficult to design, the BeagleBone Black has gained a significant following among programmers because of its extraordinary amount of computing power.

Example File Archive

All the designs, programs, and support files discussed in this book are freely available online. The archive is called eagle-book.zip and it can be downloaded from http://eagle-book.com.

1.5 More Information

One of EAGLE's greatest advantages is the staggering amount of information available. No matter what problem you face, it's likely that someone has already encountered it and found a solution. In addition to this book, here are four sources of information that I highly recommend.

1.5.1 Element14—www.element14.com

EAGLE is maintained and released by CadSoft, but in 2009, CadSoft was acquired by Premier Farnell PLC, a distributor of electronic components. That same year, Premier Farnell created element14, an online community to provide support for circuit designers. This community has grown significantly over the years, and each day its forum receives hundreds of designers asking and answering questions. In addition, it provides a library of documentation and videos related to electronic design.

EAGLE isn't the only topic discussed at element14, but the subforum devoted to EAGLE support is one of busiest places on the site. Here, users ask questions ranging from routing issues to library entries to converting file formats to those used by other tools. Richard Hammerl, a chief technician at CadSoft, frequently answers questions, which means the subforum is nearly as good as full professional support.

1.5.2 SparkFun—www.sparkfun.com

In 2003, Nathan Seidle founded SparkFun Electronics to "make electronics accessible to the average person." This site sells development tools and kits, such as Arduino boards, and it also provides articles related to electrical design. The list of tutorials includes SMT soldering, programming, robotics, and of course, EAGLE. Nathan Seidle has written a series of articles that discuss EAGLE, and SparkFun provides its own EAGLE scripts, programs, and CAM Processor jobs.

The SparkFun forum is very active and its subforums discuss topics as diverse as wireless/RF design, GPS projects, and shipping times for fabrication facilities. In the PCB Design Questions subforum, many EAGLE users submit questions and receive answers.

1.5.3 YouTube—www.youtube.com

If you search for **EAGLE** and **PCB** or **CadSoft** on YouTube, you'll find many YouTube videos devoted to explaining EAGLE usage. Some may be out of date, but taken as a whole, they provide a friendly introduction to this complicated topic.

1.5.4 CadSoft—www.cadsoftusa.com/www.cadsoft.de

Last but not least, I recommend CadSoft's main site. CadSoft provides a great deal of documentation on EAGLE, but in general, you can find the same documentation inside EAGLE's top-level doc directory. One major point of interest is the Downloads link, which makes it possible to download additional libraries, ULPs, and actual EAGLE projects.

1.6 Conclusion

I first used EAGLE around 2003, and though it had many of the same capabilities as today, it tended to crash at least three times an hour. On message forums, users railed against EAGLE's instability and exchanged workarounds for dealing with its many bugs. But CadSoft persevered in its work on EAGLE, and nearly 10 years later, the bugs and instability are gone. Instead of complaining, today's users defend the application fiercely.

I'm a devoted EAGLE user, and my goal in writing this chapter is to explain why I think the tool is so wonderful. EAGLE provides a full suite of design features, including a schematic editor, a board editor, device editors, and a CAM processor. Its libraries contain thousands and thousands of electronic components. It's stable, runs at high speed, and if I encounter issues, there are many online resources I can turn to.

One of the reasons I'm so impressed with EAGLE is its generous licensing. Users can try out the tool for free and continue using it indefinitely. If they'd like to take advantage of its advanced features, they can purchase a license without spending great sums of money.

Chapter 2

An Overview of Circuit Boards and EAGLE Design

Before you start using EAGLE, you should have a basic understanding of printed circuit boards (PCBs)—what they are and how they're constructed. This is a complex topic with many specialized concepts and terms, but I'll assume you know nothing about them. Therefore, my goal in writing this chapter is to bring you up to speed.

The first part of this chapter provides a general overview of circuit board technology. It starts by explaining what PCBs are meant to accomplish and how this is reflected in their design. It explains the basics of PCB construction for single-sided boards, double-sided boards, and multilayer boards.

The second part of this chapter explains how PCBs can be designed using EAGLE's capabilities. This process consists of five steps:

1. Create a new project.
2. Design a schematic that defines the circuit's components and their connections.
3. Create a circuit board design from the schematic and position the packages corresponding to the schematic's components.
4. Route connections between the packages.
5. Convert the board design into files that can be sent to a fabrication facility.

If you already have a solid grasp of topics like silk-screening and solder mask, feel free to skim this chapter or skip it altogether. If not, I recommend that you take the time to understand how PCBs are constructed and how EAGLE makes it possible to design them. Then you'll have no trouble with the initial circuit design presented in the next chapter.

2.1 Anatomy of a Printed Circuit Board

We all know what circuit boards look like: They're thin, rigid, and usually rectangular, with components attached to one or both surfaces. The top and bottom are generally colored in dark blue or green. Lines running between the components have a slightly different color.

In addition to the top and bottom sides, modern circuit boards have internal planes called *layers*. Internal layers don't have components but may contain metal lines that carry electricity to and from the components on the top and bottom. For example, the circuit board in the iPhone 4 handset has 10 layers.

Layers are critically important in PCB design, so circuit boards are commonly divided into three categories: single-sided, double-sided, or multilayer. This section discusses each of these categories and the manner in which the circuit boards are constructed.

But first, let's answer the question of why circuit boards exist. At the very least, a circuit board serves two purposes:

1. Provides mechanical support for a set of components
2. Provides electrical connections between the components

Given this focus on components, it's a good idea to take a brief look at the types of electric components that board designers have to deal with.

2.1.1 Electrical Components

From the dawn of circuit design to the present day, engineers have created thousands of different types of components. They can be categorized in many ways, and one common distinction involves whether they need power to operate. Components that require power, such as transistors and integrated circuits (ICs), are called *active components*. Components that don't need power, such as resistors and capacitors, are called *passive components*.

For circuit board designers, components are categorized according to their *leads*, also referred to as *terminals*. A lead is a metal extrusion that serves as the component's connection point to a circuit board. Figure 2.1 depicts three popular types of leads:

- **Through-hole**—Leads are wires that enter holes in the board.
- **Surface mount technology** (SMT)—Leads are metal tabs on the perimeter of the device.
- **Ball grid array** (BGA)—Leads are metal balls on the bottom of the device.

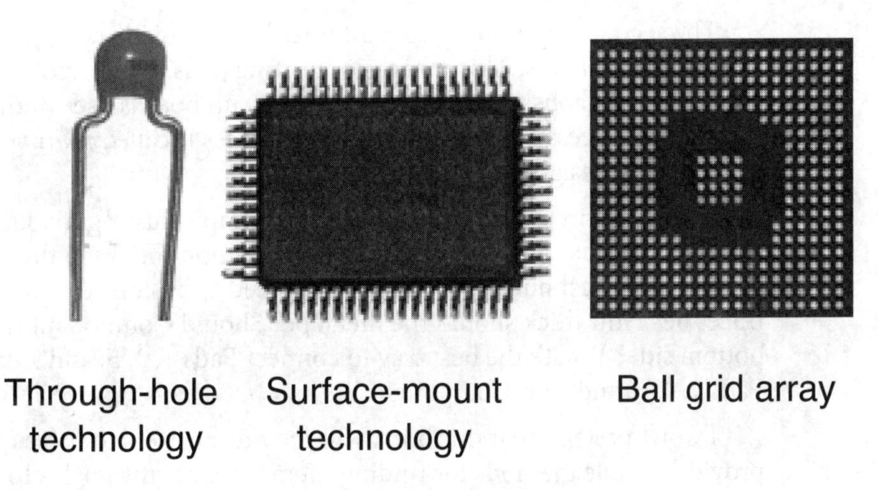

Figure 2.1: Component Leads

Through-hole components dominated the twentieth century but SMT components have become much more popular. This is because SMT components are smaller, don't require drilling, and can be automically positioned using devices called pick-and-place machines.

The drawback of SMT components is that the number of leads is limited by the component's perimeter. In contrast, the number of leads in a BGA component is limited by the component's area. Therefore, BGA components allow for many more leads.

The location on a board that comes in contact with a component's lead is called a *pad*. Leads are attached to pads using solder, and because solder connects metal to metal, pads must be metallic. The second reason that pads must be metallic is that they need to conduct electricity to or away from the lead.

In general, through-hole pads and BGA pads are round, and SMT pads are rectangular. Figure 2.2 presents typical pads for through-hole and SMT components.

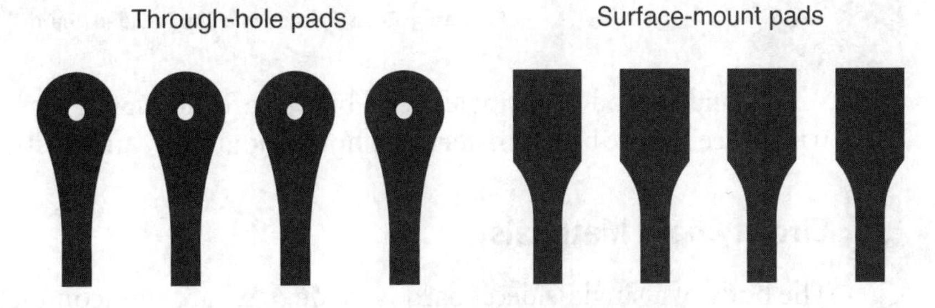

Figure 2.2: Through-Hole and SMT Pads

In all cases, the pads should be large enough so the leads can be reliably soldered to the board. But if the pads are too large, they may come in contact with other metallic surfaces.

The second function of a circuit board is to provide connections between a circuit's components. This means connecting pads to one another using a conductive material. To establish these connections, circuit boards use conductive lines called *traces*. The process of laying out a boards' traces is called *routing*, and it's a crucial part of the circuit design process.

In theory, circuit board design is fairly simple: Just figure out where the component pads should be and create connections between them. But in practice, this process has a vast number of details that need to be considered. How wide should the traces be? How thick should the metal be? Should Component A be on the top or the bottom side? What's the best way to connect Pads 1, 3, 5, and 7 on Component X to Pads 2, 4, 6, and 8 on Component Y?

I won't pretend that this book will provide answers to these questions or even provide reliable methods for finding them. Instead, my goal is to explain how EAGLE can assist you as you search for your own answers.

2.1.2 Single-Sided Boards

The simplest circuit board to understand and fabricate is the single-sided board. Figure 2.3 presents a cross-section.

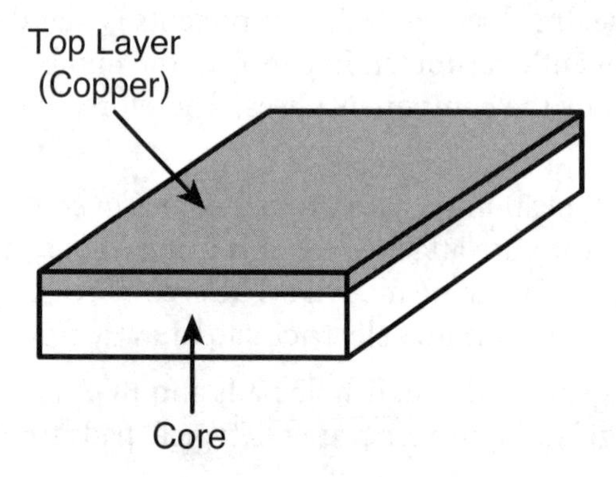

Figure 2.3: Cross-Section of a Single-Sided Board

To understand how single-sided boards are fabricated, it's important to be familiar with three topics: board materials, photolithography, and post-processing.

Circuit Board Materials

The body of a single-sided board is made of a hard, nonconductive material, typically fiberglass epoxy-resin, or FR4. Some literature refers to a board's material as substrate, but in this book, we'll refer to this material as *core*. Core thickness is given in thousandths of inches called *mils* (1 mil = 0.001 in). Standard board thicknesses are 31 mils, 39 mils, and 62 mils.

A thin layer of metal is attached (usually glued or electrodeposited) to one side of the board, and we'll call this the top side. This metal is almost always copper, which is inexpensive and provides a low-resistance path for electricity. As Table 2.1 shows, the thicker the copper layer, the lower the resistance.

Table 2.1

Copper Thickness and Sheet Resistance

Thickness in oz	Sheet Resistance in μΩ
0.5	971
1.0	486
2.0	243
3.0	162

It may seem odd that the copper thicknesses in Table 2.1 are given in ounces, but this is the common measurement used in the industry. For circuit boards, an ounce corresponds to the weight of copper per square foot. Table 2.2 relates copper ounces to the thickness of the metal in mils.

Table 2.2

Copper Thicknesses

Copper Thickness in oz	Copper Thickness in mils
0.5	0.68
1.0	1.35
2.0	2.70
3.0	3.05

Thick copper provides less resistance than thin copper and can carry greater current. But thin copper makes it easier to form small features. This is why many fabrication houses allow a smaller minimum dimension for thin copper than for thick copper.

Photolithography

To remove unwanted copper from a board, PCB fabrication houses employ photolithographic (*photo* - light, *lithos* - stone, *graphein* - write) methods. Figure 2.4 shows how this works.

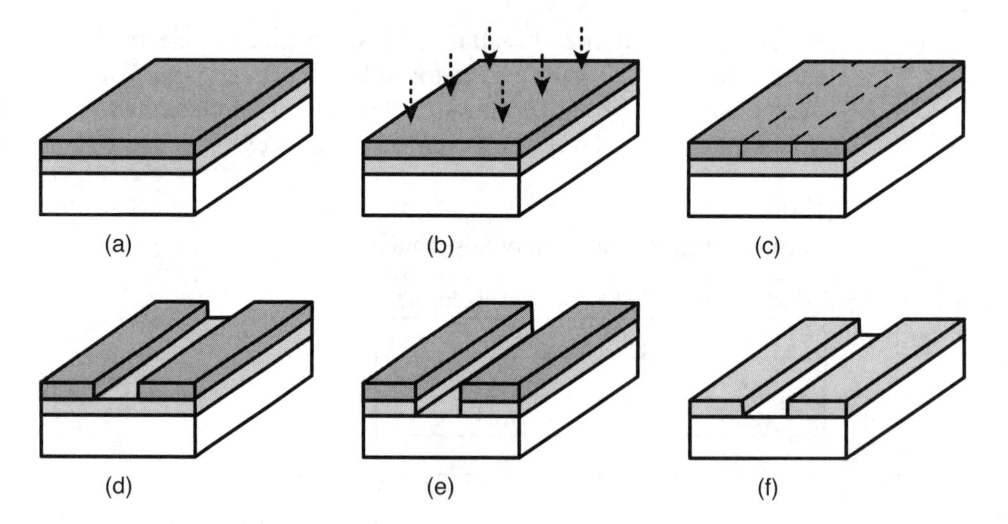

Figure 2.4: Circuit Board Photolithography

The six steps depicted in the figure are given as follows:

a. The copper layer is coated with light-sensitive material called photoresist.

b. A photoplotter uses a light source to selectively expose portions of the photoresist.

c. The photoresist's chemical properties change when exposed to light. If the photoresist is positive, the exposed photoresist softens. If the photoresist is negative, the exposed photoresist hardens. The photoresist in the figure is negative.

d. The softened photoresist is removed using a chemical called developer. The hardened photoresist remains and covers parts of the copper.

e. The uncovered copper is removed using a strong acid, such as cupric chloride. This process is called etching and the chemical is called an etchant.

f. The remaining photoresist is removed, leaving only the patterned copper on the circuit board.

Post-Processing

After photolithography, electrical components can be soldered to the board. But most board fabrication processes perform at least three steps beforehand:

1. Cover circuit board (except pads) with solder mask.
2. Apply solder paste to pads.
3. Use silk-screening to print letters and draw symbols.

To protect the copper traces, board fabrication facilities frequently cover the circuit board with a tough, nonconductive material called *solder mask*. Solder mask is usually dark green or blue, which explains why so many circuit boards have these colors. The solder mask shouldn't cover the pads because that would make it difficult to solder components to the board.

To assist in the soldering process, it's common to apply a small amount of conductive glue to SMT pads. This glue is called *solder paste*, and it is an important part of modern circuit board assembly.

On many circuit boards, you'll notice lettering and symbols printed in white. These markings may identify where components should be placed, which leads are available for testing, and which company designed the board. This printing process is referred to as *silk-screening* because silk was originally used to form the stencils that mask the printed symbols.

2.1.3 Double-Sided Boards

As the number of components on a board increases, so do the number of traces. In many instances, the full set of traces can't be connected on a single plane without intersecting one another. To solve this problem, traces need a way to jump over other traces. This jumping is made possible by adding copper to the board's bottom side. This bottom side may also support additional components. A board with copper on both sides is called a double-sided board.

The process of fabricating double-sided boards is similar to that of single-sided boards. Copper is glued to both sides of the core material and both sides are processed using similar photolithographic methods.

The important difference between double-sided and single-sided boards is the need for electrical paths between the top and bottom sides. These paths are called *vias*, and they're formed by creating tunnels through the core material and filling them with metal. Figure 2.5 depicts the cross-section of a via connecting the sides of a double-sided board.

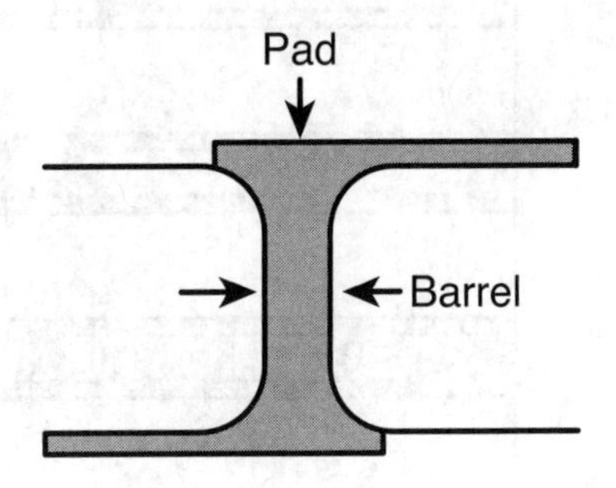

Figure 2.5: A Via Through a Double-Sided Board

The body of the via between the two layers is called the barrel. A pad is formed where the via touches either surface. A via's aspect ratio is the ratio of its height to its diameter.

NOTE A via's pad isn't necessarily a regular pad. That is, a via's pad is usually not connected to a component's lead. If a via pad is directly connected to a component's lead, the process is called *via-in-pad*.

Vias are categorized according to how their holes are created. The most popular method of creating a via is to drill through the core material and fill the hole with metal. In this case, the via is called a *plated-through hole*, or *PTH*. PTH diameters are commonly given in mils, and common via diameters are 12, 13, and 25 mils.

The second method of creating vias is to use lasers, photolithography, or plasma etching to create holes smaller than can be created by drills. These are called *microvias* and their diameters are commonly given in millimeters. Typical diameters are 0.1 and 0.3 millimeters. The holes are usually filled with solder paste.

2.1.4 Multilayer Boards

Double-sided boards enable more complex routing than single-sided boards, but sometimes two layers still aren't enough. This occurs frequently when designing circuits whose components have hundreds of closely spaced leads. It's also important for designs that require entire layers for ground or power supply.

To resolve these issues, board designers create multilayer board designs. In essence, a multilayer board is a group of double-sided boards sandwiched together using a material called *pre-preg*. Similar to glue, pre-preg is soft to begin with but hardens when heat and pressure are applied. A board's arrangement of core and pre-preg layers in a multilayer board is called its *stackup*, and Figure 2.6 shows a stackup for an eight-layer board.

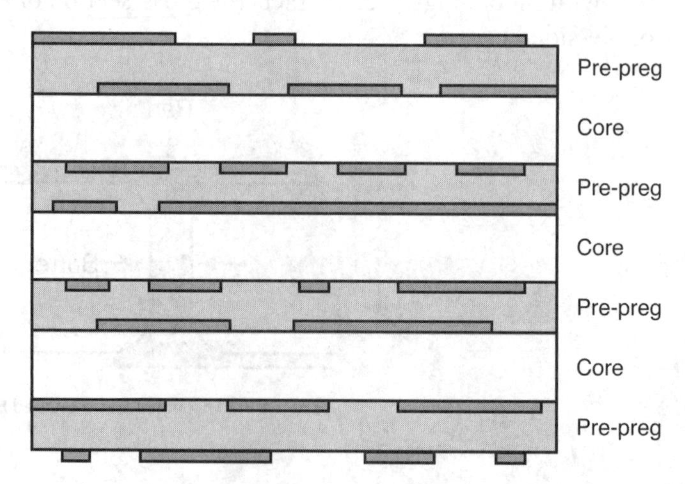

Figure 2.6: Stackup of an Eight-Layer Board

This stackup contains three layers of core material and four layers of pre-preg. But the board has eight layers because there are eight surfaces containing copper.

As with double-sided boards, vias pass electricity from layer to layer. For a multisided board, vias come in four types:

- *Through vias* connect the board's top and bottom sides. They don't come in contact with any of the internal layers.
- *Stub vias* run all the way through the board. They can connect one or both external layers to any of the internal layers.
- *Blind vias* connect an external layer to an internal layer but don't run through the entire board. They can be seen from only one side.
- *Buried vias* connect internal layers but don't touch any external layer. They can't be seen from either side of the board.

Figure 2.7 presents an example of each of these vias in a multilayer board. It's worth noting that, while many fabrication houses support different via types, the prices differ according to difficulty.

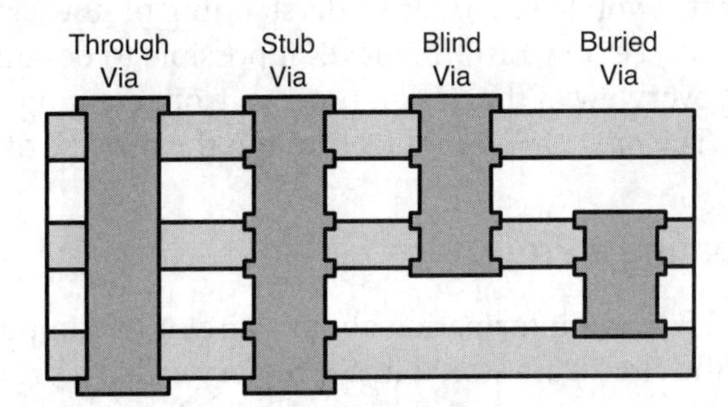

Figure 2.7: Vias in a Multilayer Board

2.1.5 Gerber and Excellon Files

Circuit board designers usually don't get involved in the fabrication process. Instead, our job is to give a fabrication facility the information it needs to construct our designs. This information consists of three parts:

- **Locations**—For each feature (such as a pad or a via), we need to identify its layer and (x, y) coordinates.
- **Dimensions**—We need to provide geometric data including the width of the traces, the spacing of the pads, and the thickness of the copper.
- **Drill holes**—For each hole to be drilled, we need to identify the hole's position and the drill diameter.

Designers provide these properties to the fabrication facility using computer files. Multiple formats exist to describe circuit boards, but at the time of this writing, the most commonly accepted format is RS-274X, which is commonly referred to as the extended Gerber format or just the Gerber format. Appendix B, "The Gerber File Format," discusses the Gerber format in detail.

Gerber files control how the photolithographic tools should be used to pattern copper on the board (see Figure 2.4). To tell the fabrication facility about the board's holes, another type of file is needed. Files containing drill information are called Excellon files.

The ultimate goal of PCB design is to create the files that accurately describe the circuit board. The next section will explain how EAGLE generates these important files automatically.

2.2 Overview of Circuit Design with EAGLE

Now that you have a basic understanding of how circuit boards are constructed, you're ready to see how EAGLE makes it possible to design them. This section provides a brief overview of the design process, from creating a project to generating the design files. The next chapter walks through the process of designing a simple circuit.

2.2.1 Creating a Project

EAGLE stores information about a circuit in computer files, and all the files for a specific circuit are stored in a directory called a project. In essence, a project is a special type of folder that stores all the files related to a single circuit design.

By default, EAGLE places each new project directory inside a folder called eagle, which is located in your Documents folder. On my Windows 7 system, the eagle directory can be found at the following path:

```
C:\Users\Matt\Documents\eagle
```

When you launch EAGLE, the Control Panel dialog box appears. The last entry in the dialog's vertical list is called Projects. Figure 2.8 presents the initial projects that ship with EAGLE 6.5.

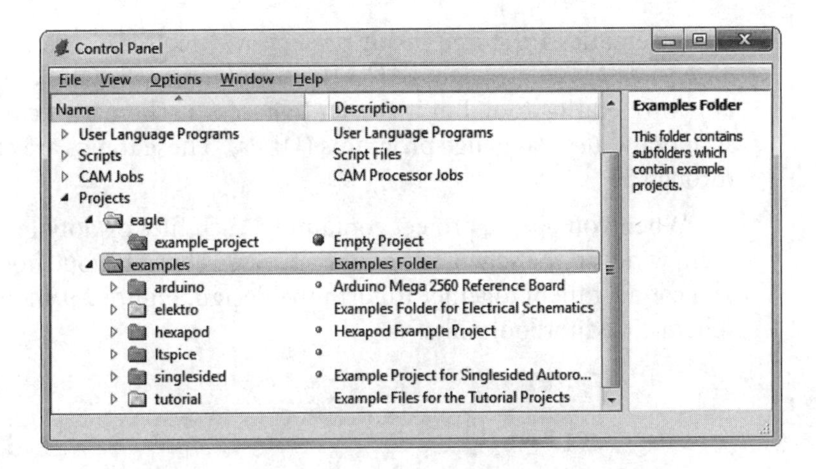

Figure 2.8: Initial Contents of EAGLE's Projects Directory

The red folders represent projects. Inside each project folder, you'll find files related to a circuit design. In the figure, the folders named arduino, hexapod, and singlesided are project folders.

The yellow folder icons correspond to regular folders. These contain project folders and other regular folders. In the figure, the folders named eagle and examples are regular folders.

If you look at the list of projects, you see that each has a circle between the project's name and description. A small circle means the project is closed. If you right-click a closed project and select Open Project, EAGLE reads the project's design files into memory and launches editors for its schematic (*.sch) and board (*.brd) files. When a project is open, its circle becomes large. Only one project can be open at a time.

Creating a new project is easy. Go to the main menu in EAGLE's Control Panel and select File > New > Project. Chapter 3, "Designing a Simple Circuit," will explain this further in its presentation of an initial circuit design.

2.2.2 Creating a Schematic Design

After you create a project, the first design file you need is a schematic. A schematic is a high-level description of a circuit's structure. It provides information about the circuit's components, the components' values (resistance, capacitance, and so on) and the manner in which they're connected.

Schematics do not provide information about the circuit's physical characteristics. That is, the schematic doesn't say anything about the board's dimensions or where the components are actually placed on the board.

NOTE Different versions of EAGLE have different limits for circuit board size, but there are never any limitations for schematic sizes.

Schematics are stored inside projects as *.sch files. Its format is based on the eXtensible Markup Language (XML), which means it can be read and modified with any XML editing tool. But it's easier to access a schematic's content by executing scripts and user langauge programs (ULPs). These topics are explored in Chapters 10 through 13.

When you open a project containing *.sch files or double-click a *.sch file in the main window, the schematic editor appears. This lets you choose circuit components and connect them together to form the design. Figure 2.9 shows what an empty schematic editor looks like.

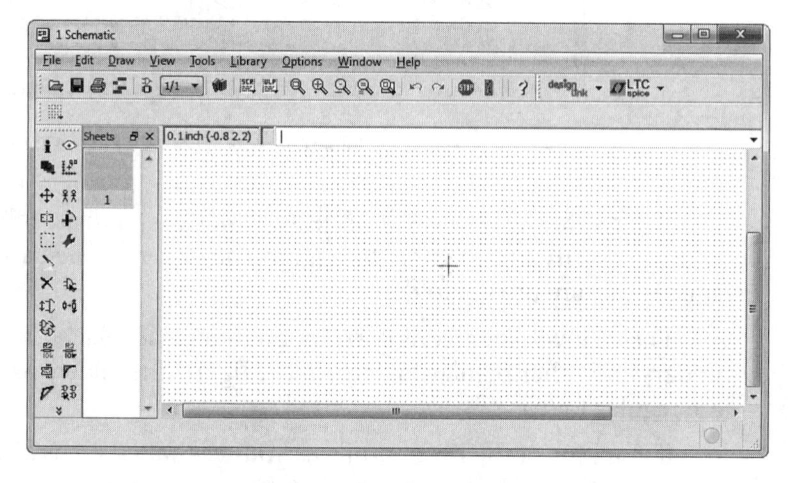

Figure 2.9: The Schematic Editor

Later chapters discuss the schematic design process in detail and the many items in the editor's toolbars and menus. For now, it's important to understand the four steps of using the schematic editor:

1. Insert components from EAGLE's library into the schematic.
2. Arrange components and set their names/values.
3. Draw connections between components.
4. Create a board file (*.brd) from the schematic.

To add a component to the schematic, you need to look through EAGLE's set of component libraries. Thankfully, this set of libraries is *huge*, so no matter what integrated circuit you're looking for, you'll probably find it or something similar. Figure 2.10 presents a small portion of the list of available components. In this case, the component to be added is the MC9S12XF512 integrated circuit from Freescale Semiconductor.

The vastness of the EAGLE libraries can be daunting, particularly if you're searching for a component without knowing the manufacturer. Thankfully, pressing Ctrl-F opens a search bar at the bottom of the window. This makes it possible to find a component using some or all the characters in its name.

If EAGLE can't provide what you're looking for, you can create custom components. This requires drawing a base shape and adding pins so that other components can connect to it. Chapter 8, "Creating Libraries and Components," explains how to create new components and add them to a library.

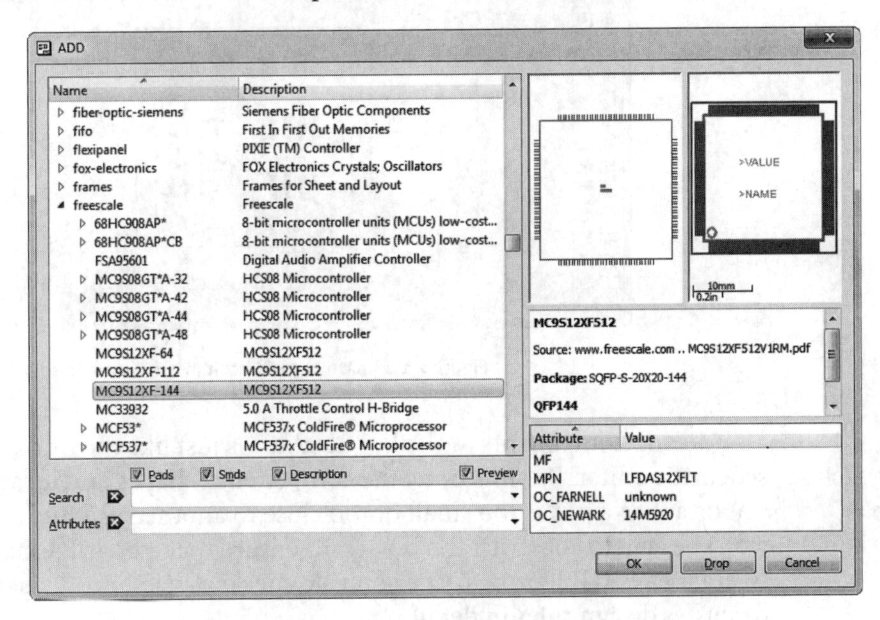

Figure 2.10: Components in an EAGLE Library

After you insert your components, it's easy to move them around, draw wires (called nets) between them, and assign values to their properties, such as resistance. When you're happy with the schematic, a single button press creates a board file.

2.2.3 Designing the Circuit Board

When EAGLE generates a new board file, it opens a new editor called the board editor. This has many of the same menu options and toolbar items as the schematic editor, but it serves a different purpose. Figure 2.11 shows what a design in the board editor looks like.

The circuit in the board editor contains the same components as those in the schematic from which it was generated. But now the shapes resemble those of physical devices. This is because the circuit in the board editor is meant to resemble the real-world circuit. Dimensions such as width, height, and thickness become critically important.

The process of designing a circuit in the board editor consists of three steps:

1. Set the size of the circuit board.
2. Position each component on the board.
3. Create connections between the components.

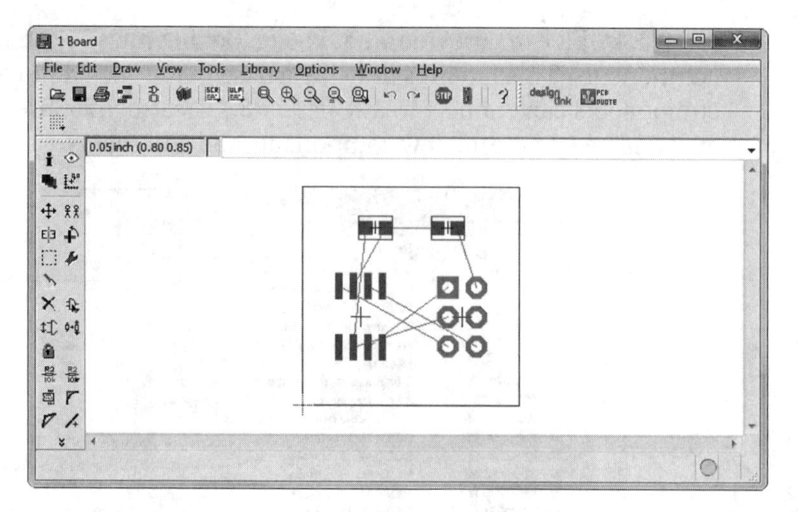

Figure 2.11: An Unrouted Circuit in the Board Editor

Moving components in the board editor is just like moving components in the schematic editor. But this time, the components' shapes and locations are vitally important. If a pad is too small or too close to another pad, the fabrication facility won't be able to construct the board. To ensure that a board design meets dimensional tolerances, EAGLE provides design rule checks. Chapter 5, "Layout and Design Rules," discusses design rules in detail.

2.2.4 Routing Connections

In Figure 2.11, the components' connections are depicted as thin lines that run from pad to pad. These lines, called *airwires*, aren't real-world connections. Their purpose is to identify which pads need to be connected with metal traces. The process of converting airwires into traces is called *routing*, and Figure 2.12 displays the circuit in Figure 2.11 after routing has been performed. Note that there is no single right way to route a circuit—the routing in Figure 2.12 is just one of many possible routings.

It's easy to manually route simple designs like the one in Figure 2.11. But as the number of components increases, it becomes more difficult to create connections between them. Thankfully, many releases of EAGLE contain an autorouter that performs most of the work for you. Chapter 6, "Routing," explains both manual routing and the process of using the autorouter.

When the components are put into position and their connections have been routed, the board design is essentially finished. To convert the board design into files for fabrication, EAGLE provides the CAM Processor.

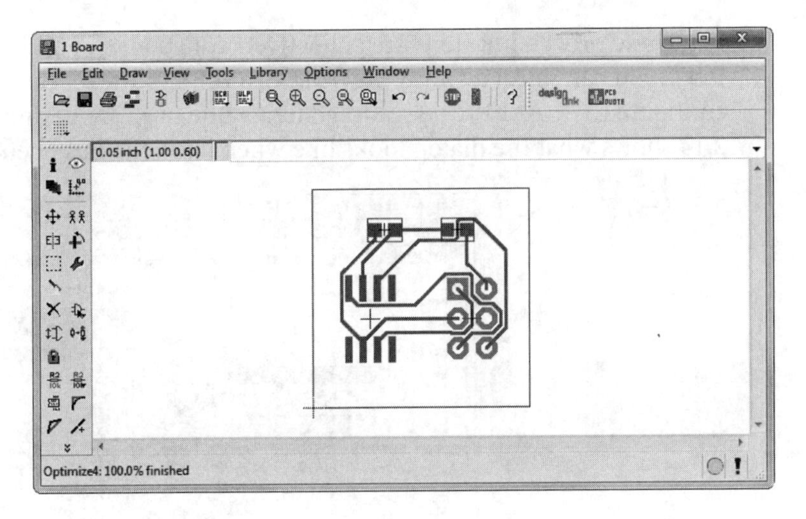

Figure 2.12: A Routed Circuit in the Board Editor

2.2.5 Generating Design Files

The last step is the simplest. There are no components or connections to worry about. All you have to do is tell EAGLE what files you want to generate and what aspects of the circuit design should be stored in the files.

Instead of a full editor, all you need is a dialog box. This is called the CAM (Computer Aided Manufacturing) Processor and Figure 2.13 shows what it looks like.

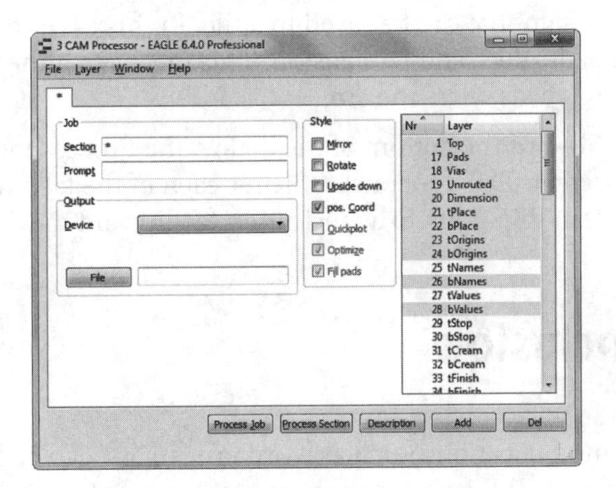

Figure 2.13: The CAM Processor (Unconfigured)

This dialog provides many options related to generating design files, including the formats of the output files and the nature of the data to be stored. Rather than enter this information manually, it's easier to read data from a file that identifies which tasks should be performed.

These processing steps are collectively called a *job*, and if you go to File > Open > Job..., you can select a *.cam file from EAGLE's top-level cam directory. These files configure the CAM Processor to generate one or more design files for a circuit. Figure 2.14 shows what the dialog looks like when a job file has been selected.

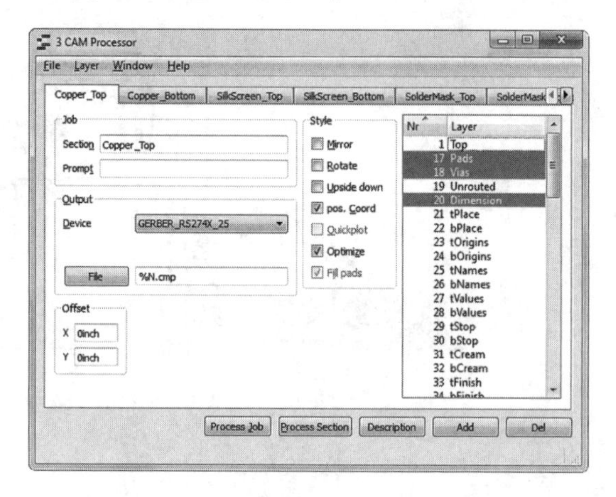

Figure 2.14: The CAM Processor Configured with a Job

Just under the main menu, each tab represents a different file to be created. The first tab, labeled Copper_Top, creates a Gerber file that contains data related to the copper pattern on the top side. The file's format is defined using the combo box next to the Device label. As shown, GERBER_RS274X_25 has been selected.

On the right of the dialog, you can see a list defining the different portions of the design that can be saved to a file. In this case, the selected aspects of the design are Pads, Vias, and Dimension. This means that the file will store data related to the board's pads, vias, and size.

Toward the bottom of the dialog, the Process Job button launches the CAM Processor. This generates a file for each of the tabs. When these files are generated, you can ship them to a fabrication facility and they'll build your board.

2.3 Conclusion

Many applications, such as word processors, can be learned quickly through trial and error. EAGLE isn't one of them. Before you start with EAGLE, you need a top-level understanding of circuit boards and how they're designed. In writing this chapter, my goal has been to provide this understanding.

The first half of the chapter discussed the technology behind circuit boards. Circuit boards provide mechanical support and electrical connections between components. Components touch the circuit board at locations called pads, which are connected to one another using copper lines called traces. As circuit designers, our goal is to determine how to pattern the board's copper to provide these pads and traces. More concretely, our goal is to generate design files that tell fabrication facilities how to construct boards with these pads and traces.

EAGLE is one of many software tools available that generate design files for a circuit. As described in the second half of this chapter, the EAGLE design process consists of five steps:

1. Create a project to hold files related to the design.
2. Design a schematic that identifies which components are present and the manner in which they're connected.
3. Create a board design and place the components within the board.
4. Set (route) connections between the components in the board design.
5. Generate design files with the CAM processor.

Now that you have a working knowledge of PCBs and their design, you're ready to get your hands dirty. Chapter 3 walks through the design of a simple circuit from creating the project to routing the connections in the board editor.

Chapter 3

Designing a Simple Circuit

When I'm working with an unfamiliar design tool, I like to start with a simple project. This is a particularly good idea when dealing with EAGLE, whose user interface takes some time to get used to. For this reason, this chapter walks through the design of a basic inverting amplifier that contains only five components. In this whirlwind tour of EAGLE, I'll explain how to create the project, design the schematic, place the components in the board editor, route the connections between them, and generate Gerber and Excellon files.

My goal isn't to make you an EAGLE expert, but to make you sufficiently comfortable so that you can tackle more interesting circuits, such as the Arduino Femtoduino presented in the next chapter. For added familiarity, you may want to work through this chapter multiple times.

3.1 An Inverting Amplifier

This chapter focuses on an inverting amplifier, and before I explain how to design the circuit, I'd like to briefly discuss how it works. The theory isn't pertinent to the design process, so if this doesn't interest you, feel free to proceed to the next section.

An operational amplifier, or op-amp, amplifies the difference between two signals: its positive and negative inputs. Denoting the positive signal as v_+ and the negative signal as v_-, the op-amp's output is $A(v_+ - v_-)$. This is shown in Figure 3.1.

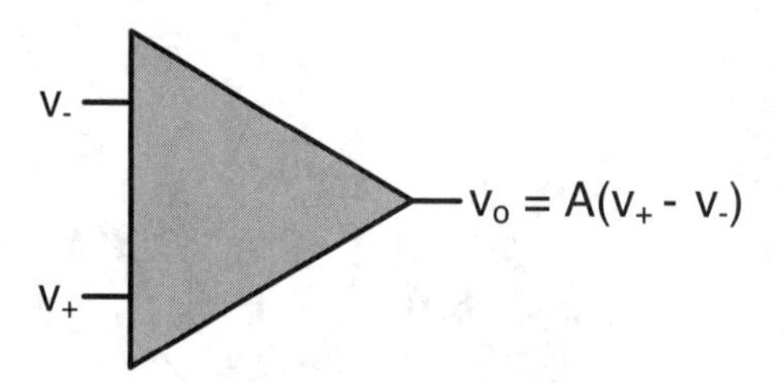

Figure 3.1: An Operational Amplifier

The result of the amplification can be unpredictably large, so circuit designers constrain the op-amp's output by feeding it back into one of its inputs. In the case of an inverting amplifier, the output is fed back into the negative input through a feedback resistor R_f. The input signal is also connected to the negative input through a resistor R_i. The op-amp's positive terminal is connected to ground. Figure 3.2 shows what the full circuit looks like.

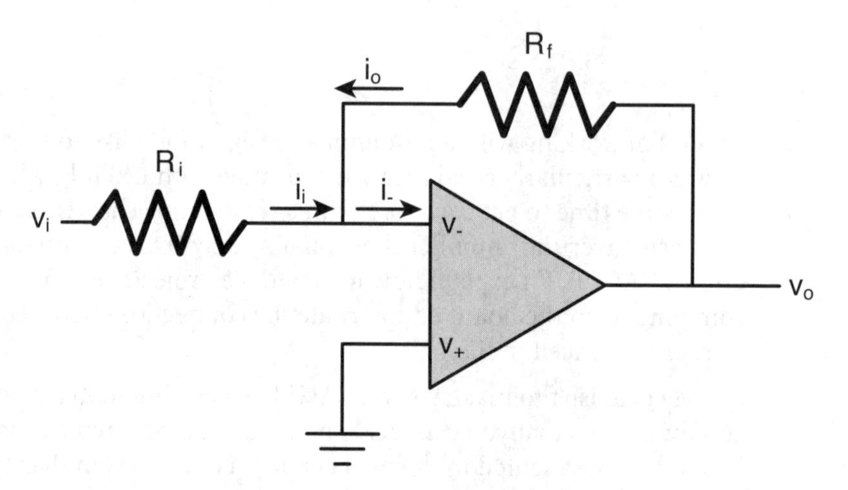

Figure 3.2: An Inverting Amplifier Circuit

Because the positive terminal is connected to ground, the op-amp's output equals Av_-. To compute v_-, we need to examine the three currents at the op-amp's negative terminal:

1. i_-—The current entering the op-amp.
2. i_{in}—The current produced by the input signal. By Ohm's law, this equals $(v_i - v_-)/R_i$.
3. i_o—The output current fed back into the input. This equals $(v_o - v_-)/R_f$.

Kirchoff's Current law states that the currents entering a junction sum to zero. Therefore, $-i_- + i_{in} + i_o = 0$. An ideal op-amp accepts zero input current, and real-world op-amps are so close to this ideal that we can set i_- to 0A. Therefore $i_{in} + i_o = 0$. Rearranging this produces $i_{in} = -i_o$, which means $(v_i - v_-)/R_i = -(v_o - v_-)/R_f$.

One more point needs to be made. Ideal op-amps adjust their output voltage to ensure that the input voltage difference, $v_+ - v_-$, is as low as possible. Therefore, it's safe to assume the two input voltages are equal. For our inverting amplifier circuit, this means $v_- = v_+ = 0$.

Rearranging the current equation leaves us with $v_i R_i = -v_o/R_f$, or $v_o = -(R_f/R_i)v_i$. The circuit is called an inverting amplifier because the output always has the opposite sign of the input. The resistors determine the degree of amplification. If R_f is greater than R_i, the output will be greater in magnitude than the input.

An amplifier's output is limited by its two voltage supplies, called voltage rails. These determine the maximum and minimum voltages that can be produced by the amplifier. For example, if the rails are set to 5V and –5V, all output from the amplifier will be clipped to +/– 5V.

3.2 Initial Steps

Before you can design the inverting amplifier, two steps are needed:

1. Install the book's EAGLE library, eagle-book.lbr.
2. Create a project and a new schematic.

The first task needs to be performed only once. The second must be performed before every new design.

3.2.1 Installing the Book's EAGLE Library

To make it easier to design the circuits presented in this book, I've included all the components in a library file called eagle-book.lbr. This file can be found in eagle_book.zip, which can be freely downloaded from http://www.eagle-book.com.

Installing this library is simple. EAGLE's installation directory contains a folder called lbr, and on my system, this can be found at C:\Program Files (x86)\EAGLE-6.5.0\lbr. Place eagle-book.lbr in this directory.

After installing a library, you need to tell EAGLE that you intend to use it. The following directions show how this can be done.

1. Launch the EAGLE application and find the Libraries option in the center of the Control Panel.
2. Open Libraries by clicking its triangle. Scroll to find the eagle-book library.
3. Click the small green dot separating eagle-book.lbr from its description. The dot will become large to indicate that the library is active. This is shown in Figure 3.3.

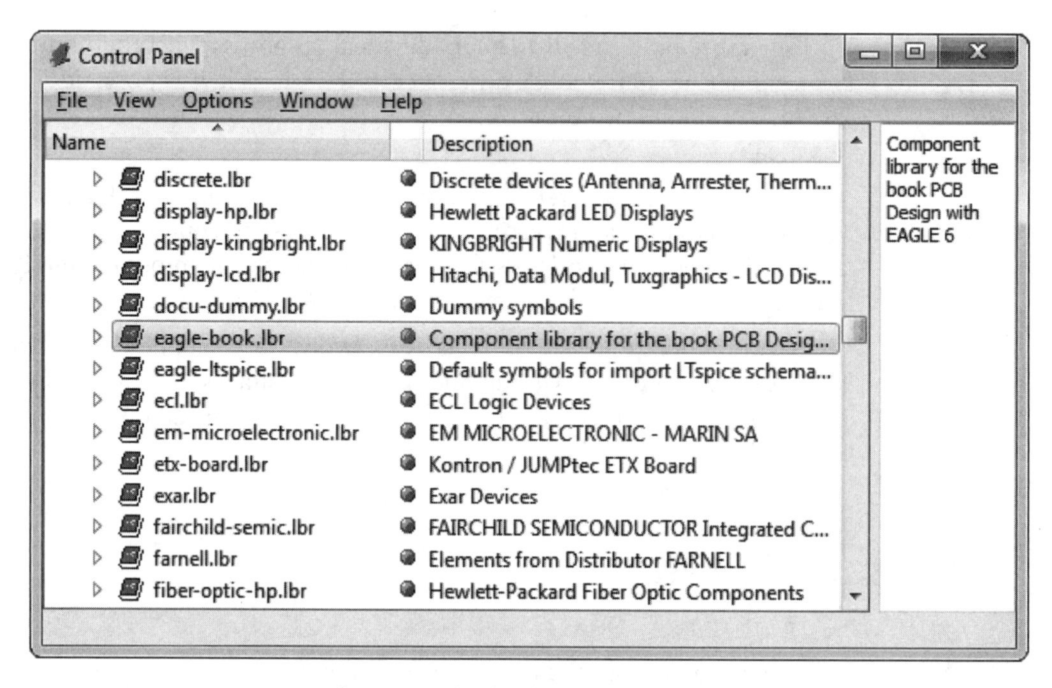

Figure 3.3: Activating the eagle-book Library

If you click the arrow to the left of a library, you can see what devices it provides. In the case of eagle-book.lbr, the devices contribute to the three example circuits in this book:

- **Inverting amplifier**—The example circuit in this chapter consists of five components that can be fully integrated in a single-layer board.
- **Femtoduino**—The Arduino-based circuit introduced in Chapter 4, "Designing the Femtoduino Schematic," serves as an example of a double-sided board.
- **BeagleBone Black**—The circuit in Chapter 14, "Schematic Design for the BeagleBone Black," contains a microprocessor, a graphics processor, and flash memory. This circuit board requires six layers.

3.2.2 Create a New Project and Schematic

As discussed in the preceding chapter, EAGLE organizes circuit designs using projects. A project is a directory that contains all the design files related to a single circuit.

To create a new project for the inverting amplifier, go to the Control Panel's main menu and select File > New > Project. EAGLE creates a new project in the eagle folder. I recommend that you give it the name **invamp**.

EAGLE automatically expands the green circle next to invamp, which makes it active, as shown in Figure 3.4.

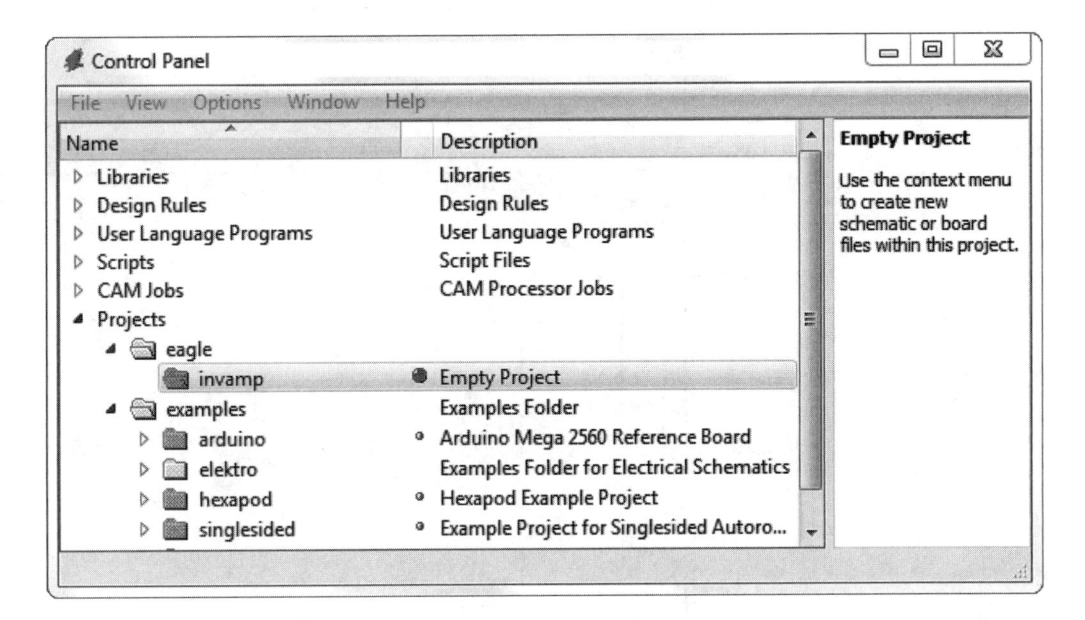

Figure 3.4: Creating a New Project

As shown in the figure, invamp is a project (red folder) and is currently active (large green dot). By default, the project's directory is created in the user's home directory. On my Windows 7 system, the invamp folder is located at C:\Users\Matt\My Documents\eagle\invamp.

If you open this folder, you'll find a file called eagle.epf. This contains settings for the project that include the user's window preferences and the project's associated files. If EAGLE gives a strange error related to projects and files, the first place you should look is eagle.epf.

At the moment, the invamp project is empty. The following sections add two new files: a schematic design (*.sch) and a board design (*.brd).

3.3 The Inverting Amplifier Schematic

To create an empty schematic for the inverting amplifier, go to the Control Panel's main menu and select File > New > Schematic. Alternatively, you can right-click the invamp project and select New > Schematic.

Figure 3.5 shows what the empty schematic editor looks like.

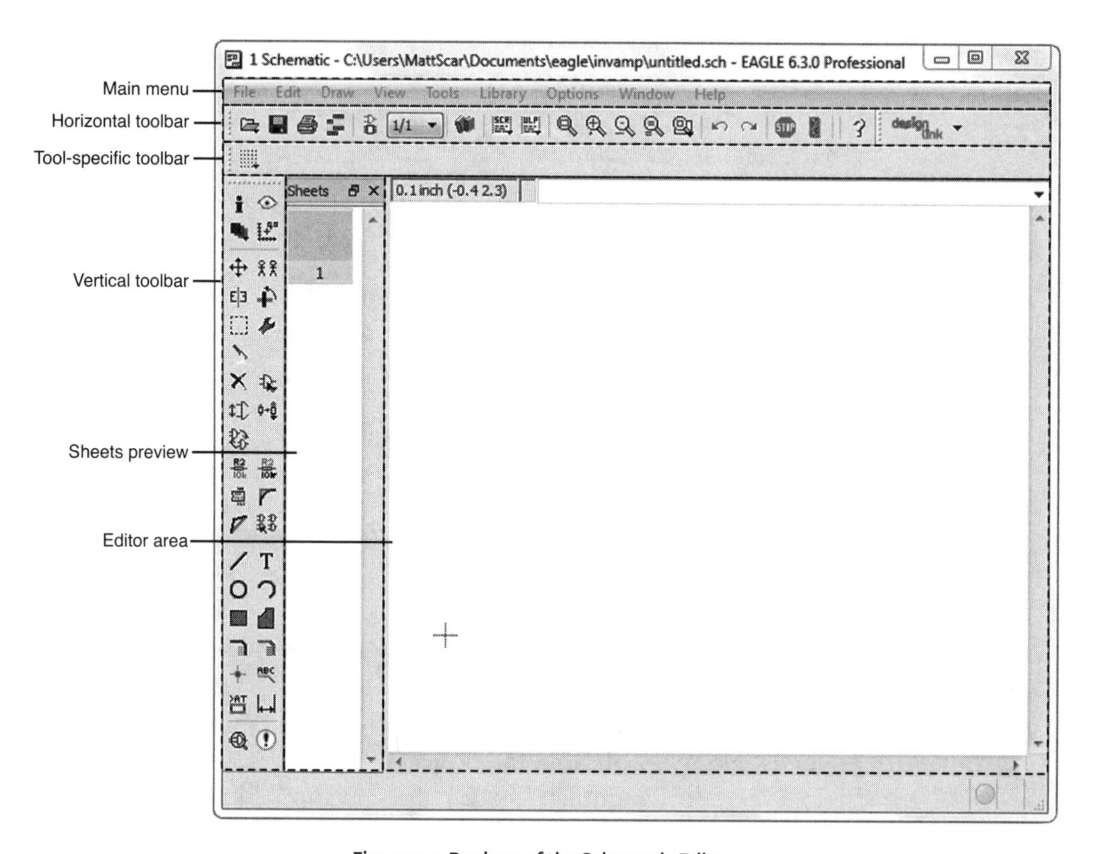

Figure 3.5: Regions of the Schematic Editor

From top to bottom and left to right, the regions of the schematic editor are as follows:

- **Main menu**—Top-level options for navigating the editor
- **Horizontal toolbar**—Basic capabilities including open file, save file, print, zoom, undo, and redo
- **Tool-specific toolbar**—Options related to the active tool
- **Vertical toolbar**—The primary tools needed to design the schematic
- **Sheets preview**—Displays the sheets in the schematic
- **Editor area**—Depicts the current schematic design

Most of these, such as the main menu and horizontal toolbar, are easy to understand. But you'll probably spend a lot of time working with the vertical toolbar, which contains the tools needed for editing.

3.3.1 Tools in the Vertical Toolbar

The number of entries in the vertical toolbar may seem overwhelming, but for designing a schematic, you really need to know only 10. Figure 3.6 shows where these tools are located.

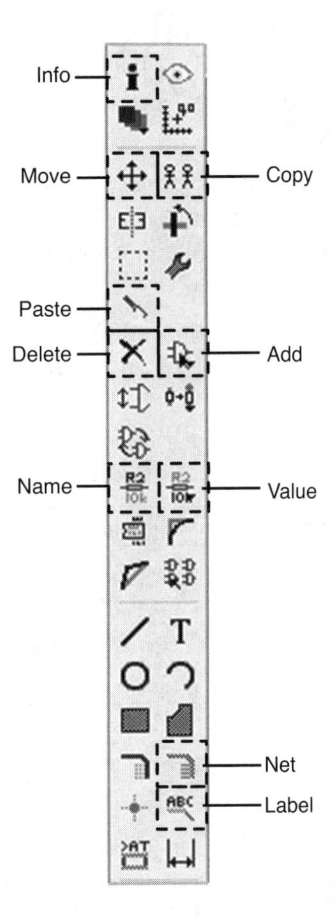

Figure 3.6: Schematic Editing Tools on the Vertical Toolbar

The functions of these tools are listed as follows:

- **Info**—Examine information about a component, net, or bus
- **Move**—Change the location of a component
- **Copy**—Store a component to the clipboard
- **Paste**—Place a component from the clipboard
- **Delete**—Remove a component from the design
- **Add**—Select a component from a list of libraries
- **Name**—Assign a name to a component

- **Value**—Assign a value to a component
- **Net**—Draw a connection between components
- **Label**—Display the name of a net or bus

Before you start editing the schematic, it's important to understand a major difference between EAGLE's user interface and those of other applications. The tools in the vertical toolbar are *modal*—when a tool becomes active, it remains active until another tool is selected.

This behavior can be frustrating, espcially when you want to cancel a drawing operation. For this reason, EAGLE provides the Stop tool. This is the red octagonal sign in the horizontal toolbar. When selected, Stop cancels the operation and deactivates the current tool.

Many of EAGLE's tools can be activated using hotkeys. Table 3.1 lists hotkeys that I use frequently in the schematic editor.

Table 3.1

Tool Hotkeys for the Schematic Editor

Name	Hotkey	Purpose
Info	Ctrl-I	Examine information for a component, net, or bus
Move	Ctrl-M	Move a component to a different location
Copy	Ctrl-Shift-C	Copy a component to the clipboard
Paste	Ctrl-V	Paste a component from the clipboard
Delete	Ctrl-D	Delete a component from the editor
Add	Ctrl-Shift-A	Select a component from a list of libraries
Name	Ctrl-Shift-N	Assign a name to a component, net, or bus
Value	Ctrl-Shift-V	Assign a value to a component, net, or bus
Net	Alt-N	Draw an electrical conductor
Undo	Ctrl-Z	Cancel preceding action
Redo	Ctrl-Y	Perform preceding action again

I've left out the Rotate tool because you don't need the hotkey or the toolbar entry. Instead, after you select a component, you can rotate it by right-clicking.

> NOTE By default, EAGLE gives all new schematics the name untitled.sch. Before proceeding further, I recommend that you change this to **invamp.sch**.

3.3.2 Adding Components to the Design

The inverting amplifier circuit presented in this chapter consists of the following components:

- An LM741 to serve as the operational amplifier
- Two resistors to set the amplification
- A six-pin external connector to provide power and signals (positive power, negative power, ground, input, output)
- A ground connection

This discussion explains how to insert these components to the schematic. If you choose to work through the design, don't be concerned about where you place the circuit's components. As long as the right components are present and they're connected in the right way, you can design the final circuit without difficulty.

Adding the LM741

This example circuit relies on the venerable LM741 to serve as the operational amplifier. The following instructions show how to insert this in a schematic:

1. Activate the Add tool by selecting the entry in the vertical toolbar or by pressing Ctrl-Shift-A.
2. When the Add dialog appears, scroll to the entry titled eagle-book.
3. Open the library and double-click the LM741 component.
4. Click inside the schematic editor to insert the op-amp in the design.
5. Click the STOP button to deactivate the Add tool.

When the op-amp appears in the schematic, the designation IC1 will appear above the figure. This serves as the component's identifier. For more information, select the Info tool (Ctrl-I) and click the device.

Adding Resistors

The circuit's amplification is set by two resistors. The following instructions show how to add them to the circuit.

1. Activate the Add tool (Ctrl-Shift-A), scroll to eagle-book, and open the entry labeled RES. Double-click the entry for RES_0603, which is a resistor whose physical package corresponds to the 0603 designation.
2. In the schematic editor, place one resistor above the op-amp.
3. Click again in the editor, placing a second resistor to the left of the op-amp.
4. Click the STOP button to deactivate the Add tool.

To set the amplifier's gain to –3, the first resistor must have three times the resistance of the second. For this example, R1's value should be set to 3.3kΩ and R2's value should be set to 1.1kΩ. The following instructions explain how this can be done:

1. Activate the Value tool (Ctrl-Shift-V) and click R1.
2. In the small New Value dialog, enter **3.3k**.
3. With the Value tool still active, click R2.
4. In the small New Value dialog, enter **1.1k**.

These values don't affect the circuit in any way. Instead, they serve as labels that will show how the resistors should be positioned in the board.

Adding the Connector and Ground

This design isn't intended to be fabricated as a real-world circuit, but if it were, it would need at least one connector to provide power and signal connections. For this reason, the design contains a connector with six pins: two rows of three pins each. The following instructions show how to insert it and the ground connection into the design:

1. Activate the Add tool, scroll to eagle-book, and double-click OPCONN.
2. In the schematic editor, place the connector to the right of the op-amp.
3. Activate the Add tool again and double-click GND.
4. In the schematic editor, place GND below the op-amp and to its left.

Now that these five components are inserted in the design, the next step is to connect them together. But first, it's important to understand EAGLE's terminology for working with schematic and board designs.

3.3.3 EAGLE Terminology

EAGLE has specific terms for circuit elements and their connections, and the better you understand them, the better you'll understand CadSoft's documentation and online discussions.

Circuit Elements

When dealing with circuit components, you should be familiar with three terms:

- **Symbol**—An element in a schematic editor
- **Package**—An element in the board editor
- **Device**—The combination of one or more symbols and a package

To clarify these terms, consider the LM741 operational amplifier. The triangular element in the schematic editor is the op-amp's symbol. In the real world, the LM741 is rectangular, so the element in the board editor (which you'll see shortly) has a rectangular shape. This is the op-amp's package.

The combination of the op-amp's symbol and package is called a device. A device doesn't have a distinct appearance: It looks like the symbol in the schematic editor and a package in the board editor.

Connections and Connection Points

Just as EAGLE has specific terms for circuit elements, it also has terms for their connections and connection points. In the schematic editor, the following terms apply:

- **Pin**—A symbol's point of connection
- **Net**—A connection representing a single electrical path
- **Bus**—A connection representing multiple electrical paths

For example, the connection between the resistors in the inverting amplifier consists of a single electrical path. In EAGLE terminology, we say that the resistors' *pins* are connected by a *net*. But if a connection between two devices contains 64 electrical paths (a 64-bit bus), the connection is referred to as a *bus*.

In the board editor, the following terms are used:

- **Pad**—A package's point of connection
- **Airwire**—An unrouted connection between two pads
- **Trace**—A routed connection between two pads

Don't be concerned if these terms aren't clear just yet. I'll explain each of them again in later chapters.

3.3.4 Connecting Symbols

The next step in the design process is to arrange the symbols in the schematic and connect their pins together. This discussion explains both tasks and then introduces the concept of signals.

Positioning the Symbols

The precise locations of the schematic's symbols aren't important, but for the sake of clarity, I recommend that you position them in a manner similar to that shown in Figure 3.7.

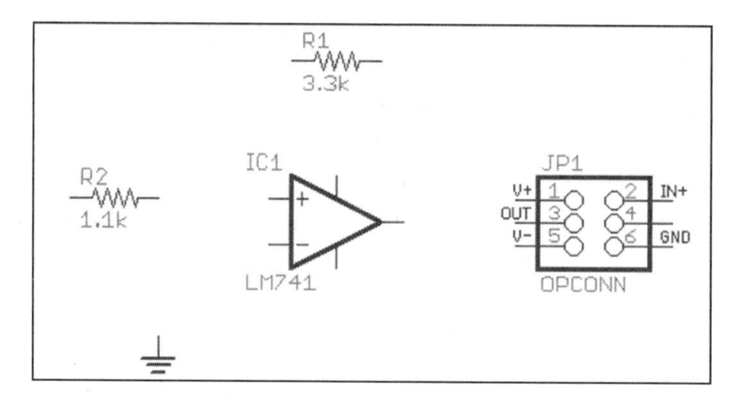

Figure 3.7: Arrangement of Symbols in the Schematic Editor

Drawing Nets

To draw a net between two pins, you need to perform three steps:

1. Activate the Net tool by selecting the entry on the toolbar or by pressing Alt-N.
2. Click the first pin. EAGLE will start drawing a net from that pin.
3. Click the second pin. EAGLE will connect the two pins with a net.

Using these steps, I recommend that you connect the schematic's pins in the manner shown in Figure 3.8.

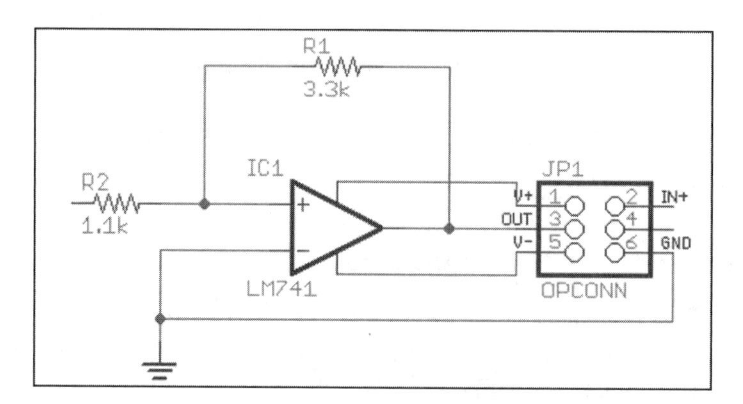

Figure 3.8: Connected Devices in the Schematic Editor

The left pin of R1 must be connected to the net between R2 and the op-amp. To make this connection, start a net from the left side of R1 and click again when the two nets touch. EAGLE will join the nets together and mark the junction with a green circle.

If two nets cross without a green circle, they're not connected. To create a connection, activate the Junction tool and click the intersection of the two nets.

There are two more points to be made. As you draw a net, you can create bendpoints by clicking in the editor. You can also terminate a net by double-clicking in the editor. This becomes important when dealing with signals, which I'll discuss next.

Signals

One more connection needs to be added: The input voltage (left side of R2) must be connected to the IN+ pin of the six-pin connector. We could draw a long net connecting the two pins, but there's a better way. If both nets are given the same name, EAGLE will know they're meant to be connected.

EAGLE assigns a default name to each net in a schematic, and they usually look like $N1 or $D5. To see this, activate the Info tool (Ctrl-I), click a net, and look at the Name field in the Properties box.

When multiple nets have the same name, EAGLE knows they should be connected. This name is called a *signal* name, and as we'll see in later chapters, signals make it possible to divide a complex circuit into simpler subcircuits.

The following instructions show how to join the input voltage to IN+ using a signal:

1. Click the connector's IN+ pin, and double-click to the right, creating a short net.
2. Activate the Name tool (Ctrl-Shift-N) and assign the name INPUT to the net.
3. Activate the Label tool, click the net, and place the text to the right.
4. On the left side of the schematic, click R2's left pin and double-click to the left, creating a short net.
5. Activate the Name tool and assign the name INPUT to the net.
6. A dialog will ask if you want to connect the net to the INPUT signal. Select Yes.
7. Activate the Label tool, click the net, and place the text to the left.

Figure 3.9 shows what the final schematic looks like.

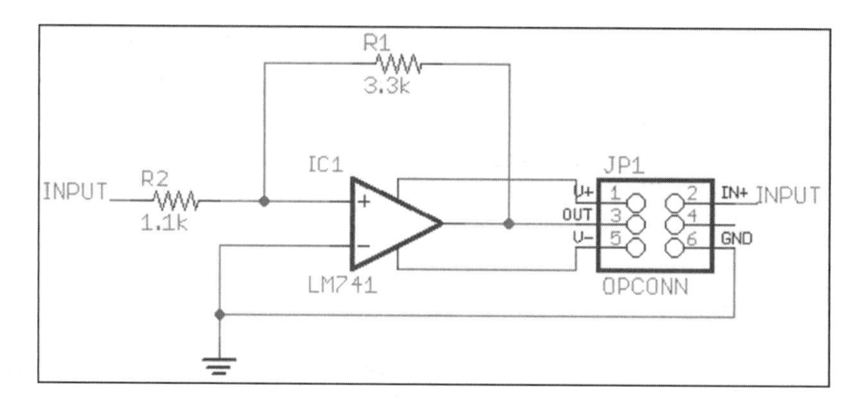

Figure 3.9: Full Schematic of the Inverting Amplifier Circuit

NOTE You can zoom in and out by scrolling with the middle mouse button. You can change the editor's position by holding down the middle mouse button and dragging.

3.4 Board Layout

Now that the schematic design is complete, the next step is to design the actual board. Unlike the symbols in the schematic editor, the packages in the board editor resemble the physical geometry of the actual components. For example, the LM741's symbol looks like a triangle, but as we'll see, its package has the same rectangular shape as the LM741 chip.

Proper orientation and placement of a circuit's packages is critical. The positions of the packages will be transferred to the PCB design, so it's important to put them exactly where they need to be. This process is called *layout* and it's the first topic discussed in this section.

The second topic focuses on creating connections between packages (or more precisely, between the packages' pads). This process is called *routing* and this section presents two methods of routing. First, I'll show how EAGLE's autorouter can create connections automatically. Then I'll explain how to route connections manually.

3.4.1 Creating the Board Design

There are two ways to create a board file (*.brd) from an open schematic file (*.sch):

- Click the Generate/Switch to board tool on the horizontal toolbar.
- In the main menu, go to File > Switch to board.

A dialog will ask if you want to create the board file from the schematic. Click Yes and EAGLE will create invamp.brd and open the board editor.

In a board design, the editor area is divided into two regions. To the left, the editor displays the packages corresponding to the symbols in the schematic. These packages are drawn to resemble the components' physical devices, and they're connected by thin, yellow lines called *airwires*.

To the right, the editor displays an empty rectangle that represents the circuit board. The goal of the layout process is to set the rectangle's dimensions and move the packages into their proper positions. Afterward, the routing process replaces the airwires with physical connections between the packages' pads.

If a circuit's schematic and board design are open in their respective editors, EAGLE will link them. That is, any changes made to the schematic will be reflected in the board design. For this reason, if you close the schematic editor while the board editor is open, EAGLE will display a ribbon in the board editor stating that F/B Annotation Has Been Severed! Despite this warning, you can continue editing the board design.

3.4.2 The Grid

One important difference between the schematic editor and the board editor is the need for precise positioning. If a package isn't in exactly the right place, the component's leads won't connect properly to the pads on the board.

To assist with positioning, EAGLE editors provide grids that enforce placement at multiples of the grid spacing. The grid's properties are configured through the Grid tool, which can be activated by clicking the button in the upper-left corner of the editor. Figure 3.10 shows what the Grid dialog looks like.

This dialog makes it possible to set four properties:

- The grid's visibility (on or off)
- The grid's appearance (lines or dots)
- The spacing and dimensions for the primary grid
- The spacing and dimensions for the alternate grid

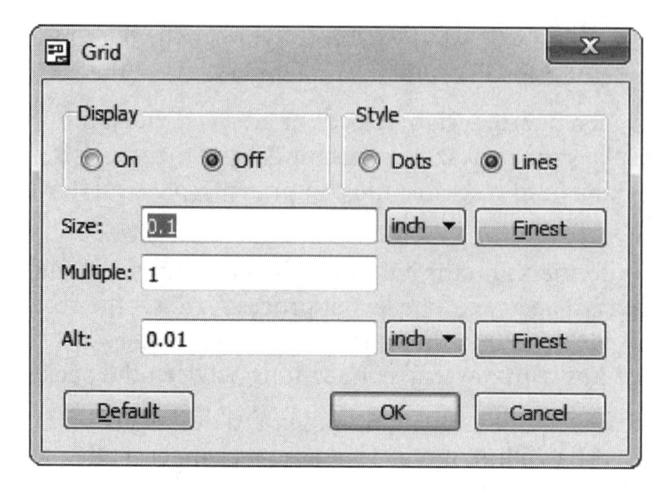

Figure 3.10: The Grid Dialog

To ensure proper routing between pads, it's important to use the same grid spacing for every package in the circuit. This common grid is called the *primary grid*, and its spacing can be set in units of inches, mils (thousandths of an inch), milimeters (mm), and microns (thousandths of a millimeter).

In addition to the primary grid, EAGLE's editors provide an alternate grid, which is generally used to position secondary features such as text. For example, when a component is inserted in an editor, it can be placed only at locations determined by the primary grid spacing. But if you hold Ctrl down while moving the component, you'll find that the grid spacing is determined by the alternate grid.

3.4.3 Board Dimensions and Origins

After the grid is configured, the next step is to set the perimeter of the circuit board. This circuit is small enough that we can fit its components in a 0.75" x 0.75" square. The following instructions explain how to configure the grid and set the perimeter in the board editor:

1. If you haven't already, open the invamp.brd design in the board editor. This can be done from the schematic editor by going to File > Switch to board on the main menu.
2. Open the Grid dialog by selecting the Grid tool on the horizontal toolbar.
3. Set the primary grid spacing to 0.05 inches and press OK.
4. Activate the Move tool and click the right line of the board's rectangle, preferably close to the center. Move the line to the left until its x-coordinate equals 0.75.

5. With the Move tool still active, click the top line of the rectangle and move it downward until its y-coordinate equals 0.75.
6. Verify that the board area is sized correctly by positioning the mouse over the board's upper-right corner. The coordinates should equal (0.75, 0.75).

The board is now ready for layout, but before we delve into that subject, I want to discuss the topic of origins. In the board's lower-left corner, you'll see that EAGLE has drawn a cross. This is the board's origin, and if you position the mouse directly over it, you'll see that the displayed coordinates are (0, 0). Every circuit element in an EAGLE editor has a similar cross that defines its origin. When a package is placed in the board, its position is given by the position of its origin relative to the board's origin. That is, if the coordinates of a package are given as (x, y), it means the package's origin is at (x, y).

3.4.4 Board Layout

Moving packages in the board editor is just like moving symbols in the schematic editor. With the Move tool active, select a package and change its position. To rotate a package, right-click while the Move tool is active or activate the Rotate tool.

If you know the intended coordinates of a package, you can quickly move it into position using the Info tool. For example, to place a package at position (x_0, y_0), activate the Info tool (Ctrl-I), click the package, and enter $\mathbf{x_0}$ and $\mathbf{y_0}$ in the boxes labeled Position. You can also enter its angular orientation (in degrees) in the box labeled Angle.

Table 3.2 lists the position and orientatation of each of the four packages in the inverting amplifier circuit.

Table 3.2

Package Placement—Inverting Amplifier Circuit

Name	Origin Position	Orientation
R1	0.25, 0.60	0
R2	0.50, 0.60	0
IC1 (LM741)	0.20, 0.30	180
JP1 (OPCONN)	0.55, 0.30	270

When the packages have the correct positions and orientations, the board area should resemble Figure 3.11.

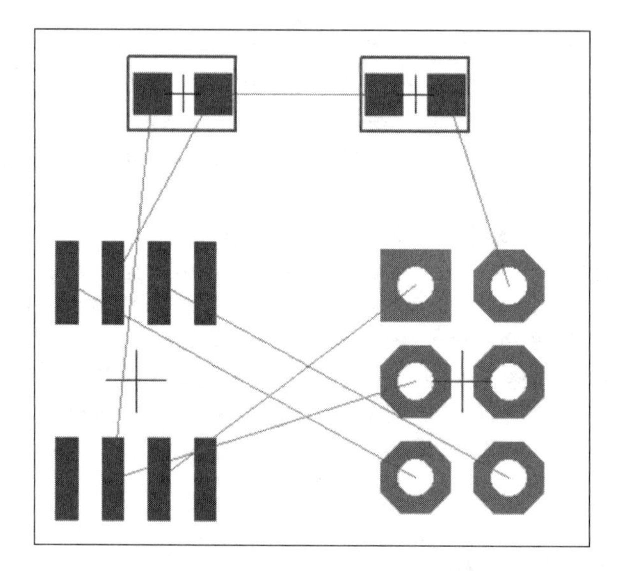

Figure 3.11: Board Layout—Inverting Amplifier Circuit

3.5 Routing

With the circuit's packages in place, the next step is to create connections between their pads. These connections are called *traces*, and in forming them, two rules must be observed:

1. A trace consists of straight lines that join at bendpoints whose angles must be a multiple of 45°.
2. No trace can intersect a trace carrying a different signal.

In Figure 3.11, there are only eight airwires connecting the four packages. Given the two rules, how would you create the traces? It might look easy, but remember that the traces can't intersect.

The goal of this section is to explain how connections can be routed in EAGLE. The process requires a different set of tools than those we've used so far. Figure 3.12 lists each of them and their names.

This section presents two methods of routing connections in EAGLE. First, I'll explain how to use the autorouter to form traces automatically. Then I'll show how to create traces manually using the Route tool.

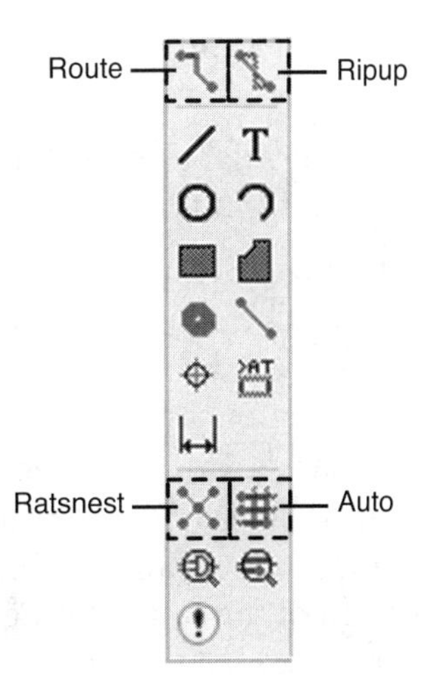

Figure 3.12: Board Layout—Inverting Amplifier Circuit

3.5.1 The Autorouter

EAGLE's autorouter is ideal for creating traces in simple circuits like the inverting amplifier circuit. To launch the autorouter, activate the Auto tool on the vertical toolbar. This brings up the Autorouter Setup dialog, as shown in Figure 3.13.

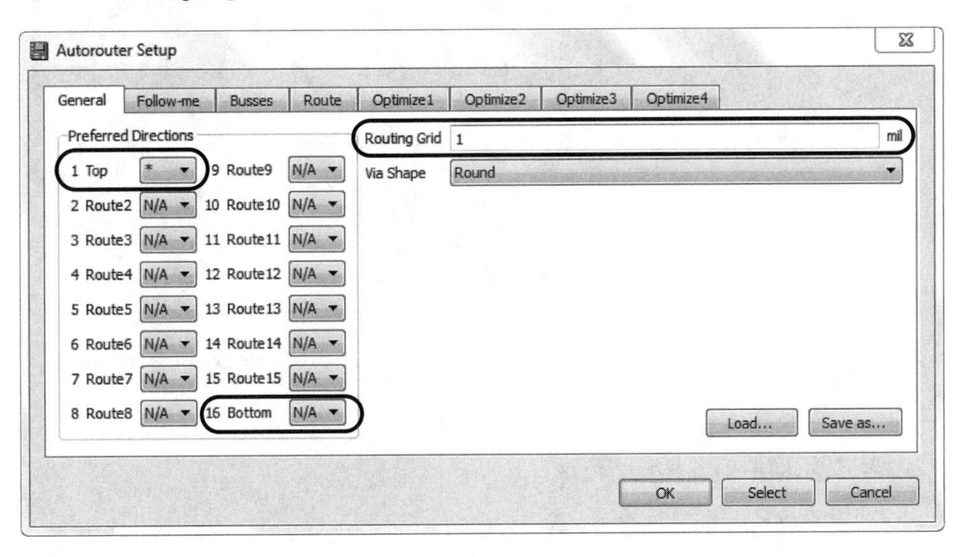

Figure 3.13: Autorouter Setup—Inverting Amplifier Circuit

A full description of the autorouter and its configuration will have to wait until Chapter 6, "Routing." For now, I recommend that you make three simple adjustments to the default settings:

1. On the left, find the combo box labeled 1 Top and choose the * setting. This tells the autorouter to form traces on the top layer in any direction.
2. Find the combo box labeled 16 Bottom and choose the N/A setting. This tells the autorouter not to form traces on the board's bottom layer.
3. On the right, find the text box labeled Routing Grid and set the autorouter's spacing to 1 mil. This reduces the spacing of the autorouter's internal grid, which means it can do a more precise job of routing connections.

Don't worry about the other tabs for now. After these changes are made, click OK at the bottom of the dialog. The autorouter will start, and as it computes the routing, you see traces drawn and redrawn in the board editor. At the bottom of the editor, a message informs you of the router's progress. When the message reads 100% finished, the routing is complete. Figure 3.14 shows what the routed design looks like on my system.

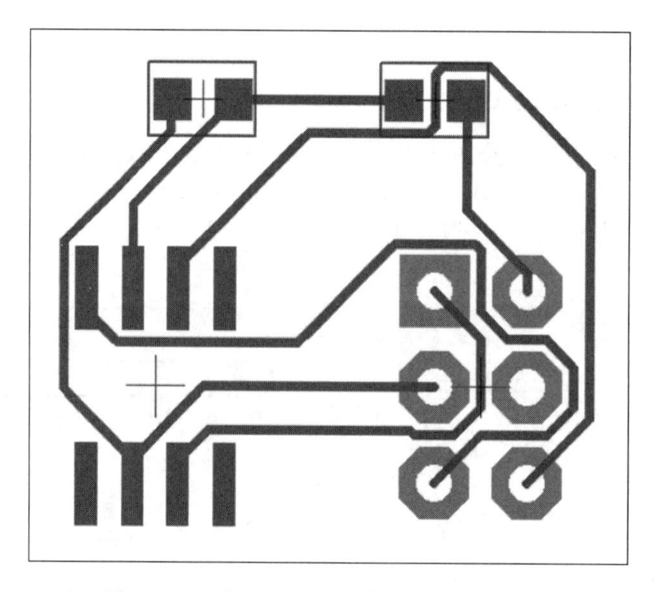

Figure 3.14: The Routed Inverting Amplifier Circuit

3.5.2 Manual Routing

EAGLE's autorouter excels at connecting simple circuits, but it isn't sufficient to route complex circuits without assistance. To provide this assistance, the designer needs to perform some of the routing on his own. The primary tool for manual routing is the Route tool. But before I discuss this, I want to briefly explain how Ripup and Ratsnest work.

Ripup and Ratsnest

The Ripup tool performs the opposite operation of the Route tool. That is, it unroutes traces and converts them back into airwires. If you activate the Ripup tool and select a trace, it transforms it and adjacent traces into unrouted airwires.

When a routed design is ripped up, the airwires maintain the shapes of the traces. Another tool, called Ratsnest, recomputes the original airwires, ensuring that each airwire follows the shortest path between its pads. I like the Ratsnest tool because it removes the clutter of post-ripup airwires and makes it easier to start manual routing.

The Route Tool

The Route tool is one of the most complex tools in the EAGLE toolbar. To see what I mean, click the Route tool and look at the new horizontal toolbar above the editor. Chapter 6 will explain what all these items mean, but for now, the only items that concern us are those depicted in Figure 3.15.

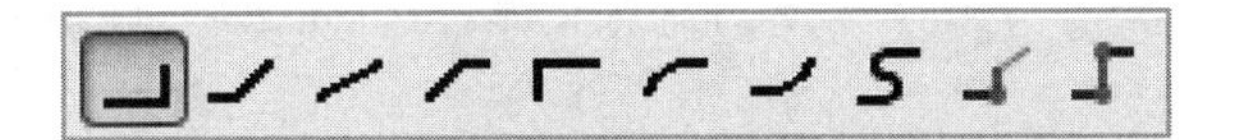

Figure 3.15: Bend Options for the Route Tool

Creating a route in EAGLE involves more than just drawing straight lines from one point to the next. As you use the Route tool, the trace can bend in various ways depending on which bend option is selected. By default, the Route tool bends traces at 90°, and this is why the left-most option in Figure 3.15 is selected.

In the layout depicted in Figure 3.14, a number of traces bend at 45° angles. This is common in many circuits. The second and fourth toolbar items cause the Route tool to create 45° bends, and I generally use them in my routing, switching between the two as needed. But I recommend that you experiment on your own to see how the bend options work.

In general, the process of manually routing a trace consists of three steps:

1. With the Route tool active, click the airwire to be routed.
2. Click the airwire's starting pad. As you move the mouse, EAGLE displays a trace in the editor.
3. Continue drawing the trace until you reach the destination pad. Click the destination pad to complete the route.

NOTE If you click an airwire close to its start pad, EAGLE may start drawing the trace automatically. In this case, Step 2 is unnecessary.

Manual routing takes practice, and a good place to start is by manually creating a trace between the resistors at the top of the circuit. If you used the autorouter earlier, I recommend that you remove this trace with the Ripup tool. Then activate the Route tool and start routing. Figure 3.16 shows how to create a trace from R1 to R2.

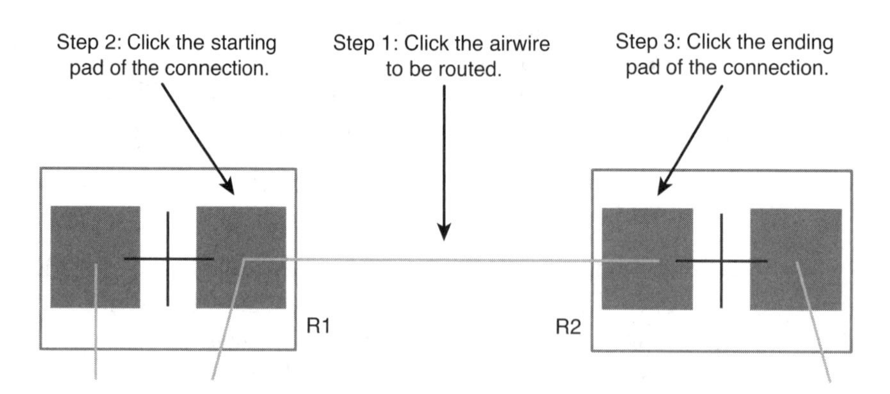

Figure 3.16: Routing a Connection Between Two Resistor Pads

After you form this connection, I recommend that you continue connecting more pads in the design. For most circuits, routing is the most difficult task in the design process, so it's important to be well-acquainted with it. But remember, this circuit isn't intended to be fabricated, so you don't have to route the entire circuit.

NOTE EAGLE provides a third method of routing called follow-me routing. Chapter 6 explains how this works.

3.6 CAM Processor

After routing, the circuit's design is complete. If a fabrication facility accepts *.brd files, you can send invamp.brd out for fabrication. Unfortunately, most businesses don't accept EAGLE files. The most common file formats are the RS-274X (Gerber) format for the copper pattern and the Excellon format for drill holes.

To generate these files in EAGLE, you need the CAM (Computer Aided Manufacturing) Processor. To launch this, select the Cam Processor button on the board editor's horizontal toolbar, or go to File > Cam Processor on the main menu. Figure 3.17 shows what the dialog box looks like.

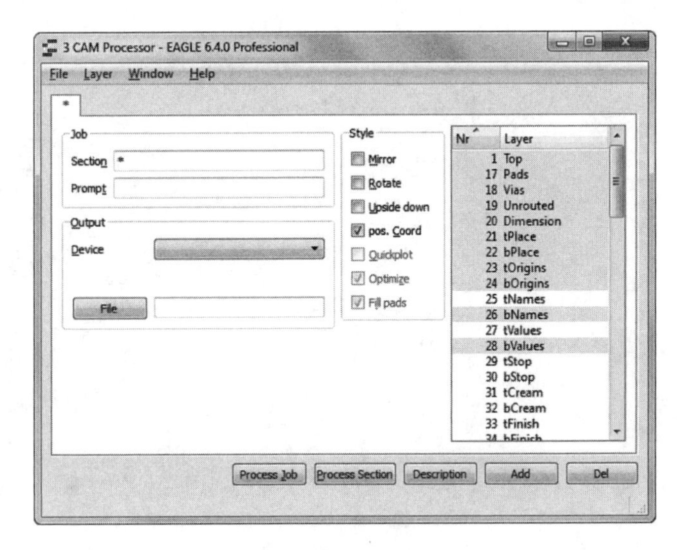

Figure 3.17: The CAM Processor

The inverting amplifier circuit has only one side (the top) and no drill holes. This means the circuit can be fully described using three Gerber files whose contents are given as follows:

- **Copper**—Defines the pattern of the copper that forms the pads and traces
- **Solder mask**—Defines where the solder mask coating should be applied to protect the circuit
- **Silk-screen**—Defines where text (usually white) should be printed to provide information about the circuit

If configured properly, the CAM Processor can generate a Gerber file for each of these three materials. The configuration process can be performed manually by changing fields in the dialog, but it's easier to read configuration data from an existing file. The example circuits in this book are provided in an archived directory, and that directory contains a folder called jobs.

A job is a sequence of tasks for the CAM Processor. If a file has the *.cam suffix, EAGLE will assume that it defines a job. If you look through the source archive for this book, you'll find a file called invamp.cam in the Ch3 folder. This defines a job that generates the three files for the inverting amplifier circuit. I recommend that you move invamp.cam to the top-level cam directory in your EAGLE installation.

To configure the CAM Processor with invamp.cam, go to File > Open > Job... on the dialog's main menu. Select invamp.cam and the dialog should display the tasks defined by the job file. Figure 3.18 shows what the updated dialog looks like.

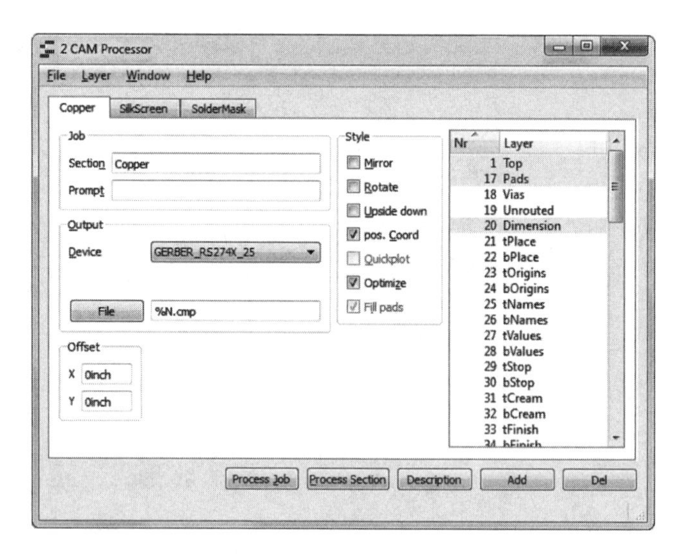

Figure 3.18: The CAM Processor Configured with Invamp.cam

The three tabs, labeled Copper, SilkScreen, and SolderMask, correspond to the three files to be created for the circuit. Each tab represents a different processing task for the CAM processor, and each provides a different set of configuration parameters.

If you look at the combo box labeled Device, you'll see that the value for each tab is set to GERBER_RS274X_25. This means that the generated file will be formatted according to the Gerber standard.

A full description of the CAM Processor's configuration fields will have to wait until Chapter 7, "Generating and Submitting Output Files." For now, all you need to know is that pressing the Process Job button will generate the files required by the job file. If you press this button and look at your project's contents in the Control Panel, you'll see three new files:

- **invamp.cmp**—Contains information for patterning copper
- **invamp.stc**—Contains information for covering the surface with solder mask
- **invamp.plc**—Contains information for setting the silk-screen (text)

The Gerber file format doesn't require a specific suffix. The suffixes of these files are arbitrarily set in the CAM Processor, but I've seen them used in a number of places. The important point is that they can be sent to a fabrication facility, which will then fabricate the circuit board for the inverting amplifier.

The Gerber file format is text-based, so you can examine these files with a text editor. But if you do this, you'll find that Gerber files are completely incomprehensible unless you understand the format standard. Appendix B, "The Gerber File Format," explains this format in great detail.

3.7 Conclusion

When I first encountered EAGLE, I needed to design a circuit containing several layers and more than 50 components. This was not a good way to begin learning how to use a design tool. The best way to become acquainted with EAGLE is to start with a simple design. This way, you gain familiarity with the application without having to deal with the added complexity of large-scale circuits.

My goal in writing this chapter has been to provide this familiarity. Inverting amplifiers aren't particularly interesting, but if you've worked through this example, you should be comfortable with how EAGLE works. If you're new to circuit design, I hope you've acquired a deeper understanding of how a schematic design becomes a board design, and how a board design becomes a fabricated circuit board.

This chapter has presented an overview of the EAGLE design process, from schematic design to automatic routing. If you haven't absorbed it all just yet, don't worry. The next chapter presents the schematic design of a nontrivial circuit, the Arduino Femtoduino. During the discussion, I'll review much of the material presented in this chapter.

Chapter 4

Designing the Femtoduino Schematic

As discussed in the preceding chapter, the first step in designing a circuit is to create a schematic. This identifies which components are used in the design and the manner in which they're connected.

In EAGLE, schematics are created using the Schematic Editor. In this chapter, the goal is to explore the editor in depth by showing how a practical circuit is designed. The next chapter explains how to design the physical board and Chapter 7, "Generating and Submitting Output Files," shows how to convert the board design into a set of Gerber files.

In selecting the example circuit, I based my decision on six criteria:

- **Simplicity**—The circuit should be simple enough that a newcomer to EAGLE doesn't need to spend hours designing the schematic.
- **Testability**—It should be straightforward to determine if the circuit works without an oscilloscope or logic analyzer.
- **Utility**—The circuit should have a practical use or be easily integrated into a larger, useful circuit. Further, designing the circuit should broaden the designer's understanding of digital technology.
- **Low-cost components**—All the components in the circuit should cost less than $50.00.
- **Small size**—The board design must be workable on the free version of EAGLE as well as the paid versions.
- **Straightforward routing**—The connections between components should be routable using only two layers, and it mustn't be too complicated for amateurs to route the board manually.

For these reasons, I've selected the Arduino Femtoduino as the first nontrivial circuit in this book. Of the many Arduino-compatible designs available (Uno, Due, Nano, Lilypad, and so on), this is the smallest and simplest. Yet it's complex enough to support the functions of the Arduino programming language.

The circuit board connects to a personal computer through a six-pin ICSP connector and allows the user to program the Femtoduino's microcontroller, an Atmel ATmega328p. Figure 4.1 shows a diagram of what the final circuit board looks like.

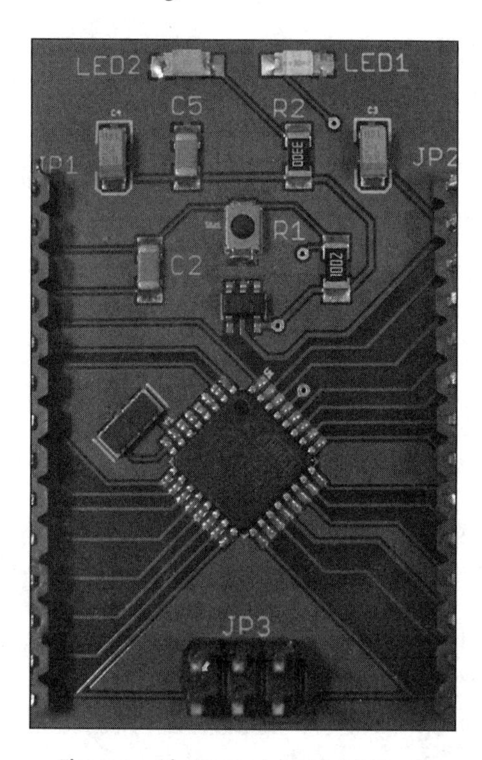

Figure 4.1: The Femtoduino Circuit Board

In addition to the ICSP connection, the board has headers that make it easy to plug it into breadboards or other electronic projects. A good slogan for Arduino might be "Simple enough for the newcomer, powerful enough for the pro." That also might be a good slogan for EAGLE.

The designer of the Femtoduino, Fabio Varesano, has made the circuit design available for free. I've included the schematic's PDF in the Ch4 folder in this book's archive, along with my EAGLE files. So if you choose to work through the Femtoduino design, you can compare your work with my designs and those of the original designer.

This chapter walks through the schematic design of the Femtoduino, proceeding from one component to the next. But before you start connecting components, there are a few preliminary tasks that need to be performed.

4.1 Initial Steps

In the preceding chapter, Section 3.2 explained how to install this book's example library, eagle-book.lbr. This needs to be performed only once, and if you haven't installed the library, I recommend that you go back to Chapter 3, "Designing a Simple Circuit."

The preceding chapter also explained how to create new projects and schematics. The first part of this section reviews this topic and the second part explains how to configure the schematic editor's grid.

4.1.1 Creating the New Project and Schematic

Chapter 3 explained how EAGLE organizes design files inside projects. To create a project for the Femtoduino, start EAGLE if you haven't already, and go to File > New > Project in the command panel. For the project name, I recommend that you enter **femtoduino**. This is shown in Figure 4.2.

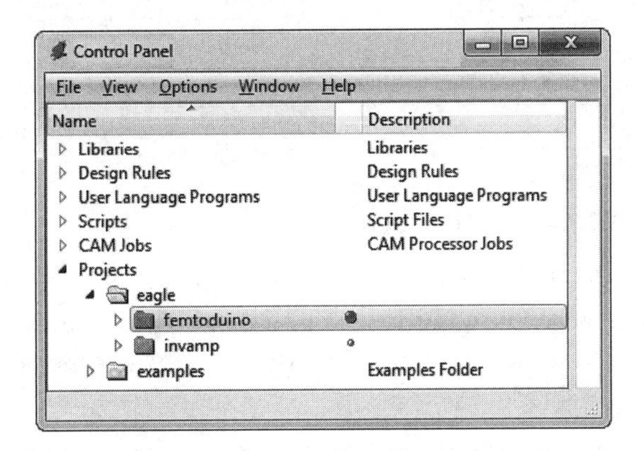

Figure 4.2: Creating the Femtoduino Project

Make sure the circle in the second column of the panel is green. This means the project is active.

After the project is created, right-click its name and choose New > Schematic. This creates a new file called untitled.sch and opens the schematic editor. I recommend that you save the blank design with a better name, such as **femtoduino.sch**.

Chapter 3 discussed many of EAGLE's toolbar items and their usage. Table 4.1 provides a reminder of many of the hotkeys available.

Table 4.1

Useful Hotkeys in the Schematic Editor

Name	Hotkey	Purpose
Add	Ctrl-A	Select a component from a list of libraries
Info	Ctrl-I	Examine information for a component, net, or bus
Move	Ctrl-M	Move a component to a different location
Copy	Ctrl-Shift-C	Copy a component to the clipboard
Paste	Ctrl-V	Paste an element from the clipboard
Delete	Ctrl-D	Delete a component from the editor
Name	Ctrl-Shift-N	Assign a name to a component, net, or bus
Value	Ctrl-Shift-V	Assign a value to a component, net, or bus
Net	Alt-N	Draw an electrical conductor
Smash	Ctrl-Shift-S	Detach a component's name and value
Undo	Ctrl-Z	Cancel preceding action
Redo	Ctrl-Y	Perform preceding action again
Zoom In	F3	Zoom in
Zoom Out	F4	Zoom out
Zoom to Fit	Alt-F2	Zoom to fit circuit in window
Find	Ctrl-F	Search for a component by name

4.1.2 Configuring the Grid

EAGLE editors rely on a grid to set the location of components. If a component is placed according to the grid and origin is located at coordinates (x, y), x and y will be multiples of the grid interval. By default (in America, at least), the grid interval is set to 0.1 inches.

Understanding grid configuration is very important, so I recommend that you open the Grid dialog by clicking the Grid item in the upper-left part of the editor. Figure 4.3 shows what the default grid settings look like on my system.

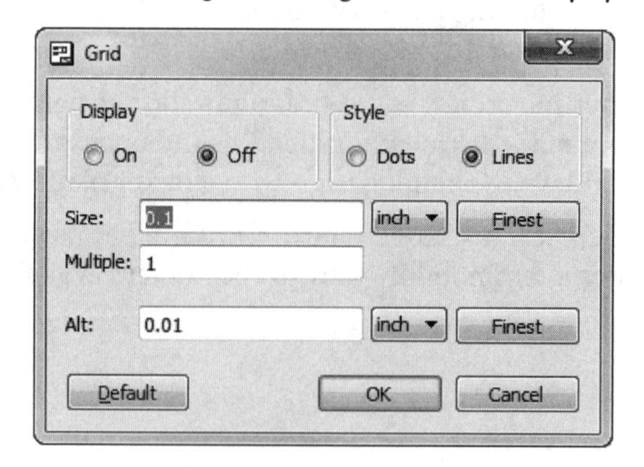

Figure 4.3: The Grid Dialog

Starting at the top, the Display radio buttons set the grid's visibility and the Style buttons set whether it should be displayed using dots or intersecting lines. I prefer to make the grid visible (dots) in my designs, but you may keep or change these settings as you see fit.

Do not change the Size value, which configures the interval between grid points. The default value, 0.1", is the agreed-upon distance between pins in the schematic editor, and it's important to leave this unchanged. If a pin is placed at a location that isn't a multiple of 0.1, such as (0.15, 0.72), you won't be able to connect a net to it.

To the right of the Size box is a combo box that allows you to set the units of the grid distance. This can be set to inch, mm, mil (thousandth of an inch), and mic (thousandth of a millimeter). Leave this set to inch.

Below the Size box is a box that allows you to set a multiple of the grid interval. This changes how the grid looks in the editor but doesn't change where you can place components. For example, if you set the multiple to 3, the grid lines/dots will be spaced 3x apart, but you can still place components at regular positions.

Beneath Multiple, the Alt box makes it possible to configure settings of the editor's alternative grid. The alternative grid can't be displayed, but it can be used to place items in the editor at a finer scale. In Figure 4.3, the alternative grid interval is set to 0.01 inches, which means that if the alternative grid is used, components can be placed at coordinates like (8.89, 6.74) instead of (8.8, 6.7).

To use the alternative grid, press the Alt key while moving an item in the editor. This makes it convenient to place nonessential items like labels and measurements but should not be used to position electrical components. The need for regular component placement will become clear when you start inserting devices into your design.

4.2 The Reset Switch

In the world of software development, complex programs are divided into subroutines to make them more manageable. In circuit design, complex circuits are divided into subcircuits. The Femtoduino consists of four subcircuits, and the simplest of them contains the reset switch. This subcircuit contains the following components:

- A single-pole, single-through (SPST) switch
- A 10k resistor
- A 0.1uF capacitor (non-polarized)
- A connection to +5V and a connection to GND

Figure 4.4 presents the schematic for the reset switch subcircuit.

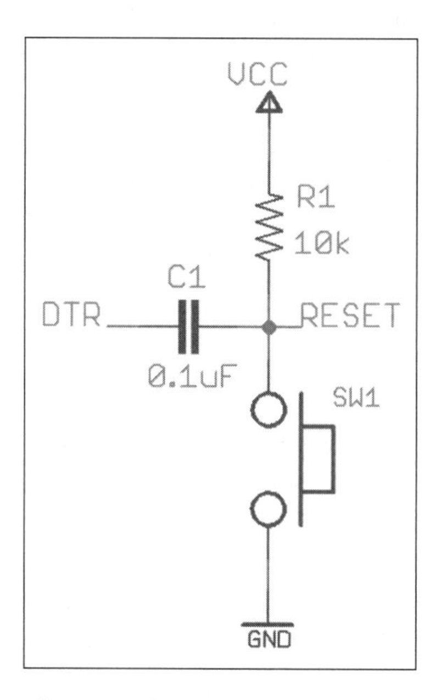

Figure 4.4: The Reset Switch Subcircuit

With the schematic editor open, you can create this circuit with the following steps:

1. Open the Add dialog by pressing Ctrl-Shift-A or by clicking the toolbar entry. Scroll and find the eagle-book library. Open this library and double-click the VCC entry.

When you select a device from a library and return to the editor, the cursor will take the shape of the selected device. Before you place the device in the editor, I recommend that you practice rotating it by right-clicking.

2. Move the mouse toward the upper-left part of the editor and left-click. This places the VCC supply in the schematic.
3. EAGLE assumes you want to add more copies of the device, but you need only one. Open the Add dialog again by pressing Ctrl-Shift-A or by clicking the toolbar entry. Scroll to eagle-book, open the RES entry, and add a RES_0603 to the design. This will be referred to as R1.
4. Rotate R1 until the >VALUE label is on the left. Place it beneath the VCC supply.
5. Repeat the insertion process and add three components to the schematic: a SWITCH beneath the resistor, a GND beneath the SWITCH, and a CAP_0603 (under the CAP entry) to the left of the SWITCH. Rotate this capacitor, referred to as C1, to make it horizontal.

Before proceeding further, I'd like to briefly review EAGLE's terminology as it applies to connections in the schematic editor:

- **Pin**—The point on a symbol at which connections are made.
- **Net**—A line representing a single electrical connection. Nets connect to symbols at the symbols' pins.
- **Bus**—A line representing multiple electrical connections. Buses connect to nets with specifically named pins.
- **Wire**—A graphical line that does not represent an electrical connection.

Be sure to understand the differences between these terms. Personally, I've always found it strange that a net defines an electrical connection but a wire doesn't.

6. To create an electrical connection, click the Net entry on the toolbar or press Alt-N. Connect the upper pin of R1 to VCC. EAGLE will draw a solid green line representing the net.
7. With the Net tool still active, connect R1's lower pin to the switch's upper pin. Draw another net from the switch's lower pin to ground.
8. Activate the Move tool by pressing Ctrl-M or by clicking the toolbar entry. Move C1 so that its right pin is vertically positioned between the bottom of R1 and the top of the switch.
9. Activate the Net tool (Alt-N) and draw a net from the capacitor's right pin to the net connecting the resistor and the switch. EAGLE will draw a green dot at the connection of the two nets. This represents a junction.

At this point, the design should look somewhat similar to that shown in Figure 4.5. You don't have to place your components in the same positions, but their pins should be connected in the depicted manner.

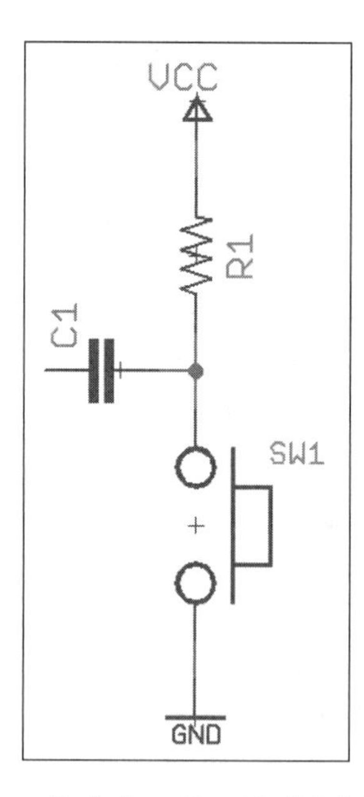

Figure 4.5: Preliminary Reset Switch Subcircuit

Now let's set values for the resistor and capacitor. Remember that these values aren't used by EAGLE during the design process.

10. Activate the Value tool (Ctrl-Shift-V) and click R1. Assign the value to 10k.
11. With the Value tool still active, click C1 and assign the value to 0.1uF.

We can improve the circuit's appearance by repositioning the name and value of the resistor and capacitor. This is made possible by the Smash tool, which graphically separates a component's name and value from the symbol. This is shown in the following instructions.

12. Activate the Smash tool (Ctrl-Shift-S or the toolbar entry) and click the resistor and the capacitor. Notice that the components' names and values now have their own origins.
13. Using the Move and Rotate tools, position the names and values to make the schematic more readable. As you place the text, remember that pressing Alt will use the alternate grid instead of the primary grid.

The last step involves signals. If you activate the Info tool (Ctrl-I) and click a net, a Properties dialog will appear. Beneath the Net heading, you'll see two boxes: one labeled Name and one labeled Net Class. I'll explain net classes toward the end of this chapter. Right now, I want to discuss net names.

EAGLE assigns a name to every net in the editor, and by default, each name is unique. But designers can set a custom name for a net by using the Name tool (Ctrl-Shift-N). If two nets have the same name, EAGLE presumes they're electrically connected, even if they're not connected graphically. In EAGLE terminology, multiple nets with the same name are said to carry the same *signal.*

Using signals, designers can create subcircuits that are electrically connected to one another without being connected by lines in the schematic. This not only makes the schematic easier to read, but also it makes it possible for different designers to work on portions of the same circuit without knowing the positions of other subcircuits.

In addition to setting a net's name, a designer can also set a net's label. This displays the net's name on the schematic and has no other significance.

To connect the reset switch subcircuit to other subcircuits, two named nets must be added: RESET and DTR. The following directions show how these nets can be named and associated with labels.

14. Activate the Net tool (Alt-N), click on C1's left pin, and double-click a point two or three spaces to its left. This creates a short net.
15. Activate the Name tool (Ctrl-Shift-N), click the net, and set its name to DTR. Activate the Label tool, click the net, and position the label to its left.
16. Activate the Net tool and click the junction (the green dot) joining the switch, resistor, and capacitor. Double-click the point two or three grid spaces to the right. This creates another short net.
17. Activate the Name tool, click the net, and set its name to RESET. Activate the Label tool, click the net, and position the label to its right.

At this point, the subcircuit should look similar to that shown in Figure 4.4. Again, precise positions aren't important but the components' pins should be connected in the given manner. Also, the named nets (DTR and RESET) must have the same names and locations as those in the figure.

NOTE In addition to having correct names for nets, it's crucial to have the correct names for components, such as R1 and C2. Check to make sure your components have the same names than those shown in the figures. Otherwise, the board design process in the next chapter will present errors.

4.3 Voltage Regulation

The microcontroller's supply voltage (VCC) must be set between 1.8 and 5.5 volts. In the Femtoduino, VCC is provided by a voltage regulator that accepts the input voltage, VIN, and regulates its output to 5 volts. The regulator is an MIC5205 low-noise, low dropout regulator from Micrel. Figure 4.6 shows what its subcircuit looks like.

NOTE The Femtoduino documentation also accepts a VCC of 3.3V, but the resonator's frequency must be lowered from 16 Mhz to 8 Mhz.

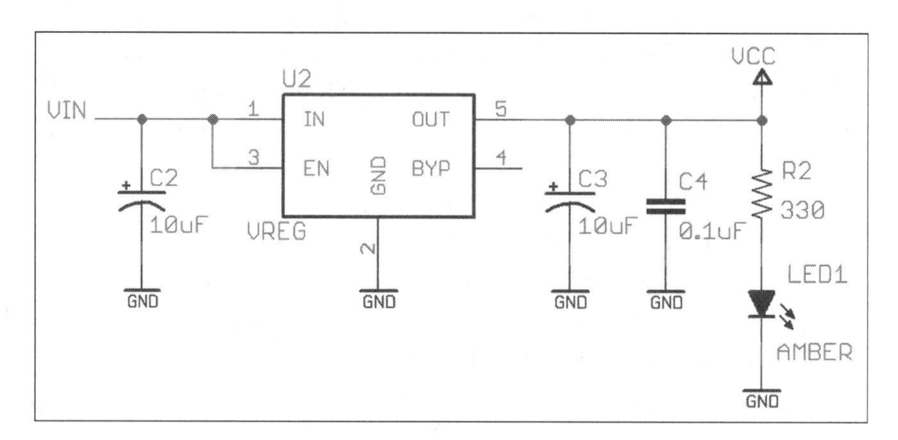

Figure 4.6: The Voltage Regulation Subcircuit

The following steps explain how to draw the regulator subcircuit in the schematic editor. The directions proceed from left to right.

1. In EAGLE, open the schematic file (*.sch) containing the subcircuit from the preceding section. Activate the Add tool and open the eagle-book library.
2. Insert the following components into the design: one MIC5205, five GNDs, one VCC, one RES_0603 (R2), one LED_0805 (LED1), two CAP_TANTs (C2 and C3), and one CAP_0603 (C4).
3. Place the MIC5205 to the right of the reset switch subcircuit, leaving plenty of space. Draw a net extending left from the MIC5205's IN pin. Assign the signal VIN to the net and place a label to its left.
4. Place C2 beneath the VIN net and a GND beneath C2. Connect C2's upper pin to the VIN net and its lower pin to GND.
5. Connect the MIC5205's EN pin to the VIN net. Move a GND beneath the MIC5205 and connect it to the MIC5205's GND pin.
6. Position C3 and C4 to the right of MIC5205 in parallel. Draw a net extending right from the MIC5205's OUT pin. Connect this net to the upper pins of C3 and C4.
7. Place two GND components beneath C3 and C4. Connect the GNDs to the lower pins of C3 and C4.
8. To the right of C4, place VCC, R2, LED1, and a GND in series. Draw nets to connect them in series. Make sure the net connecting VCC and R2 forms a junction with the net extending from the MIC5205's OUT pin.

9. Assign the following values to the components: 10uF to C2 and C3, 0.1uF to C4, 330 to R2, and AMBER to LED1.
10. Use the Smash tool to position the components' names/values as you see fit.

C2 and C3 are polarized, so make sure their positive terminals (the straight lines) are always connected to a higher voltage than their negative terminals (the curved lines).

4.4 The ATmega328P Microcontroller

Every Arduino circuit, including the Femtoduino, relies on an Atmel microcontroller to serve as its brain. Microcontrollers are similar to microprocessors, but in addition to having a processor, a microcontroller contains its own RAM, ROM, and I/O circuitry, including analog-to-digital converters (ADCs). This makes circuits with microcontrollers easier to design and fabricate than circuits with microprocessors.

Atmel is one of the leading manufacturers of microcontrollers, and the ATmega328p is one of their most powerful 8-bit microcontrollers. Here is a subset of its features:

- 32kB of flash RAM, 2kB of SRAM, and 1kB of EEPROM
- 20MHz CPU
- 8 analog-to-digital (ADC) channels
- 23 I/O pins, 32 pins total

One important advantage of using Atmel microcontrollers is the availability of free software. Atmel provides the Atmel Studio for free, and it can be downloaded from http://www.atmel.com/tools/atmelstudio.aspx.

4.4.1 Pins of the ATmega328P

The names of the ATmega328p's 32 pins can be confusing to deal with. First of all, the 23 I/O pins are divided into three ports:

- **Port B**—Contains eight pins, PB0–PB7
- **Port C**—Contains seven pins, PC0–PC6
- **Port D**—Contains eight pins, PD0–PD7

Most of the device's pins serve multiple roles, and with each role, each pin takes a different name. For example, Pin PB2 may take three different names depending on how Port B is configured:

- **SS**—If configured as the SPI (Serial Peripheral Interface) Bus Master Slave Select
- **OC1B**—If configured as the Timer/Counter1 Output Compare Match B Output
- **PCINT2**—If configured as Pin Change Interrupt 2

Thankfully, you don't have to understand these roles or even how to configure the ports—the Arduino framework handles these details for you. But you should understand that the pin names used in this schematic may be different than those used in other schematics containing the ATmega328p.

Table 4.2 lists each of the microcontroller's 32 pins. Each I/O pin is identified by its port, position, and role, which is given in parentheses. For example, the fifth pin in Port B is named PB4 and it serves the role of Master In/Slave Out, so its designation in the table is PB4 (MISO).

Table 4.2

ATmega328p Pins

Name	Purpose
PB0 (ICP)	Timer/Counter1 Input Capture Input
PB1 (OC1A)	Timer/Counter1 Output Compare Match A Output
PB2 (SS/OC1B)	Timer/Counter1 Output Compare Match B Output
PB3 (MOSI)	Master Out/Slave In pin for SPI communication
PB4 (MISO)	Master In/Slave Out pin for SPI communication
PB5 (SCK)	Serial clock used for SPI communication
PB6 (XTAL1)	Connects to ceramic resonator to provide timing
PB7 (XTAL2)	Connects to ceramic resonator to provide timing
PC0 (ADC0) - PC5 (ADC5)	Six pins of the analog-to-digital converter
PC6 (RESET)	Used to reset the device
ADC6	Seventh pin of the analog-to-digital converter
ADC7	Eighth pin of the analog-to-digital converter
PD0 (RXD)	USART input pin
PD1 (TXD)	USART output pin
PD2 (INT0)	External Interrupt 0 Input
PD3 (INT1)	External Interrupt 1 Input
PD4 (XCK/T0)	Timer/Counter 0 External Counter Input
PD5 (T1)	Timer/Counter 1 External Counter Input
PD6 (AIN0)	Analog Comparator Positive Input
PD7 (AIN1)	Analog Comparator Negative Input

AREF	Analog reference voltage for the analog-to-digital converter
AVCC	Supply voltage for the analog-to-digital converter
VCC, VCC	Supply voltages for the digital circuit
GND, GND, GND, GND	Ground signals

These pin names are based on Atmel's naming convention. The Arduino framework assigns its own names to the microcontroller's pins. For this schematic, the ATmega328p's pins will take Atmel's names but the nets connecting the pins will take the names given by Arduino.

An example will help make this clear. The pin labeled PC6(RESET) will be connected to a net whose signal is named RESET. This signal connects the ATmega328p to the reset subcircuit designed earlier.

4.4.2 Drawing the Schematic—ATmega328p

In the Femtoduino, most of the microcontroller's pins are connected to headers that can be plugged into a breadboard. A handful of pins are connected to the the reset circuit, the Serial Peripheral Interface (SPI), the power supply, and ground. Figure 4.7 depicts the subcircuit containing the ATmega328p.

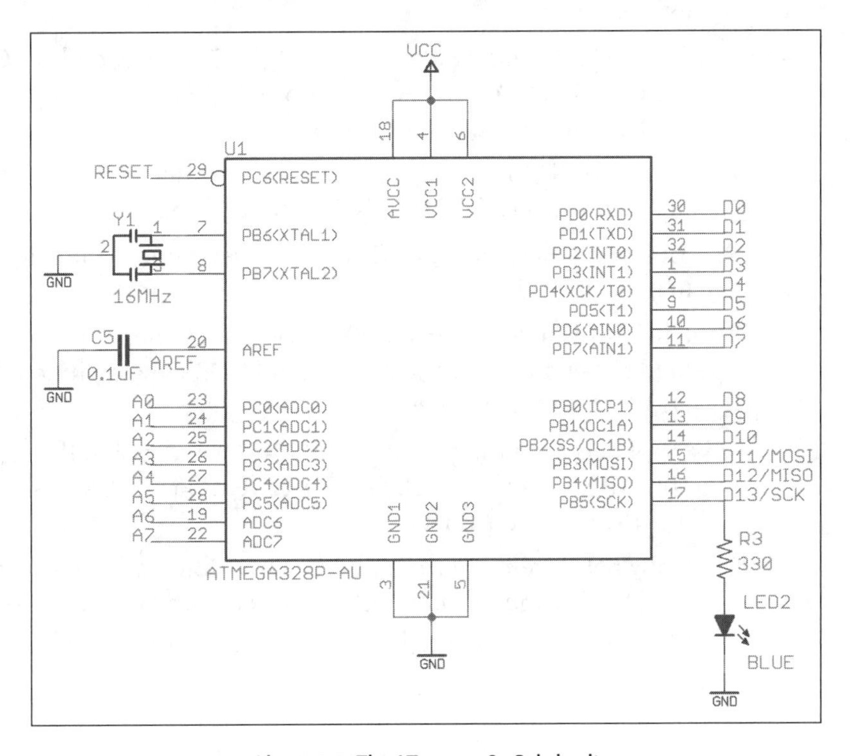

Figure 4.7: The ATmega328p Subcircuit

The following instructions explain how to design this subcircuit in EAGLE. They start from the device's upper-left pins and proceed counterclockwise.

1. In EAGLE, open the schematic file (*.sch) containing the subcircuits from the preceding sections. Activate the Add tool and open the eagle-book library.
2. Select the ATMEGA328P and place it to the right of the voltage regulation subcircuit, leaving plenty of surrounding space.
3. Using the Add tool, insert the following components into the design: one VCC, four GNDs, one LED_0805 (LED2), one RES_0603 (R3), one CAP_0603 (C5), and one CSTCE (Y1).
4. Draw a net outward from the RESET pin and assign it to the RESET signal. This connects the microcontroller to the reset switch presented earlier. Create a label for the signal and place it left of the net.
5. Move Y1 and a GND to the left of Pins PB6(XTAL1) and PB7(XTAL2). Connect Pin 1 of Y1 to PB6(XTAL1) and connect Pin 3 of Y1 to PB7(XTAL2). Connect Pin 2 of Y1 to the GND.
6. Move C5 and a GND to the left of the AREF pin. Rotate C5 to make it horizontal. Draw two nets: one from AREF to C5's right pin and one from C5's left pin to ground. Assign the AREF signal to the net connecting AREF to C5. Create a label for the signal.
7. Draw nets extending from pins PC0-PC5, ADC6, and ADC7, and assign the signals A0, A1, A2, A3, A4, A5, A6, and A7, respectively. For each signal, create a label and move it to the left of the net.
8. Move a GND to the bottom center of the microcontroller. Connect it to the microcontroller's GND1, GND2, and GND3 pins.
9. Move R3, LED2, and a GND to the lower-right of the microcontroller. Rotate R3 to make it vertical.
10. Draw three nets: one from PB5(SCK) to R3's upper pin, one from R3's lower pin to LED2's upper pin, and a net from LED2's lower pin to GND. Assign the signal D13/SCK to the first net.
11. Draw nets extending from pins PB0-PB4 and assign the signals D8, D9, D10, D11/MOSI, and D12/MISO, respectively. For each signal, create a label and move it to the right of the net.
12. Draw nets extending from pins PD0-PD7 and assign the signals D0, D1, D2, D3, D4, D5, D6, and D7. For each signal, create a label and move it to the right of the net.

13. Move the VCC to the top center of the microcontroller. Connect this to the microcontroller's AVCC, VCC1, and VCC2 pins.
14. Assign the following values to the circuit components: 16MHz to Y1, 0.1uF to C5, 330 to R3, and BLUE to LED2. Use the Smash tool to reposition the names and values.

4.5 Header Connections

The Femtoduino board can communicate with external circuits in one of two ways:

- **Breadboard header**—Allows the Femtoduino to be plugged into a breadboard
- **ICSP header**—Allows the Femtoduino to receive programming through the six-pin AVR protocol

This section discusses both headers and explains how to design the respective subcircuits in the schematic editor.

4.5.1 Breadboard Headers

One of the advantages of the Femtoduino (and Arduino circuits in general) is that it can be plugged into a breadboard. To make this possible, the Arduino must provide pins spaced 0.1 inches apart in single columns.

The circuit elements that provide these pins are called *headers*. The Femtoduino has two headers with 14 pins each. Unlike the other components we've encountered so far, these are connected to the circuit board using through-hole soldering, which means holes must be drilled in the board.

The following instructions show how to design the subcircuit for these headers:

1. In EAGLE, open the schematic file (*.sch) containing the subcircuits from the preceding sections. Use the Add operation and open the eagle-book library.
2. Insert two HEADER1X14 symbols into the schematic.
3. For the first header, JP1, extend a net from each pin and assign the name shown on the left side of Figure 4.8.

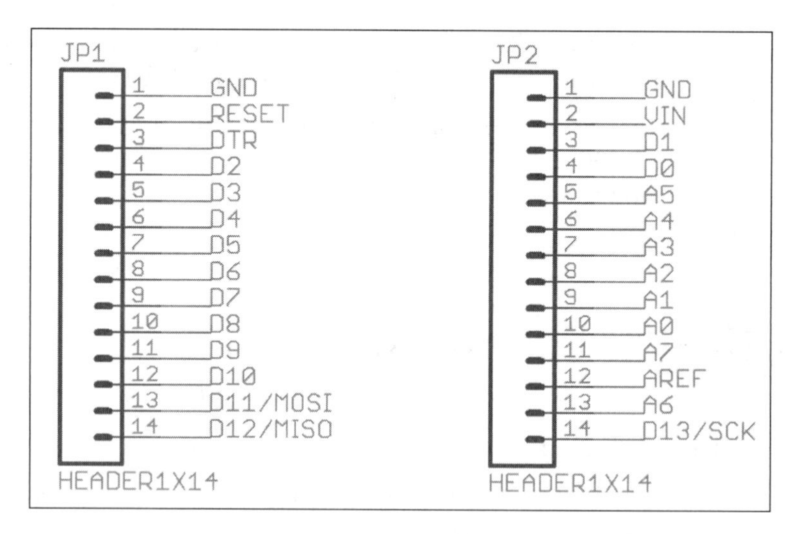

Figure 4.8: The Breadboard Headers

4. For the second header, JP2, draw nets from each pin and assign the names shown on the right side of Figure 4.8.

It's important to understand why these pins are named the way they are. Pins D0-D13 correspond to digital data and A0–A7 correspond to analog data. These designations are used by all Arduino boards.

4.5.2 AVR's In-Circuit Serial Programming (ICSP) Header

The Femtoduino is programmed through the In-Circuit Serial Programming (ICSP) header on the top side of the board. The data transfer protocol is the Serial Peripheral Interface (SPI), which provides synchronous, full-duplex communication between a master and slave. In this case, the programmer is the master and the Femtoduino is the slave.

SPI communication relies mainly on three signals. The master initiates communication by activating the clock (SCK) signal. It transfers data one bit at at time through the Master-Output Slave-Input (MOSI) signal. When data is required from the slave, the slave transmits data on the Master-Input Slave-Output (MISO) pin.

In the schematic, the D11 pin carries the MOSI signal, D12 carries the MISO signal, and D13 carries the SCK signal. This is shown in Figure 4.9, which depicts the schematic containing the ICSP header.

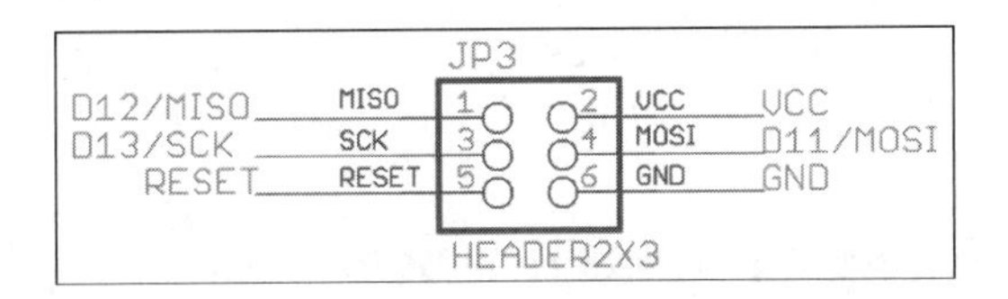

Figure 4.9: The ICSP Header

To design this subcircuit in EAGLE, simply add a HEADER2X3 to the design and create the nets as shown in Figure 4.9. When you finish, the schematic will be complete and you'll be ready to start the board design.

4.6 Net Classes

As the board design continues, the nets in the schematic will be converted into copper traces. By default, these connections all have the same width. To ensure dense routing, this width is usually set to be as small as possible (~5–8 mil). However, tracks that carry significant current need greater width. The relationship between current and a track's minimum width is beyond the scope of this book, but you can find more information in the design standards released by IPC.

To tell EAGLE that some nets have different properties than others, designers create net classes. This is accomplished by going to Edit > Net classes... in the schematic editor. Figure 4.10 shows what the resulting dialog looks like.

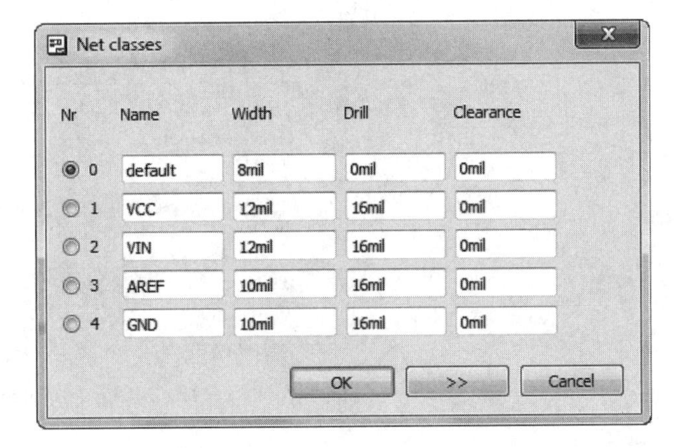

Figure 4.10: Net Classes of the Femtoduino

As shown, the default net class has a width of 8 mil. The VCC, VIN, AREF, and GND nets carry more current than other nets, so their net classes have greater widths. The following instructions explain how to set these net classes for the Femtoduino circuit.

1. Open the Net classes dialog by going to Edit > Net classes... on the main menu. Set the width of the default class to 8 mil, the drill to 0 mil, and the clearance to 0 mil.
2. For net class 1, set the name to VCC, the width to 12 mil, the drill to 16 mil, and the clearance to 0 mil.
3. For net class 2, set the name to VIN, the width to 12 mil, the drill to 16 mil, and the clearance to 0 mil.
4. For net class 3, set the name to AREF, the width to 10 mil, the drill to 16 mil, and the clearance to 0 mil.
5. For net class 4, set the name to GND, the width to 10 mil, the drill to 16 mil, and the clearance to 0 mil.
6. Click OK to close the Net Classes dialog.

4.7 Electrical Rule Check

Before a board design can be generated from a schematic, an electrical rule check (ERC) should be performed. The process is simple. In the schematic editor, click the ERC tool item, which can be found in the bottom-left corner of the toolbar. When you click this item, EAGLE will analyze the design. If the analysis produces errors or warnings, they display in the ERC Errors dialog. Figure 4.11 shows what this looks like.

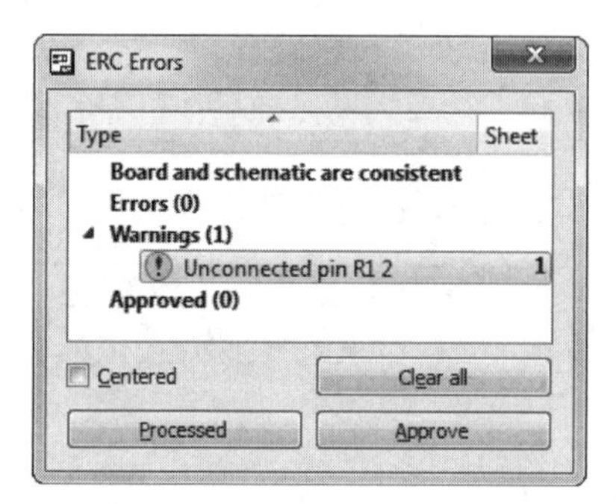

Figure 4.11: The ERC Errors Dialog

If you double-click an entry in this dialog, the editor will focus on the part of the circuit causing the error or warning. Some errors and warnings can be tolerated, but most can't. If an error or warning is tolerable, you can press the Accept button in the dialog to place it in the list of accepted errors and warnings.

Every pin can take one of nine directions (in, out, io, nc, oc, pwr, pas, hiz, and sup). Many warnings and errors are the result of a mismatch between a pin's direction and its usage in the circuit. This section presents the errors and warnings I've encountered and their underlying causes.

NOTE The electrical rule check also tests the consistency between the schematic design and the board design from which it was generated. If you haven't generated a board design, the consistency check won't be made.

4.7.1 ERC Warnings

Here are a number of warnings I've encountered in my EAGLE schematics:

- **<direction> pin <name> connected to...**—The pin's direction conflicts with its usage in the circuit.
- **Only one pin on net...**—A net has only one connection. This may mean that a net isn't fully connected or that a name was misspelled in the circuit.
- **Net __ overlaps pin...**—A net is drawn over a pin connection. This may result from a net that is too long, or it may mean the pin is oriented in the opposite direction.
- **Unconnected pin...**—A component's pin has no net attached to it. Particularly common with passive elements, such as resistors and capacitors.
- **Close but unconnected wires...**—Nets or buses intersect in the circuit without a junction (green dot). Therefore, there is no electrical connection.
- **Part __ has no value...**—A component can take a user-defined value, such as a resistance or capacitance. But the designer hasn't given it a value in the circuit.

4.7.2 ERC Errors

If EAGLE is certain that an aspect of the circuit represents an error, the ERC will present an error condition in the dialog. I can't present every possible error, but I'll list the ones I've encountered in order of decreasing frequency:

- **Unconnected INPUT/Unconnected OUTPUT**—EAGLE feels that every pin capable of transmitting or receiving a signal should be connected. This isn't the case in many circuits, so this error can usually be disregarded. However, it's always a good idea to make sure that the unconnected pins are meant to be unconnected.
- **SUPPLY pin __ overwritten with more than one signal...**—EAGLE assigns its own names to nets for pins connected to supplies and ground terminals. To fix this, remove the name of the corresponding net or change the property of the corresponding pin.

- **OUTPUT and SUPPLY pins mixed on net __...**—A pin whose direction equals out has been connected to a supply or a ground. To fix this, change the pin's direction or change its connection. Similar errors can be fixed in the same way.

4.8 Generating the Board Design

Now that you've drawn the schematic and performed an electrical rule check, you're ready to create the board design. This replaces the symbols of the circuit's devices with their physical packages.

To generate the board design, click the Generate/Switch to Board item, which is located fifth from the left in the horizontal toolbar. Alternatively, you can select File > Switch to Board in the main menu.

If this is the first time you've created a board for the schematic, EAGLE displays a dialog asking if this is your intention. Click Yes, and the board editor appears containing the connected device packages for the circuit. The next chapter explains how to position these packages in the Arduino Femtoduino.

4.9 Framing the Schematic

After the components are connected and the errors are checked, the last step involves drawing a frame around the schematic's perimeter. This is completely optional and I've seen plenty of EAGLE schematics without frames. But they provide three main advantages:

1. They make it possible to identify the designer and date of the finished design.
2. They provide a coordinate system for identifying sections of the schematic.
3. They serve as esthetically pleasing borders for the circuit design.

To create a frame, go to Draw > Frame on the main menu. When this tool is active, two options appear in the horizontal toolbar: one that sets the number of rows and one that sets the number of columns.

To draw the frame, click in the upper-left corner of the schematic. The frame's dimensions will change as you move the mouse. When you click again, the frame will bound the selected region. The horizontal border is labeled with numbers and the vertical border uses letters. Figure 4.12 shows what a frame looks like with six columns and two rows.

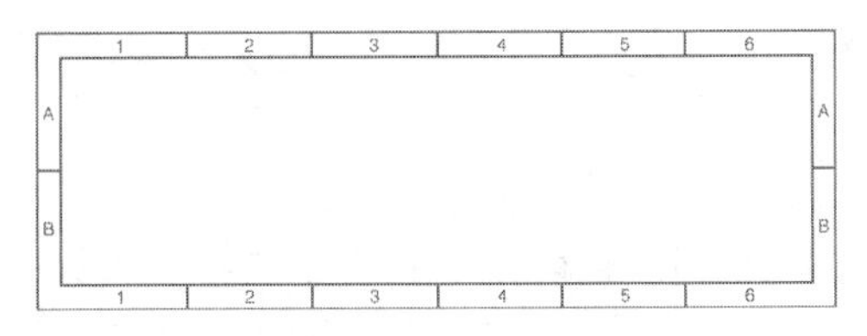

Figure 4.12: A 6x2 Frame

In addition to the frame, I like to put a title block in the lower-right corner. This provides four pieces of information:

- The schematic's title
- The designer's name
- The document number and revision number
- The date

To add the title block to the schematic, activate the Add tool and look through the frames library. Select the DOCFIELD entry and click in the editor as though adding a regular symbol.

4.10 Attributes and Assembly Variants

Before leaving the topic of schematic design, there are two last points I want to mention. First, the schematic editor makes it possible to associate name/value pairs with the design and each of its components. You can also modify aspects of the design's components so that they can be assembled by different fabrication facilities.

4.10.1 Global Variables

A global variable consists of a name-value pair similar to the values used for components, such as C = 10uF. Global variables are helpful when you need to store information such as a design's version/revision number or the original source of the schematic design.

Global variables can be set by going to the Edit > Global attributes... entry in the editor's main menu. This opens the Global Attributes dialog shown in Figure 4.13.

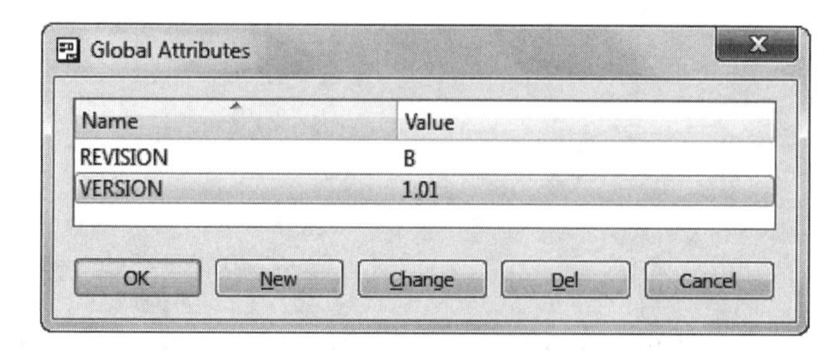

Figure 4.13: The Global Attributes Dialog

Pressing the New button opens a dialog that lets you enter a name and value to be associated with the design. This information won't be displayed in the schematic, but it will be accessible through User Language programs. Chapters 11 through 13 explain the User Language in detail.

4.10.2 Component-Specific Attributes

Now that you've seen how to set global variables, you may wonder how to associate data with the individual parts. This can be very useful, particularly if you want to create a proper Bill of Materials. The following three steps show how this can be done.

1. Enter `attribute` in the text box just above the editor area.
2. Select a component in the schematic.
3. In the Attributes box (which closely resembles Figure 4.13), enter a name/value pair to be associated with the selected part.

It's important to note that these attributes will be associated only with the specific parts in the schematic. That is, if you create a new schematic and add the part, the attribute won't be present. To make an attribute permanent, you need to modify the library containing the component.

The attribute command is part of EAGLE's command language. Chapter 10, "Editor Commands," discusses this command and many more.

4.10.3 Assembly Variants

If you're an individual designer, you're probably going to design a single schematic and use it to create a single board design. But if you're a large company, you may want to reuse the same circuit in multiple products, but with small variations. The need for new variations may be caused by differing voltage requirements, I/O compatibility, or updated component technology, such as moving from surface-mount to BGA.

Rather than force you to redesign the circuit for each variation, EAGLE makes it possible to create assembly variants. Here, an assembly refers to a specific combination of connected components. When you create an assembly variant, you're essentially copying an existing schematic, but in such a way as to easily modify properties of the original.

To create a new variant, go to Edit > Assembly variants... on the editor's main menu. The Assembly Variants dialog will appear and Figure 4.14 shows what it looks like.

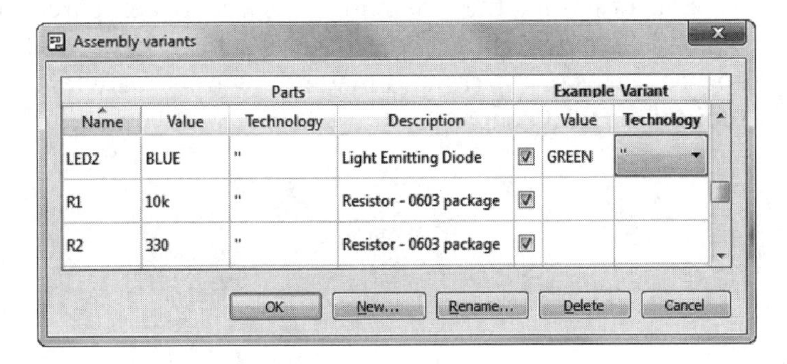

Figure 4.14: The Assembly Variants Dialog

The rightmost column in the dialog identifies a new variant called Example_Variant. By default, this contains all the components of the original schematic. The dialog provides three ways to customize the variant:

- **Populate**—If the checkbox for a part is checked, the part will be included in the schematic. If not, the part won't be present.
- **Value**—Changes the part's value for the variant. For example, the original capacitor may have a value of 10uF. The variant's value could be set to 47uF.
- **Technology**—Changes the technology of the part's package. For example, the original resistor may be surface-mount and the variant contains a BGA part.

When a new variant is created, it can be viewed and edited in the editor. To switch to a variant, find the combo box in the horizontal toolbar. (It will be initially blank because the default variant has no name.) When you select a different variant, it will be displayed in the editor. Any component not present in the variant will be displayed with a red X.

4.11 Conclusion

Drawing the schematic is my favorite part of the PCB design process. Most of the symbols are just rectangles, and it doesn't matter where you place them. Drawing nets is easy, and it doesn't matter where they're positioned so long as they connect the right pins. If the design has been properly partitioned into subcircuits, the full circuit can be grasped quickly.

But I'll admit that schematic design can be tedious at times. Creating nets and labels for all the header pins can be tiresome. Searching through EAGLE's vast component list gets annoying after a while. Worse still, the components' names and values are never where they should be, so I constantly use Smash, Move, and Rotate to get them into position.

Still, designing the schematic is a pleasure compared to designing a physical board. When designing a board, every detail takes on extraordinary significance. The consequences of an error, no matter how minor, are far more harrowing. This is the topic of the next chapter, so turn the page—if you dare.

Chapter 5

Layout and Design Rules

The preceding chapter explained how to insert symbols into a schematic editor and connect their pins. This chapter presents the next step: laying out the actual circuit board with the devices' physical representations, called packages. The first part of this chapter explains how to lay out packages in the Femtoduino design.

The second part of this chapter discusses the topic of design rules. Just as schematic designs can be validated with Electrical Rule Checks (ERCs), board designs can be validated with Design Rule Checks (DRCs). The main difference is that DRCs are much more involved and the design rules are set by the designer.

This chapter focuses on EAGLE's board editor, but before you start moving packages, it's important to understand how layers work. I'll cover this topic first.

5.1 Layers

Chapter 4, "Designing the Femtoduino Schematic," briefly explained how to access layers in the schematic editor, but the layers used by the board editor are markedly different. This is because, in addition to setting the visibility of graphical elements, the layers determine where circuit elements are placed on the physical board.

For example, when you position a device on the board, it's crucial to specify whether it should be on the top or bottom. Similarly, when you route a connection, you need to specify whether it should be on the top, bottom, or any of the routing layers in between.

The board editor provides two important commands for dealing with layers:

- **Layer settings...**—Opens the Display dialog, which controls how layers are displayed in the editor
- **Mirror**—Flips a package from top to bottom or vice-versa

5.1.1 The Display Command

If you've used Adobe Photoshop, the GNU Image Manipulation Program (GIMP), or a similar graphical tool, you're probably familiar with layers. When a group of graphical elements are placed in the same layer, the entire group can be made visible or invisible. In the board editor, layer visibility is controlled through the Display dialog, which is opened through the Layer settings... item on the toolbar. Figure 5.1 shows what this looks like.

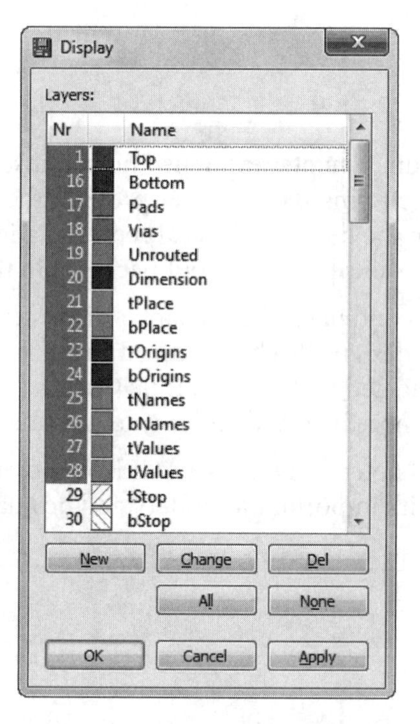

Figure 5.1: The Display Dialog

Unlike the layers in Photoshop or GIMP, EAGLE's layers have predefined purposes. As shown in the figure, each layer has a name and number, and each contains information related to a specific aspect of the circuit board.

Table 5.1 lists most of the layers available in the board editor. Many of them come in pairs: Layers whose name start with "t" denote the top of the board and layers whose name start with "b" refer to the bottom.

Table 5.1

Important Layers in the Board Editor

Number	Name	Purpose
1	Top	Contains connections on top layer
2-15	Inner Layers	Contains traces for layers between top and bottom
16	Bottom	Contains connections on bottom layer
17	Pads	Through-hole pads
18	Vias	Through-hole vias
19	Unrouted	Unrouted components (airwires)
20	Dimension	Board outline
21/22	tPlace/bPlace	Contains outlines of devices—printed in silk-screen
23/24	tOrigins/bOrigins	Required to move and rotate components
25/26	tNames/bNames	Contains component names—printed in silk-screen
27/28	tValues/bValues	Contains component values—printed in silk-screen
29/30	tStop/bStop	Stops application of soldermask (used with vias)
31/32	tCream/bCream	Specify where cutouts should be made for solder paste
33/34	tFinish/bFinish	Masks for finishing material (e.g. gold contacts)
35/36	tGlue/bGlue	Glue mask
37/38	tTest/bTest	Provides additional information
39/40	tKeepout/bKeepout	Restricted regions for components
41/42/43	tRestrict/bRestrict/vRestrict	Restricted regions for copper
44	Drills	Through-holes (conducting)
45	Holes	Through-holes (not conducting)
46	Milling	Draws contours for the milling machine
47	Measures	Dimension labels
48	Document	Documentation for the board—printed
49	Reference	Reference marks used for alignment
51/52	tDocu/bDocu	Documentation for the board—not printed

The most important layers are 1 through 16, which represent board layers that may contain copper. Layer 1 (the top layer) and Layer 16 (the bottom) may also have components associated with them. A double-sided board like the Femtoduino doesn't have internal layers, so the Femtoduino circuit has no use for Layers 2–15.

Most of the other layers are straightforward, but Layers 25, 26, 27, 28, 51, and 52 merit attention. These contain descriptive text for the board. In the last chapter, I explained the difference between a symbol's name (for example, R1, R2) and its value (for example, 30k). In EAGLE, Layers 25 (tNames) and 26 (bNames) contain the names of the board's packages and Layers 27 (tValues) and 28 (bValues) contain their values. Any text associated with these layers will be printed in silk-screen on the actual board.

But if text is associated with Layer 51 (tDocu) or Layer 52 (bDocu), the text won't be printed in silk-screen. Therefore, if there's any text that should be printed in the board design but not on the board, it should be associated with Layer 51 or 52.

EAGLE allows for additional layers whose numbers go up to the 200s, but most designs won't even use the majority of the layers in Table 5.1. I'll describe specific layers further as they become necessary in laying out the Femtoduino.

Before proceeding, I recommend that you remove the visibility of Layer 19 (Unrouted). To do this, open the Display dialog, find the row for Layer 19, click the cell in the first column, and press OK. This removes the highlighting for the cell, which means the corresponding layer won't be displayed in the editor.

Layer 19 contains the unrouted wires (airwires) in the diagram. They will play an important role later on, but for the moment, our only concern is the circuit's components.

5.1.2 The Mirror Command

No matter how many layers a PCB has, components can be soldered only to the top layer (Layer 1) or the bottom layer (Layer 16). By default, EAGLE assumes that surface-mount components will always be soldered to the top side. This is why, when the board editor is first opened for a design, all the pads are red.

The Mirror tool can be activated by clicking the entry in the vertical toolbar. When this tool is active, clicking a component flips it from the top layer (Layer 1) to the bottom layer (Layer 16) or the reverse. When a component moves from top to bottom, its shape is mirrored horizontally and EAGLE colors it blue instead of red.

It's important to note that the blue/red coloring applies only to surface-mount components. Pads of through-hole components are colored green because holes affect both the top and bottom layers. If you mirror a through-hole component, its shape will be reversed but the color of its pads won't change.

5.2 Board Layout

In the preceding chapter, Section 4.8 explained how to generate an unrouted board design from the Femtoduino's schematic. Figure 5.2 presents the resulting content in EAGLE's board editor. The circuit components are grouped together on the left and the empty board (reduced size) is displayed on the right. The unrouted wires aren't displayed because Layer 19's visibility has been set to off.

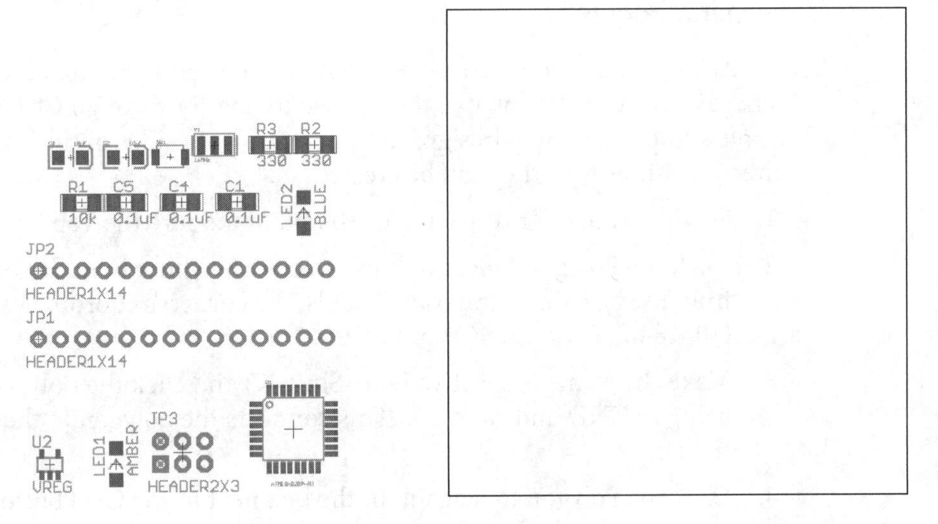

Figure 5.2: The Initial Board Editor

To design the Femtoduino board, the board's perimeter must be set and the components must be positioned inside of it. There is no single correct way to do this, but in this section, I'll explain the method that has worked successfully for me. But first, it's important to configure the properties of the overall board.

5.2.1 Preparing the Board

The following steps explain how to set up the board for the circuit:

1. Set the grid interval to 0.05". To do this, click the Grid item in the toolbar, set the units to inches, and enter 0.05 in the box labeled Size.
2. Using the Move command (Ctrl-M), drag the boundaries of the circuit region until the width equals 1.2" and the height equals 1.85". (You may want to use the Info tool to set the dimensions exactly.) This box serves as the boundary of the Femtoduino circuit board.

5.2.2 Ground Planes and Polygons

By default, all the copper that isn't part of a trace, via, or pad is etched away from a board's surface. But in many cases, designers set aside an area to be filled with copper. This area is called a *copper pour*.

If a copper pour is connected to ground and it takes up an entire layer (or most of the layer), the layer is called a *ground plane*. Ground planes provide electrical advantages involving board capacitance and electromagnetic interference (EMI) reduction, but for the Femtoduino, the primary advantage is that they simplify the routing of ground connections. If you look at the last chapter, you can see that each subdesign contains multiple ground connections, and ground planes make it easy to connect them together.

Forming Polygons

In EAGLE, copper pours are created with the Polygon command, which can be activated through the entry in the vertical toolbar or through Ctrl-Shift-P. The Femtoduino design requires ground planes on both sides of the board. The following directions show how they can be created:

1. Set the primary grid spacing to 10 mil. Make sure the Top layer is visible.
2. Click the Polygon tool and draw a rectangle whose boundaries are 10 mil away from the board's boundaries. That is, the corners's coordinates are (10, 10), (10, 1840), (1190, 1840), and (1190, 10).
3. Make the Name tool active (Ctrl-Shift-N) and click the polygon. Set the polygon's name to GND and click OK. This connects the polygon to the existing GND signal.
4. Click the Polygon tool again. In the box next to the Grid button, make sure the Bottom layer is selected.
5. Draw a polygon on the bottom layer with the same characteristics (dimensions and name) as the polygon on the top layer.

Configuring Polygons

There's no need to change the default characteristics of the polygons in this design, but this can be done by activating the Info tool and selecting the polygon. This brings up the Properties dialog, whose lower half is depicted in Figure 5.3. Six properties are listed as follows:

- **Polygon Pour**—This configures the copper's pattern in the region. If set to Solid, the region's copper will be a solid sheet. If set to Hatch, the copper will be etched in a hatched pattern.

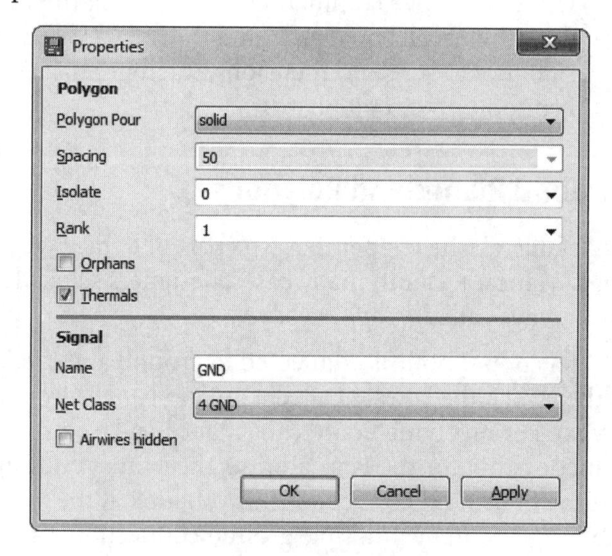

Figure 5.3: Properties of a Polygon

- **Spacing**—Sets the distance between hatch lines if Polygon Pour is set to Hatch.
- **Isolate**—This value defines the separation (in inches) between the polygon and unconnected elements inside of it.
- **Rank**—Determines how EAGLE assigns priority to overlapping polygons. The lower the rank, the higher the priority. If Polygons A and B overlap and A has a lower rank than B, B's area will be subtracted to remove the overlapping region.
- **Orphans**—If checked, unconnected portions of the region (islands) will be filled with copper. If not, the unconnected copper will be etched away.
- **Thermals**—If checked, pads in the polygon will be connected using thermal symbols.

These properties have design trade-offs that must be considered. Solid-patterned polygons are easier to fabricate but hatched polygons provide less capacitance. A low value for Isolate provides more room for routing, but makes the circuit more vulnerable to fabrication defects.

NOTE In Figure 5.3, the value of Isolate is set to 0. This does not mean that polygons come into contact with external elements. It means that the isolation value is set by the board's design rules, which will be discussed shortly.

When a regular pad is connected to a polygon, soldering can be difficult because the heat dissipates quickly through the copper. This is why designers use special pads that connect to the polygon using narrow connections on each side. These pads are called thermals.

Orphans are regions of copper within a polygon's area that aren't connected to a signal. If the Orphans box in the dialog is checked, these copper regions will remain on the board, which means less etchant will be needed during fabrication. If the box is left unchecked, the orphans will be removed, which reduces the potential of bridging caused by fabrication errors.

5.2.3 Placing Devices in the Board Region

When positioning devices on a board, the first step is to determine which should be on top and which should be on the bottom. In the case of the Femtoduino, the vast majority of the devices can be positioned on the board's top surface.

EAGLE provides two main ways to place devices in the board region. When working with a new design, I use the Move and Rotate tools. But if I already know where the devices should be placed, the Info tool simplifies the process. When this tool is active, clicking a device opens the Properties dialog, which is shown in Figure 5.4.

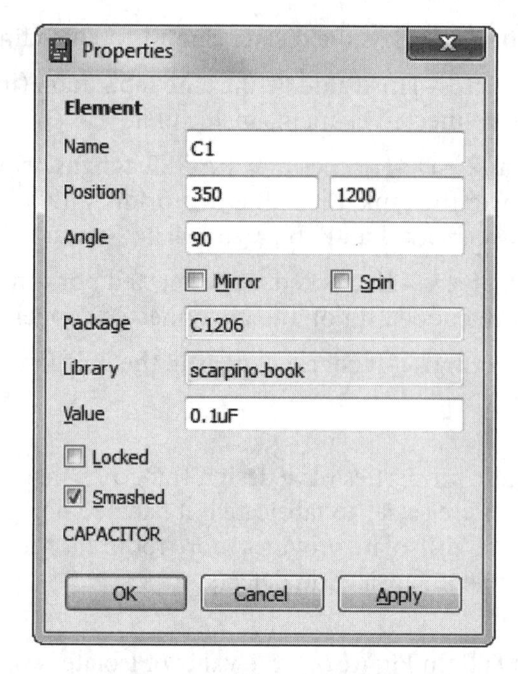

Figure 5.4: Device Properties

To set a package's position, it's easier to enter values in the dialog's Position and Angle boxes than it is to use the Move and Rotate tools. Of course, this assumes that EAGLE has assigned the same initial orientation to your packages as it does to mine. As you move components, take time to check for proper orientation.

Packages on the Top Face

Table 5.2 lists the positions and angles of each package on the Femtoduino's top layer. Note that, for resistors and nonpolarized capacitors, the angular orientation may be misaligned by 180 degrees. Before you begin placement, I recommend that you set the grid spacing to 10 mils and turn off the visibility of the bottom layer (16).

Table 5.2

Positions and Angles of Top-Side Components

Name	Position (X) in mils	Position (Y) in mils	Angle
R1	850	1200	270
R2	750	1500	90
C1	350	1200	90
C2	950	1500	270
C3	250	1500	270
C4	450	1500	90

LED1	750	1750	270
LED2	450	1750	90
U1	600	1100	180
U2	600	700	315
SW1	600	1300	270
G1	300	800	225
JP1	50	1400	270
JP2	1150	1400	270
JP3	600	150	0

After placing the packages in the board region, you may want to reposition their names and values. Just as in the schematic editor, this can be done by using the Smash tool, selecting a package, and then using Move/Rotate on the name/value text.

Packages on the Bottom Face

The Femtoduino has only two devices on its bottom layer: R3 and C5. Before you place them on the board, be sure to tell EAGLE that they're meant to be on the bottom. This can be done by activating the Mirror tool and selecting both devices. The pads of R3 and C5 will turn blue to indicate that they're meant to be soldered to the bottom side.

Before placing the bottom-side devices, I recommend that you turn on visibility of the bottom layer (16). Then, using the Info tool and the Properties dialog, set the positions and angles of the two devices according to the parameters listed in Table 5.3. Due to angular differences in the schematic, these angles may be misaligned by 180 degrees.

Table 5.3

Positions and Angles of Bottom-Side Components

Name	Position (X) in mils	Position (Y) in mils	Angle
R3	850	900	270
C5	1000	900	270

When you finish positioning the components on the bottom side of the board, the layout should look similar to that shown in Figure 5.5.

NOTE The Femtoduino circuit could be instantiated on a single side, but I want to explain aspects of double-sided boards, such as the Mirror tool, vias, and so on.

In this book's example archive at http://www.eagle-book.com, the Ch5 folder contains a file called femtoduino_centroid.csv. This file, called a centroid file, combines Tables 5.1 and 5.2. Many fabrication facilities require centroid files to identify the layers, locations, and orientations of a board's devices.

Now that you've placed the components, you're ready to create connections. But first, you should understand how to set and check design rules.

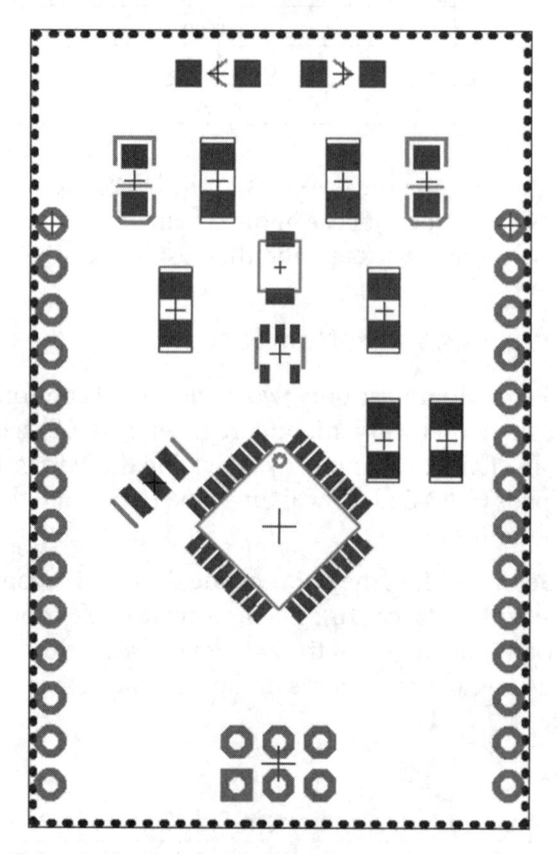

Figure 5.5: The Femtoduino Board After Layout

5.3 Design Rule Check

In an ideal world, signal traces on a circuit board would be infinitely thin, allowing for easy routing and ultra-dense layout. In the real world, traces have a minimum width and must be separated by a minimum distance. These minimum dimensions, usually given in thousandths of an inch or mils, are determined by the board fabrication facility.

For example, Advanced Circuits prefers that PCB traces be at most 5 mils thick with 5 mils separating one trace from another. These and other dimensional considerations, collectively called design rules, should be decided before the routing process starts.

EAGLE provides a design rule check, or DRC, to validate a circuit design against these design rules. This is like the electrical rule check (ERC) discussed in Chapter 4. Both checks can be launched through the vertical toolbar and both examine the current design for errors. If the DRC finds an error, it displays a dialog box like that created by the ERC.

There are two significant differences between the DRC and ERC:

- The DRC is primarily concerned with dimensional errors, such as whether a via is too close to a pad or drill hole. The ERC doesn't care about dimensions.
- The DRC criteria can be customized by the user through the Design Rules dialog. The ERC criteria is determined by EAGLE.

In my circuit designs, I use design rules in a four-step procedure:

1. Before I design a board, I decide where I'm going to go to get the board fabricated. I'll refer to this organization as the fabrication facility.
2. I check the facility's web site to determine what its processing capabilities are. This provides information such as the minimum trace width, minimum drill hole diameter, and so on.
3. I enter the facility's dimensional requirements into EAGLE's Design Rules dialog.
4. As I design the board, I launch design rule checks (DRCs) to make sure I'm not breaking any rules.

The design rule check is optional, but if you send Gerber files to a facility for a circuit that it can't fabricate, you're going to waste time and money. For this reason, I consider the DRC to be a necessary part of the design process.

To start, I recommend that you open the Design Rules dialog by clicking the DRC entry at the bottom of the vertical toolbar or by going to Edit > Design Rules... on the board editor's main menu. Figure 5.6 shows what the dialog looks like.

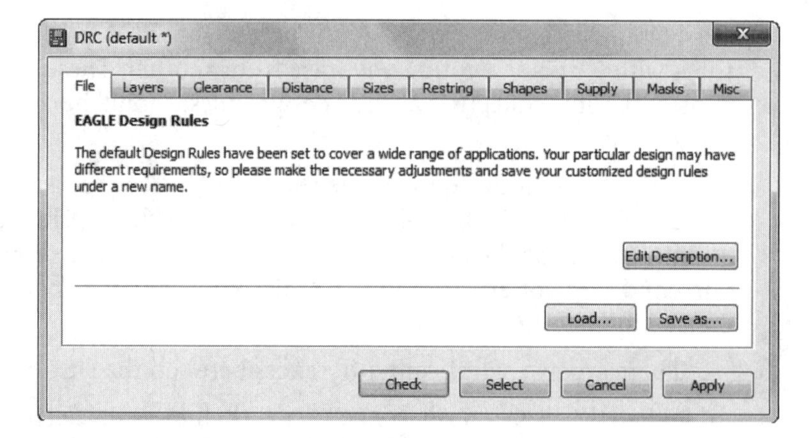

Figure 5.6: The File Tab of the Design Rules Dialog

This dialog has 10 tabs, and this section explains each of them in turn. The first tab is particularly important because it allows you to access the book's design rules for the Femtoduino circuit.

5.3.1 The File Tab

EAGLE accesses design rules through *.dru files. These are located in the top-level dru folder and the default rules are in the file called default.dru. The File tab of the Design Rules dialog, displayed in Figure 5.6, makes it possible to load existing rules files or create new ones.

To load rules from an existing file, click the Load... button on the dialog's File tab. This opens EAGLE's top level dru directory and allows you to select a *.dru file. After you choose a file, its design rules will be displayed throughout the dialog. Similarly, the Save As... button makes it possible to store the current design rules to a file.

The Ch5 folder in this book's example file archive contains a file called femtoduino.dru. This defines the design rules I used to design the Femtoduino. To use these rules in your design, click the Load... button in the File tab, find femtoduino.dru, and click OK.

NOTE The book's design rules constrain the board's features to a minimum of 8 mils. This would be considered difficult 5 years ago, but it's pretty tame for today's processing. Modern state-of-the-art features can be created at sizes of 4-5 mils.

5.3.2 The Layers Tab

The Layers tab makes it possible to set the number of signal layers (layers containing copper) for a board design and configure the layers' properties. The maximum number of layers and board size depends on which version of EAGLE you use, and Table 5.4 lists the available options.

Table 5.4

EAGLE Releases and Layer Properties

EAGLE Release	Max # of Layers	Maximum Area
EAGLE Hobbyist	6	160 x 100 mm^2
EAGLE Light	2	160 x 80 mm^2
EAGLE Standard	6	160 x 100 mm^2
EAGLE Professional	16	4 x 4 m^2

The expression in the text box labeled Setup determines which layers will be present in the board and the nature of the material separating them. By default, EAGLE assumes your design consists of two layers. This is shown in Figure 5.7.

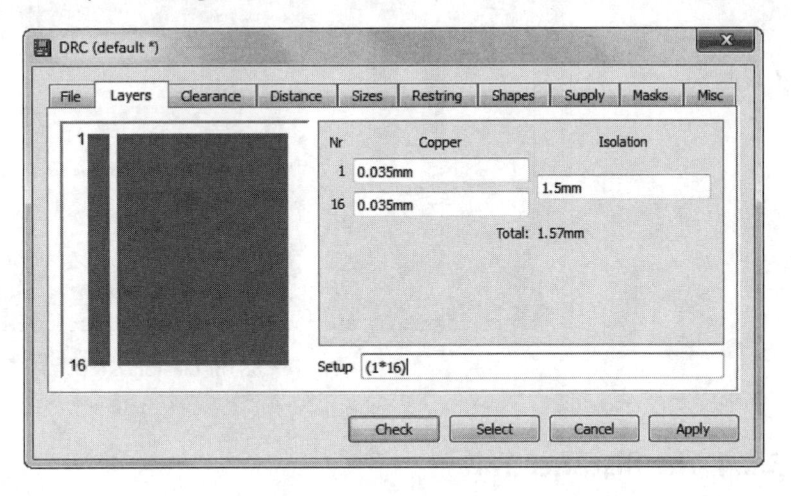

Figure 5.7: The Layers Tab

The default expression is (1*16). This provides three pieces of information:

1. The numbers identify which layers are present. The numbers 1 and 16 identify the top and bottom layers, so the board will be double-sided. For a single-sided board, this value should be set to 1.
2. The parentheses indicate that Layer 1 and Layer 16 can be connected using vias. As discussed in Chapter 2, "An Overview of Circuit Boards and EAGLE Design," a via is a hole through PCB layers that is filled with conductive material.
3. The asterisk identifies that the layers will be separated by core material. A plus sign identifies that the layers will be separated by prepreg. For the Femtoduino, all we need is two layers separated by core, so the value of (1*16) will be sufficient.

After you specify which layers are present, you can set their thicknesses. EAGLE's default thicknesses can be accepted in most cases.

5.3.3 The Clearance Tab

The minimum spacing between a board's conductive elements (traces, vias, pads, and devices) is called *clearance*, and the Clearance tab sets these values for the current design. Figure 5.8 shows what it looks like.

The clearance values in the dialog depend on the elements' types and whether they carry the same signal. In this example design rules, all the clearance values are set to 8 mils. This means every pad and trace in the design must be at least 8 mils away from every other pad and trace.

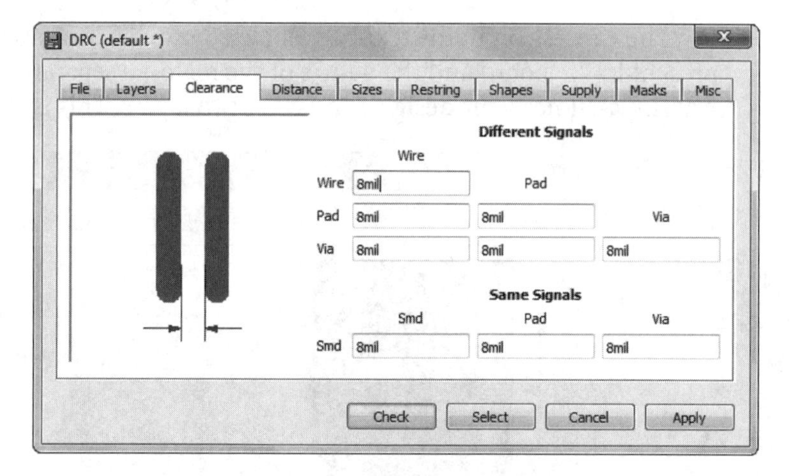

Figure 5.8: The Clearance Tab

5.3.4 The Distance Tab

Distance is similar to clearance, but instead of defining the minimum spacing between electrical elements, it defines two spacing parameters:

- The minimum spacing between electrical elements and the edge of the board
- The minimum spacing between drill holes

As with the clearance values, this book's design rules set the distance values to 8 mils. This is shown in Figure 5.9.

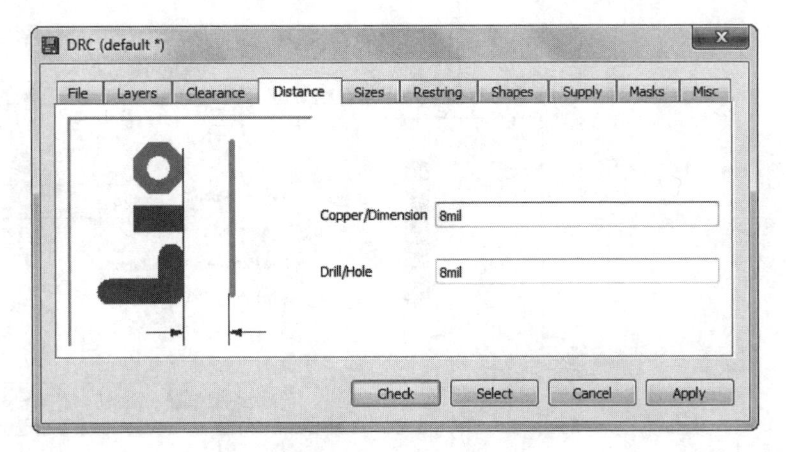

Figure 5.9: The Distance Tab

5.3.5 The Sizes Tab

The Sizes tab accepts four minimum dimensions:

- Minimum trace width
- Minimum diameter of a drilled hole
- Minimum diameter of a microvia
- Minimum ratio of a blind via

As discussed in Chapter 2, a microvia is a via formed using advanced processes such as lasers or plasma etching. They're usually much smaller than regular vias and usually come with a higher price. In this dialog tab, the microvia diameter refers to a blind via that is one layer deep.

If the minimum microvia diameter is larger than that of a drilled hole, EAGLE will assume microvias are unavailable. This is why the default value for microvia diameters is 9.99mm (393.3 mil), which is much larger than the minimum drilled hole diameter, 16 mil.

The last dimension refers to the minimum drill diameter needed for a given layer thickness. The ratio of layer thickness (t) to drill diameter (d) is usually given by fabrication facilities as t : d. In this tab, t is normalized to 1, so only the d dimension is needed. In Figure 5.10, the value is set to 0.5, which means the diameter of a blind via must be at least one-half the size of the layer's thickness.

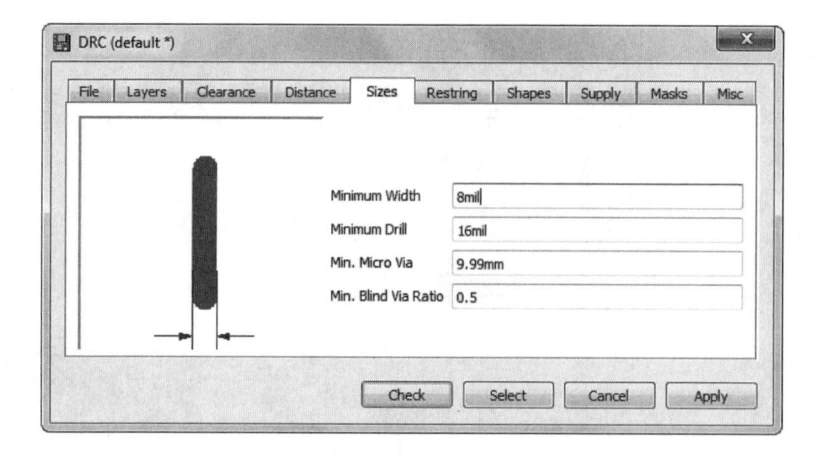

Figure 5.10: The Sizes Tab

5.3.6 The Restring Tab

In EAGLE terminology, a restring is the layer of copper that remains after a through-hole pad or via has been drilled. The Restring tab lets you set the restring width as a percentage of the hole's diameter. The width can be set for pads, vias, and microvias, and different values can be entered for different layers. This is shown in Figure 5.11.

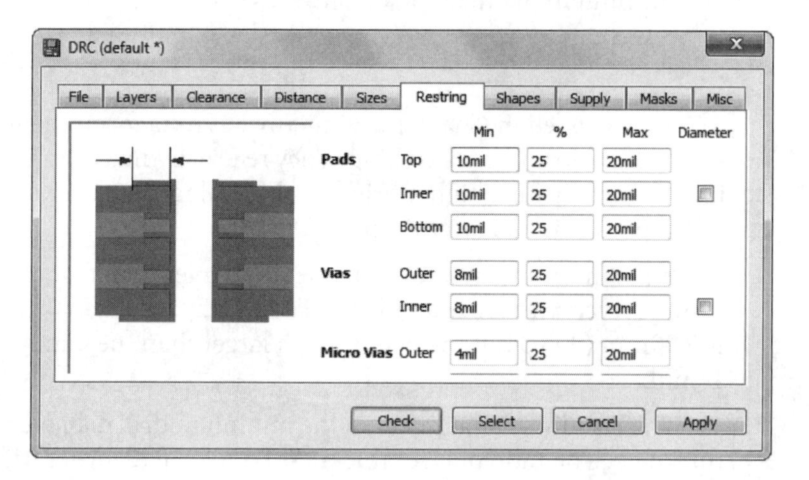

Figure 5.11: The Restring Tab

As shown in the second column, the standard width of restring is 25% of the hole diameter. But for small holes like microvias, this may be technically unfeasible. In this case, minimum and maximum restring values should be provided. For example, if a via has a diameter of 12 mils, a restring of 3 mils would be difficult to fabricate. In this case, it would be a good idea to use the default minimum restring width of 8 mils.

5.3.7 The Shapes Tab

The Shapes tab configures the geometry of pads in the design. Pads are rectangular by default, but this tab makes it possible to form pads with rounded corners. The rounding value ranges from 0 (rectangular) to 100 (fully rounded corners). For example, if an SMD pad is square and its rounding value is set to 100, its final appearance will be a circle. Figure 5.12 shows what the Shapes tab looks like.

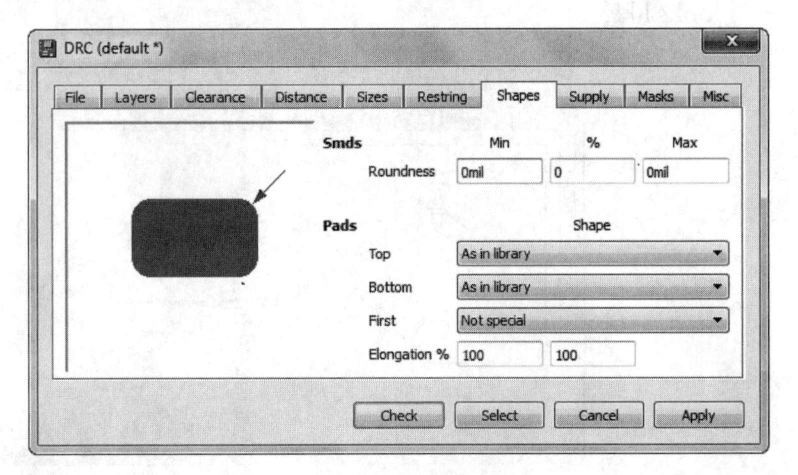

Figure 5.12: The Shapes Tab

For other pads, such as those for vias and through-hole connections, the Shapes tab makes it possible to configure the shape depending on whether it's on the top or bottom:

- **As in Library**—The pad's shape is determined by the component library.
- **Square**—The pad's shape will be square.
- **Round**—The pad's corners will be rounded.
- **Octagon**—The pad will take an octagonal shape.

If a component has many pads, it may help to give the first pad a special shape. The combo box called First makes it possible to configure a component's first pad to be square, round, or octagonal.

The last two text boxes in the Shapes tab relate to elongation, or the lengthwise stretching of the pad's shape. In both cases, the elongation is given as a percentage (greater than 100%) of the pad's diameter. The first box accepts a percentage that stretches the pad symmetrically about its center. The percentage in the second box determines the extent of the stretching in a single dimension.

5.3.8 The Supply Tab

In some cases, a trace may connect a pad to a large amount of copper (called a copper pour). If a pad is connected directly to the copper pour, the copper will draw away heat, and this makes it difficult to solder a lead to the pad. For this reason, PCB designers use special pads called thermal relief pads. These connect to copper pours using multiple narrow traces, and the left side of Figure 5.13 shows what this looks like.

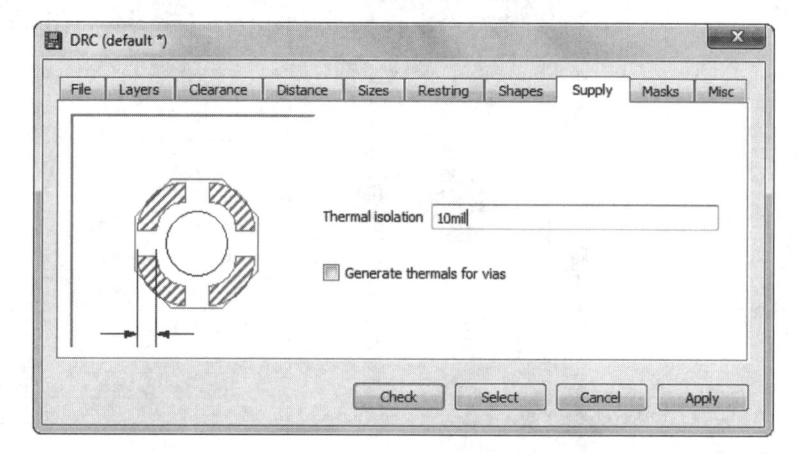

Figure 5.13: The Supply Tab

The Supply tab sets whether the design should use thermal relief pads. If so, the Thermal Isolation setting sets the spacing between the pads and the copper pour.

5.3.9 The Masks Tab

Solder mask is a protective covering (frequently green) for traces on a circuit board. This insulates the metal and prevents shorting between connections. The solder mask should never cover a pad because this prevents soldering. In addition, for testing purposes, you may want to keep vias from being covered with solder mask. Therefore, it's a good idea to set a minimum spacing between pads, vias, and solder mask.

In EAGLE, a board's top solder mask is represented by the tStop layer and the bottom solder mask is represented by the bStop layer. The properties of both layers are configured using the Masks tab, which is shown in Figure 5.14.

The Masks tab identifies the minimum spacing between the solder mask and a pad. EAGLE's default value is 4 mils, but this can be modified by setting values in the Min and Max boxes. The box marked % defines the mask spacing in terms of the smallest dimension of the pads and vias.

In EAGLE, the term cream refers to solder paste, which is a special glue used to cover SMD pads to make soldering easier. In a way, the cream layer is the opposite of the solder mask layer because you want paste on the board's SMD pads and nowhere else.

In EAGLE, the solder paste layers are tCream and bCream. By default, the dimensions set for solder paste features exactly match those of the underlying pads. For mask/cream dimensions, this book's example design rules exactly match EAGLE's default rules.

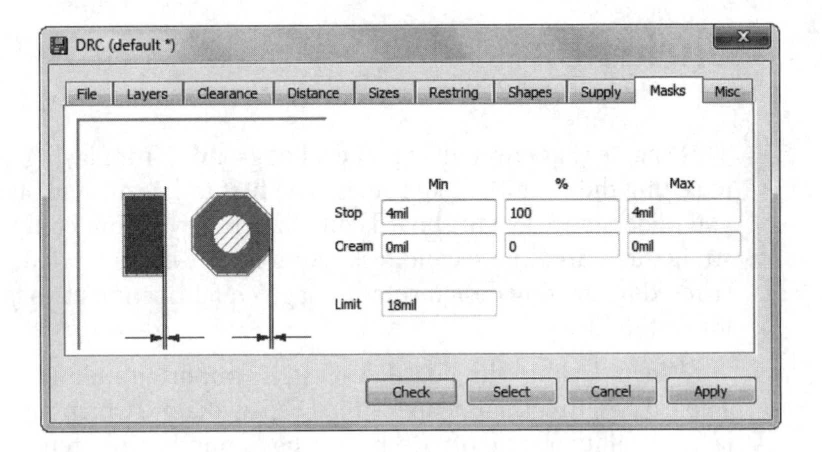

Figure 5.14: The Masks Tab

5.3.10 The Misc Tab

The final tab in the Design Rules dialog has two sets of configuration parameters. The first determines the types of checks performed during the design rule checking process. The four options are the following:

- **Check grid**—Verify that the elements in the signal layers belong to the current grid.
- **Check angle**—Verify that all wires' angles are multiples of 45 degrees.
- **Check font**—Verify that all text is written using vector fonts.
- **Check restrict**—Verify that no copper is allowed within a restricted area. Restricted areas are identified by tRestrict on the top layer and bRestrict on the bottom layer.

In my opinion, the first two checks aren't necessary. For this reason, the example design rules disable the first two checks and enable the second two.

The last two options in the Misc tab relate to differential pairs, which Chapter 14, "Schematic Design for the BeagleBone Black," discusses in detail. A differential pair is a pair of traces that carry complementary signals—when one trace carries a voltage of +V, the other's voltage is –V.

In theory, the traces that make up a differential pair should always be the same length. But this can be difficult in practice. The Misc tab allows you to set a maximum length difference between them, and in the example rules, this value is set to 10 mil.

The last option in the Misc tab relates to the gap factor between traces in a differential pair. The value in the example rules matches that of EAGLE's default gap factor: 2.5.

5.4 Conclusion

This chapter has covered a great deal of ground, from laying out a board design to setting the circuit's design rules. The first task is difficult at first, but when you understand how the board editor works, laying out components becomes straightforward. The second task requires knowing a lot of details, but when you succeed in creating design rules for a given fabrication facility, you can use those rules for further designs.

When dealing with board designs, it's important to understand how layers work. The Layer settings... tool opens the Display dialog, which makes it possible to turn layer visibility on and off. Each layer has a number and represents a different aspect of the design. The top layer of the circuit board corresponds to Layer 1, and the bottom layer of the board corresponds to Layer 16. The Mirror tool reflects components from the top layer to the bottom layer, and from the bottom to the top.

Design rules make it possible to validate a circuit against a set of dimensional criteria. EAGLE doesn't require that you use the Design Rule Check (DRC), but I highly recommend it to ensure that your design can be suitably fabricated. Most of the DRC criteria involves feature sizes and the distance between features. In the example design rules file provided by this book, the minimum dimensions are usually 8 mils.

Just as this chapter explained how to place components in the board area, the next chapter explains how to create connections between them. This process is called routing and EAGLE provides a number of ways to accomplish it.

Chapter 6

Routing

In Chapter 3, "Designing a Simple Circuit," the op-amp circuit was so simple that its connections could be routed on a single layer. In contrast, the Femtoduino circuit has many more connections and these connections span two layers. Large-scale circuit designs may require thousands of connections and more layers than just the top and bottom. As an example, the BeagleBone Black in Chapter 14, "Schematic Design for the BeagleBone Black," and Chapter 15, "Board Design for the BeagleBone Black," requires six layers.

The process of forming these connections between the packages' pads is called routing, and in my designs, I spend more time routing than in any other activity. If you read reviews of circuit design applications, you'll see that routing is a major priority. The more assistance a tool provides in the routing process, the less time is required and the fewer errors you'll end up with (in theory).

Thankfully, EAGLE provides three methods for routing traces in a design:

1. **Manual routing**—The designer routes traces by herself.
2. **Follow-me routing**—The designer selects pads to be routed and EAGLE determines how best to connect them.
3. **Automatic routing**—EAGLE attempts to route all the unrouted connections in the design.

This chapter discusses each of these methods and shows how they can be performed in the board editor. Of course, the main focus is to route connections for the Femtoduino.

NOTE This chapter's board design makes use of the design rules set through the DRC dialog discussed in the preceding chapter.

6.1 An Overview of Routing

Before writing this chapter, I searched the Internet for opinions on the best method for routing traces on a circuit board. I couldn't find a consensus—some designers use manual routing all the time and others use automatic routing whenever they can.

Because I can't give you an industry-accepted method for routing circuits, you'll have to settle for my method. Here are the five steps I use when routing circuits with EAGLE:

1. Manually route connections that are obvious, high-priority, or high-frequency.
2. Configure the autorouter using the Autorouter Setup dialog.
3. Execute the autorouter to create connections for the rest of the board.
4. Remove (ripup) poorly routed connections and make adjustments as needed. These adjustments usually involve manual routing or moving components on the board.
5. Repeat Steps 3 and 4 until the entire board is routed.

This chapter explains how these tasks can be accomplished in EAGLE's board editor. Six commands are required and Table 6.1 lists each of them.

Table 6.1

EAGLE Commands for Connection Routing

Name	Purpose
Ratsnest	Computes shortest paths for airwires and smallest polygons
Route	Creates routed trace from airwire (manual routing)
Via	Places via in design
Ripup	Removes routed trace
Signal	Defines airwire
Auto	Executes/configures autorouter

Each command has a corresponding entry in the board editor's vertical toolbar. Figure 6.1 shows where they're located.

The following section will show how they're used to route connections in the Femtoduino. The first topic involves manual routing with the Route command. I'll explain how to route initial connections using manual routing, and later, you can use the autorouter to connect the rest of the traces.

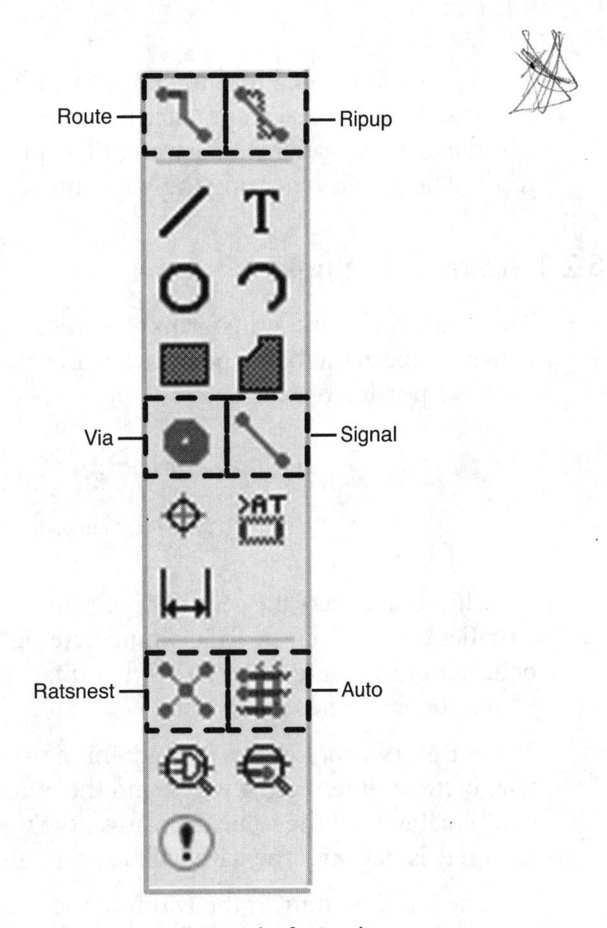

Figure 6.1: Toolbar Entries for Routing

6.2 Manual Routing

Before launching the autorouter, it's a good idea to make routing decisions of your own. If a trace carries high current or a high frequency signal, you should make sure that its path proceeds directly from start to finish. This is also true for clock signals, which need to reach their destination as quickly as possible.

The Femtoduino circuit is straightforward to route, and in this subsection, I'll explain how to make simple initial connections. But first, I recommend the following steps:

1. Set the grid markers to be closely spaced. In the Grid dialog, set Display to On and Size to 1 mil.
2. In the main menu, go to Options > Set... and click the Misc tab. Make sure the box labeled "Auto set route width and drill" is checked. This ensures that the trace width will match the width set by the net classes that were configured in Chapter 4, "Designing the Femtoduino Schematic."

To explain manual routing, I'll start with a simple example of using the Route command. Then you can practice by routing connections between the microcontroller and the board's headers. Afterward, I'll explain how to create connections between the top and bottom layers using the Via command.

6.2.1 A Simple Example

The Route command transforms airwires into routed connections. When Route is active, a special toolbar appears below the editor's horizontal toolbar. Figure 6.2 depicts a portion of it.

Figure 6.2: Configuring the Routing Process

This toolbar sets the geometry of routed wires. On the far left, the combo box specifies the wire's layer. To its immediate right, a series of buttons configure the behavior of the routed wire at bend points. I prefer the second and fourth options, which add 45° bends to the wire.

In EAGLE, *miter* refers to the joining two wires at an angle. EAGLE can insert a wire between intersecting wires, and the miter value determines the properties of this additional wire. If the value is positive, the wire will be an arc with the given radius. If the value is negative, the wire will be a straight line.

The last box controls the width of the routed wires. In the figure, the width is set to 10 mils. This value should be set to 8 mils for the Femtoduino routing.

After you configure the Route tool, you're ready to make connections. This process consists of four steps:

1. Find an airwire connecting two components.
2. Click a point on the airwire close to where it touches the first component.
3. Move the mouse to change the trace's shape. Click to form bend points.
4. Click the pad where the airwire connects to the second component.

As an example, Pad 7 of the microcontroller (U2) needs to be connected to Pad 1 of the resonator (Y1). The left side of Figure 6.3 presents the two pads and the airwire between them. The right side presents the routed wire.

To make this connection on your own, follow these steps:

1. Press the Route button in the vertical toolbar to the left.
2. On the horizontal toolbar above the editor, select the second wire bend option.
3. Click the airwire between the two pads near where it touches the microcontroller pad. As you move the mouse off the pad, the routing tool will follow.
4. Click the resonator pad to complete routing the connection.

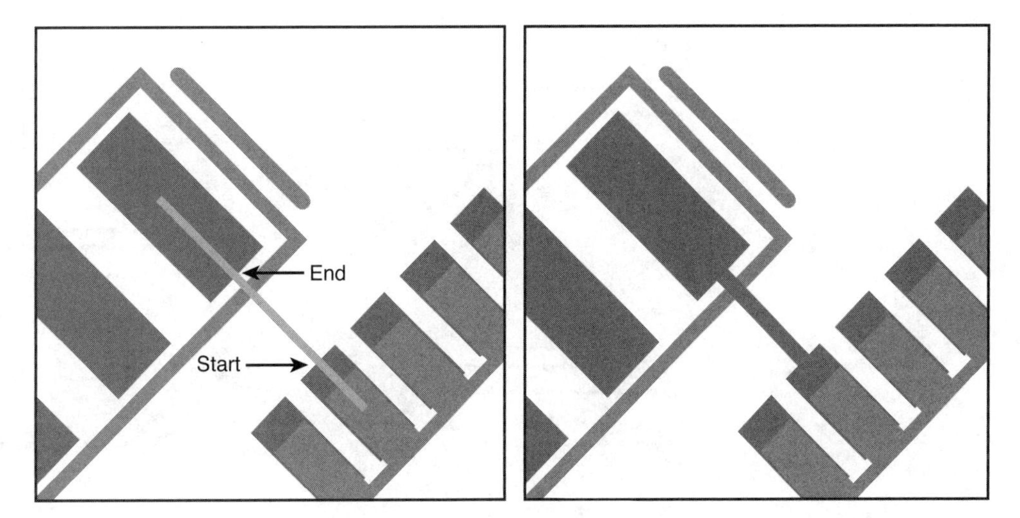

Figure 6.3: A Manual Routing Example

If you make a mistake or want to improve upon a trace, select the Ripup item in the toolbar and click the routed wire. This converts a segment of a trace into an airwire. If you double-click a routed wire with Ripup active, the entire trace becomes an airwire.

6.2.2 Connecting the Microcontroller to the Headers

Now that you've created a connection between the microcontroller and the resonator, you're ready for more practice. Most of the microcontroller's other pads can be directly connected to the long headers on either side of the board. Figure 6.4 shows what this looks like.

The more connections you make on your own, the better the autorouter will be able to do its job. Therefore, I recommend that you route as many of the displayed connections as you can. Don't be concerned if the routed wires don't look exactly like those depicted in the figure. For a low-frequency device like the ATmega328p, small differences in wire length won't cause a noticeable difference in the circuit's operation.

As you route more connections, you'll become more familiar with the tool. Here are some useful tricks:

- Right-clicking changes the wire bend style.
- Center-clicking changes the current layer.
- Holding Shift and left-clicking forms a via.
- Holding Ctrl and left-clicking starts routing a wire at any point along a wire or via.

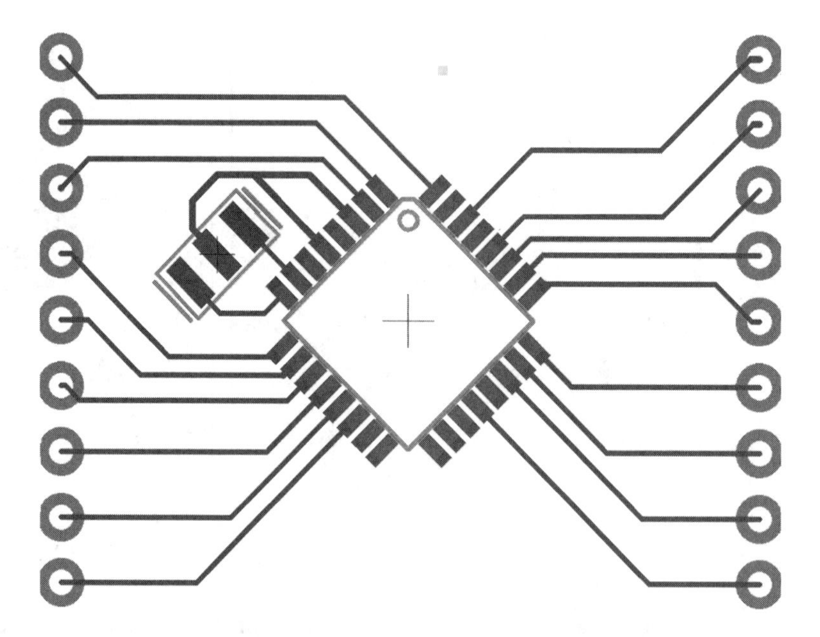

Figure 6.4: Connections Between the Microcontroller and the Headers

6.2.3 Creating Vias

Pads 4, 6, and 18 of the microcontroller need to be connected to VCC, and so does Pad 2 of the ICSP header. Connecting the microcontroller pads on the top side is easy, but connecting them to the header isn't. For this reason, we're going to create a conductive tunnel to the bottom side and route a trace to the header. As explained in Chapter 2, "An Overview of Circuit Boards and EAGLE Design," this conductive tunnel is called a via.

EAGLE provides a Via item in the vertical toolbar, but I rarely use it. I find it more convenient to activate the Route command, route a wire up to the point where I want to place the via, press Shift-left to form the via, and then use the mouse's middle button to select the layer that continues the connection. It takes practice, but this method makes it easy to route signals across multiple layers.

The geometry of a via is configured like that of a route. If the Via tool is active, a toolbar appears below the main toolbar, as shown in Figure 6.5.

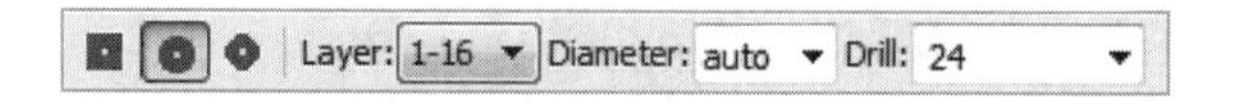

Figure 6.5: Via Configuration Toolbar

If the Route command is active, the items in Figure 6.5 will be to the far right of the toolbar. The default via shape is square, but I prefer circular vias.

Figure 6.6 shows what the connections between the microcontroller and the ICSP header look like. Pins 4, 6, and 18 are connected to one another on the top side, and a via connects them to Pin 2 of the ICSP header.

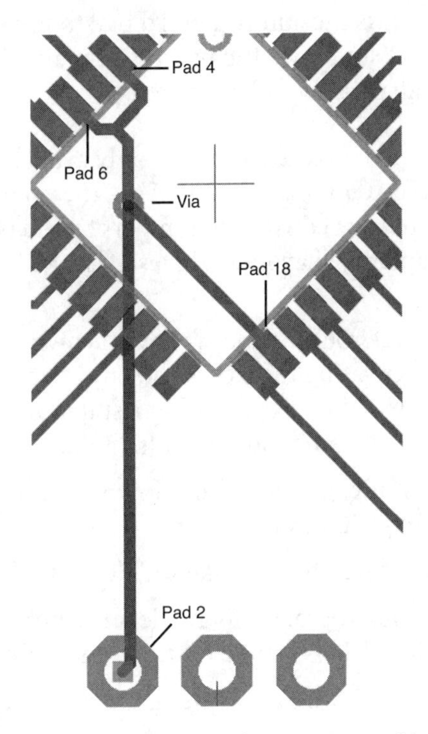

Figure 6.6: Connecting Top and Bottom Traces with a Via

The following steps explain how to create a similar set of connections in the board editor:

1. Click the Route tool and start drawing a trace from Pad 18 of the microcontroller.
2. When the trace reaches a good position for a via, press Shift and left-click.
3. Press the mouse's middle button to switch to the bottom layer. Check the current layer in the upper-left part of the editor to be sure.
4. On the bottom layer, draw a trace from the via to Pad 2 of the ICSP header.
5. Returning to the top layer, connect Pads 4 and 6 to one another and to the via.

Note that the via and header pads are both green, which means traces can connect to them on either the top or bottom layer. Also, keep a close eye on trace width and via diameter in the board editor. Occasionally, EAGLE fails to assign the dimensions given by the net classes and uses its own default values instead. Thankfully, the Change tool makes it easy to update properties of features in the editor.

6.3 Follow-Me Routing

Before I discuss the autorouter, I'd like to present a fascinating feature called follow-me routing. This simplifies the process of manual routing by using EAGLE to decide how a selected airwire should be routed.

> **NOTE** Follow-me routing is only available if the autorouter has been licensed for the EAGLE installation. Also, the operation of the follow-me router (and the autorouter) is constrained in part by the board's design rules, which are discussed in Chapter 5, "Layout and Design Rules."

Follow-me routing doesn't have its own entry in the vertical toolbar. Instead, it's started by activating the Route tool and selecting one of the two items to the right of the regular wire bend styles. The first item's tooltip reads "Follow-me (from one side)" and the second item's tooltip reads "Follow-me (from both sides)."

The best way to understand follow-me routing is to give it a try. The following four steps explain how this works:

1. In the board editor, make the Route tool active (Ctrl-R).
2. In the horizontal toolbar, select the item whose tooltip reads "Follow-me (from one side)." This can be found to the right of the wire bend styles.
3. Select an airwire in the design. Left-click the airwire near its start or ending pad.
4. Slowly move the mouse toward the opposite pad. EAGLE will do its best to draw a trace from the starting pad to the mouse's position. If there's no way to route the airwire, the cursor will take the shape of a red circle with a diagonal line.

The difference between the two follow-me router items involves how the route is computed. The first tool, whose tool-tip reads "Follow-me (from one side)," forms a trace from the initial pad to the mouse pointer. The second tool, whose tool-tip reads "Follow-me (from both sides)," routes the airwire from both pads to the mouse pointer.

I think follow-me routing is wonderful in theory, but I rarely use it. There are two reasons for this.

1. Follow-me routing is significantly slower than regular manual routing. The time lag, which increases with board complexity, gets annoying.
2. From what I've seen, anyone with moderate experience can route connections as well as the follow-me router.

6.4 The Autorouter

One of EAGLE's selling points is its autorouter, and its presence may lead you to believe that EAGLE can route an entire board's connections all by itself. But for many circuits, this isn't the case. The autorouter can be helpful if used properly, but a better name might be "routing assistant" instead of "autorouter."

For this reason, I launch the autorouter only after I've manually routed a number of connections. If it succeeds in routing the remaining signals, I consider the design complete. If not, I rip up some or all the autorouter's traces, continue manual routing, and launch the autorouter again. I continue switching between manual and automatic routing until all the traces are routed.

In the vertical toolbar, the Autorouter item can be found to the left of Ratsnest. If you click this, the Autorouter Setup dialog will appear. This is shown in Figure 6.7. Using the autorouter is simple: Press OK and the automatic routing process will begin.

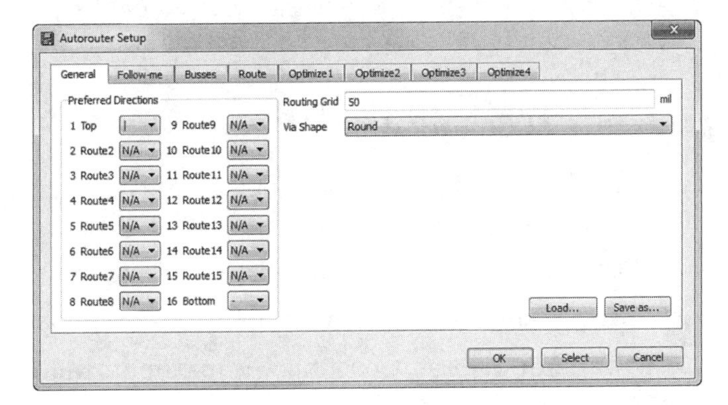

Figure 6.7: The Autorouter Setup Dialog

When I first used the autorouter, I didn't look at the setup dialog or its options. I simply pressed OK, and when the autorouter failed to route my design, I blamed CadSoft. I've learned a lot since then, and my goal in writing this section is to share what I've learned so that you can get the best performance out of EAGLE's autorouter.

6.4.1 General Configuration

The first tab presents three configuration options:

- **Preferred directions**—Preferred routing direction of traces in a given layer
- **Routing grid**—The grid spacing used by the autorouter
- **Via shape**—Geometry of the interlayer connections

The preferred direction options specify whether the autorouter should prefer horizontal routing, vertical routing, or diagonal routing for a given layer. This becomes an issue for high-frequency circuits, in which wires running in the same direction can induce voltages in one another through mutual inductance. This isn't a concern for the Femtoduino, so the preferred directions of Layer 1 and Layer 16 can be set to *. This tells the autorouter not to prefer any direction over another.

The second option is vitally important. When routing wires, the autorouter uses this value as the minimum spacing between routed wires. As this value decreases, the autorouter can place more routes in a given region. The drawback is that the autorouter requires more time and more memory to execute.

It's important to see that the autorouter's spacing distance must divide evenly into the pad's spacing distance. Otherwise, the autorouter can't connect routed wires to pads, and EAGLE will provide messages such as "Unreachable SMD at a b." For my designs, I set the routing grid to **1 mil**. On my 2-year-old laptop, a 60-component design autoroutes in about 5 minutes.

The last option specifies whether vias should be circular or octagonal in shape. There's no significant difference between the two shapes, so you can choose either.

6.4.2 Advanced Configuration Options

After the General tab, all the other tabs identify constraints the autorouter should take into account when routing paths. These constraints can be divided into three groups:

1. **Layer costs**—Constrain which layers the autorouter should favor or avoid.
2. **General costs**—Constrain a number of aspects of the circuit routing.
3. **Maximum values**—Constrain the maximum number of different types of features.

In the first two groups, the cost factors are given as positive integers. The higher the integer, the greater the cost and the more the autorouter will avoid that option. For example, if Layer X has a cost factor of 1 and Layer Y has a cost factor of 0, the autorouter will prefer routing on Layer Y to routing on Layer X.

In the second group, the names of the cost factors aren't immediately comprehensible. Table 6.2 lists each of them along with its range and a description of its constraint.

Table 6.2

General Cost Parameters for the Autorouter

Name	Range	Description
Via	0–99	Sets whether many vias should be used (low) or few vias (high)
NonPref	0–10	Sets whether the preferred direction should be a high priority (high) or a low priority (low)
ChangeDIr	0–25	Sets whether the traces should be straight (high) or have bends (low)
OrthStep	—	The cost of using orthogonal traces versus the hypotenuse (see after this table)
DiagStep	—	The cost of using orthogonal traces versus the hypotenuse (see after this table)
ExtdStep	0–30	Allows traces that run at 45° to the preferred direction, splitting the design (low) or avoids such traces (high)
BonusStep	1–3	Differentiates between good and bad layout areas (see after this table)
MalusStep	1–3	Differentiates between good and bad layout areas (see after this table)
PadImpact	0–10	Differentiates between good and bad layout areas (see after this table)
SmdImpact	0–10	Differentiates between good and bad layout areas (see after this table)
BusImpact	0–10	Differentiates between good and bad layout areas (see after this table)
Hugging	0–5	Sets whether traces should be close to one another (high) or distributed throughout the layout region (low)
Polygon	0–30	Sets whether traces should be drawn inside polygons (low) or kept outside of polygons whenever possible (high)
Avoid	0–10	After traces are ripped up, the router strongly avoids (high) or weakly avoids (low) the region

Suppose the autorouter needs to connect points P and Q. The OrthStep and DiagStep parameters tell the autorouter whether it should prefer a direct trace from P to Q or whether it should use two traces joined at a right angle. If OrthStep is low, the autorouter will prefer traces joined at right angles (orthogonal traces). If DiagStep is low, the autorouter will prefer diagonal traces.

If a layer has a preferred direction, EAGLE will do its best to route traces in that direction, especially traces departing from pads. For this reason, the region away from a pad in the preferred direction is considered *good* and the regions perpendicular to the preferred direction are considered *bad*. The BonusStep and MalusStep options tell the autorouter how seriously it should care about good and bad regions.

The PadImpact and SmdImpact options identify how much space should separate the good and bad regions from through-hole pads and SMD pads. If these options are set to low cost, the autorouter will do its best to route traces from pads as far as possible in the preferred direction. If set to high cost, the autorouter will pay less attention to preferred direction when routing traces from pads.

The last group of configuration options relate to the maximum number of features allowed in the design. Table 6.3 lists the different features available.

Table 6.3

Features Whose Maximum Can Be Configured for the Autorouter

Name	Description
Via	Controls the maximum number of vias
Segments	Controls the maximum number of trace segments
ExtdSteps	Constrains how many segments can be routed 45° to the preferred direction
RipupLevel	Maximum number of previously routed traces ripped up with each failed route
RipupSteps	Maximum number of ripup sequences that require rerouting
RipupTotal	Maximum number of traces that can be ripped up at once

The first three entries in the table are straightforward and don't require explanation. The last three entries are only available in the Route tab, and to understand their purpose, you have to understand how the autorouter does its job. The following discussion introduces this topic and explains what these dialog options represent.

6.4.3 Operation of the Autorouter

At a top level, EAGLE's autorouter works in three main stages:

1. **Bus routing**—Routes straight connections running in the x and y directions
2. **Routing pass**—Routes all other connections in the design
3. **Optimization**—Makes improvements in the routing result

These three stages correspond to the dialog's tabs: Busses, Route, Optimize1, Optimize2, Optimize3, and Optimize4. The settings on each tab constrain the autorouter's behavior in the associated stage.

The autorouter attempts to route one connection, then another, and another until all the connections are routed. If a connection can't be routed, it rips up a number of previously routed traces and tries again. This number of ripped-up traces corresponds to the RipupLevel value in the Route tab.

As the autorouter reroutes traces after a ripup, it may encounter an unroutable trace. This may produce overlapping ripup operations, and the autorouter will store a number of such overlapping operations. This is determined by the RipupSteps value in the Route tab. Similarly, RipupTotal determines how many traces can be ripped up at the same time.

By default, RipupLevel is set to 10, and RipupSteps and RipupTotal are set to 100. If any of these values is increased, the autorouter will do a better job but take a greater amount of time to finish.

6.5 Home PCB Fabrication

After you've finished the routing, the board design is complete. At this point, there are two options available for fabricating the board. First, you can generate Gerber/Excellon files using EAGLE and send them to a fabrication facility that will build the boards and ship them to you. The next chapter explains how to generate these files and provides information about five fabrication services.

The second option is to fabricate the board yourself. This requires a fair amount of work, but has a number of advantages including low cost and rapid turnaround. Another advantage is that you don't need Gerber files. The only required file is EAGLE's board design file (*.brd).

One popular method of fabricating circuit boards uses laser toner as photoresist. This is called the toner transfer method. It has worked well for me and it doesn't require any special equipment beyond a laser printer and an iron.

Before I present the method, I feel obligated to provide a warning. This method makes use of ferric chloride to remove copper. This substance is poisonous if swallowed, corrosive to the eyes, and irritates the skin. It also stains everything it touches and requires special steps for disposal.

6.5.1 Overview

A good way to understand the toner transfer method is to compare it to the fabrication method presented in Chapter 2. In that chapter, I explained how fabrication facilities use photolithography to pattern copper on circuit boards. The overall process consists of six steps:

a. The copper layer is coated with light-sensitive material called photoresist.
b. A photoplotter uses a light source to selectively expose portions of the photoresist.
c. The photoresist's chemical properties change when exposed to light. If the photoresist is positive, the exposed photoresist softens. If the photoresist is negative, the exposed photoresist hardens.
d. The softened photoresist is removed using a chemical called developer. The hardened photoresist remains and covers parts of the copper.
e. The uncovered copper is removed using a strong acid, such as cupric chloride. This process is called etching and the chemical is called an etchant.
f. The remaining photoresist is removed, leaving only the patterned copper on the circuit board.

Figure 6.8 depicts each of the steps involved for a negative photoresist. If thc photoresist were positive, the copper under the exposed region would be weakened and the copper beneath it would be etched away.

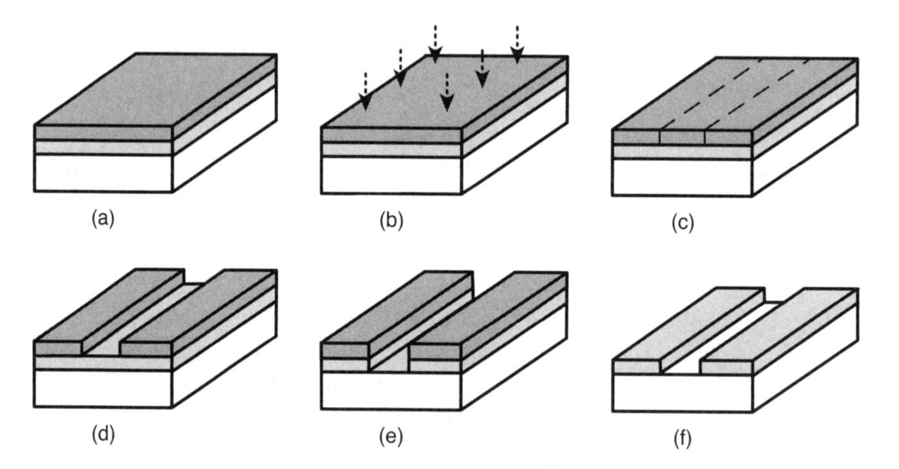

Figure 6.8: PCB Fabrication with Photolithography (Negative Photoresist)

The toner transfer method skips Steps a through c. The first step doesn't cover the entire board with photoresist, but places it only on top of the copper features that should remain on the board.

This method is made possible by one vital fact: Laser toner can serve the same purpose as hardened photoresist. When laser toner is applied to portions of a board, it's strong enough to protect the copper underneath from being etched away.

Printing toner directly to a circuit board is beyond the abilities of most printers. For this reason, this method recommends printing the design onto a sheet of paper and using an iron to transfer the toner to the board.

Figure 6.9 presents the steps graphically. The board design is printed on paper and the toner is transferred to the circuit board using an iron. When the transfer is complete, the uncovered copper is removed with ferric chloride and the toner is removed with a solvent.

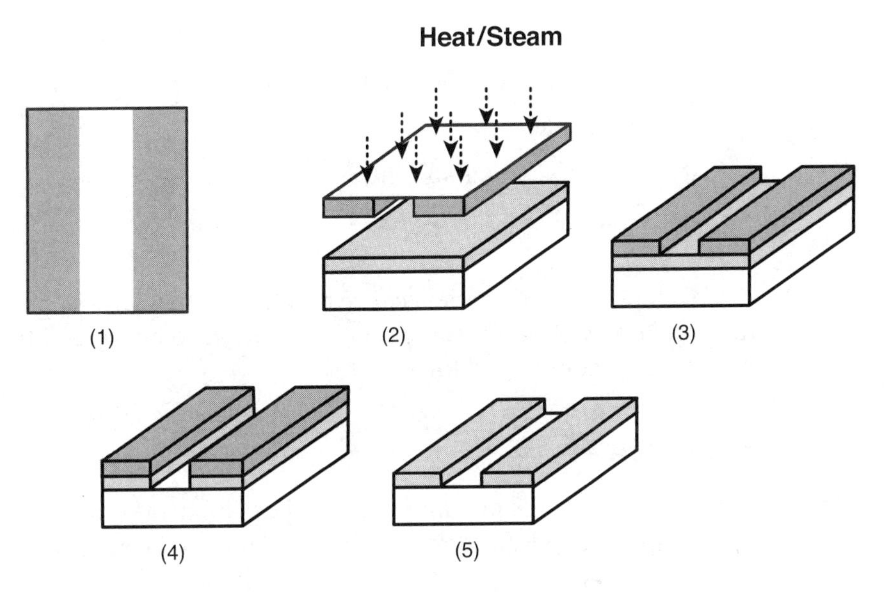

Figure 6.9: PCB Fabrication with the Toner Transfer Method

The final result in Figures 6.8 and 6.9 is the same: patterned copper on the circuit board. Photolithography produces designs with better resolution but the toner transfer method is more suitable for home fabrication. The following discussion presents the toner transfer process in detail.

6.5.2 The Toner Transfer Method

To fabricate PCBs with the toner transfer method, you need the following items:

- Copper-clad circuit board
- Laser printer (the more dots per inch, or dpi, the better)
- Paper (preferably glossy laser paper)
- Regular household iron

- Abrasive sponge, such as a scrub sponge or steel wool
- Plastic container large enough to hold the board
- Ferric chloride
- Acetone (or a similar solvent)
- Safety glasses and safety gloves

For the sake of convenience, I've divided the process's steps into two parts. The first part applies the laser toner on the copper-clad board and the second part removes the unwanted copper and toner.

Toner Transfer

The first step involves printing the board design using a laser printer and transferring toner from the paper to the circuit board. This can be accomplished with the following steps:

1. In EAGLE's board editor, activate the Layer settings... tool and deselect all layers that shouldn't be transferred to the circuit board. I recommend keeping Layer 20 (Dimension) visible so that the board's outline will be printed.
2. Go to File > Print... in the main menu. In the Print dialog, make sure Black is checked, the scale factor is set to 1, and the destination printer is a laser printer. Press OK to print the board design.
3. To prepare the board, scrub the copper surface with the abrasive sponge and wipe it clean with a washcloth.
4. Place the paper with the printed design on top of the board so that the printed side of the paper is in contact with the copper surface. Align the paper using the printed outline. If possible, use tape to secure the paper to the board.
5. Set the iron to its highest heat setting. Keeping the paper in contact with the board, press the paper with the iron. Move the iron to ensure the heat is evenly distributed, but don't let the paper move.
6. After 4 minutes, take away the iron. Put warm water in the container and immerse the board and paper. After 5 minutes, remove the board and slowly peel the paper from its surface. Carefully remove any paper stuck to the board with a toothbrush or your fingers.

At this point, you should check to see how well the board's design transferred to the copper. Excess toner can be removed with a modeling knife. Insufficient toner can be fixed with a marker. If there are major errors, you can remove the toner with acetone and start over.

Removing Copper and Toner

With the toner pattern on the board, the next step is to etch the uncovered copper and then strip the toner. Etching copper is a delicate operation and it may require multiple attempts to find a process that works for you. Here are the steps that have proven useful to me:

1. Put on the safety glasses and safety gloves.
2. Mix equal parts of water and ferric chloride in the plastic container. There should be just enough to completely immerse the board. Heat the mixture to about 100°F (38°C).
3. Immerse the board in the liquid, making sure to cover the copper surface.
4. The etching process usually takes between 15 and 20 minutes. Gently agitate the container and check the progress of the etching every few minutes. If it seems to be taking a great deal of time, you can increase the temperature slightly or add slight amounts of ferric chloride.
5. When you're satisfied with the etching, remove the board and clean it with water.
6. When the board is dry, use the acetone (or other solvent) to remove the toner. Aftewrard, clean the board again with water.

When the circuit board is cleaned and dry, you can drill holes and solder components as needed. It's a good idea to perform these tasks soon after the etching is completed. If you wait too long, the surface will form a layer of copper oxide, which makes soldering difficult.

The etchant can be used multiple times. Unfortunately, the copper remaining in the etchant is toxic, so you can't simply pour it down the drain. The resulting solution needs to be taken to a hazardous waste facility for disposal.

A popular alternative to ferric chloride is cupric chloride. Unfortunately, this requires mixing hydrochloric acid and hydrogren peroxide. I've read many accounts of success with this etchant and the resulting solution appears to be less damaging to the environment than ferric chloride. But I have no experience with using it. Also, while ferric chloride will irritate skin and stain clothing, hydrochloric acid will eat through skin and clothing. If you intend to use this etchant, be sure you have the right equipment and understand the risks involved.

6.6 Conclusion

As more components are added to a design, the routing complexity increases dramatically. Still, I prefer to begin the routing process by manually routing traces. After the important traces are routed, I let the autorouter do its best to form traces on its own. This is an iterative process, so if the autorouter doesn't finish the job, I rearrange components and manually route traces as needed.

Thankfully, EAGLE makes manual routing easy. All you need is the Route tool. After you select the bend type and trace width, you can form connections between pads with simple points and clicks. To create a via, press Shift and left-click. Then the middle mouse button lets you switch between layers of the design.

The follow-me router is also easy to use. After you click on a start pad, it will automatically route a connection up to the position of your mouse. This is a fascinating tool, but it tends to be slow, so I generally avoid it.

EAGLE's autorouter can provide significant assistance in routing a design, but only if you know how to configure it properly. Of the many configuration options available, the most significant involves grid spacing. A small spacing value means connections can be made densely at the expense of higher processing time. In my experience, the increased precision is worth the additional computation time, so I set the autorouter's spacing to 1 mil.

The autorouter's setup dialog provides many more configuration options, including cost factors for different types of features. It's a good idea to know what options are available, but in general, I've been satisfied with EAGLE's default values.

The last section in this chapter discussed the toner transfer method of fabricating PCBs. I think the first part of the process, which involves ironing toner onto copper, is ingenious. Unfortunately, the process of removing the uncovered copper presents difficulties no matter which etchant you use. If you use ferric chloride, the resulting solution must be disposed of at a hazardous waste center. If you use cupric chloride, you have to face the risks of working with hydrochloric acid.

Chapter 7

Generating and Submitting Output Files

Unless you construct circuit boards yourself, you'll need to convert EAGLE's board design into files that a fabrication facility can understand. The most common file format for PCB designs is the RS-274X format, also called the Gerber format. Drill information is stored separately in files formatted using the Excellon format.

To generate fabrication files, EAGLE provides a separate tool called the CAM (Computer Aided Manufacturing) Processor. The goal of this chapter is to explain how to configure the processor and use it to generate Gerber and Excellon files. Later, I'll explain how to view the Gerber files using a free tool called gerbv.

The final section of this chapter leaves the topic of EAGLE and discusses fabrication. That is, I'll discuss five fabrication facilities and explain how to submit design files (Gerber, Excellon, and others) to get a circuit board fabricated.

7.1 Jobs and the CAM Processor

To understand how EAGLE creates output files, it's important to know what a job is. A *job* is a group of tasks where each task reads layer information from a board design and creates a file. For example, the first task of a job may be to read Layers 18–20, process the data, and produce file_a.txt. The next task may read data from Layers 25, 29, and 34, process the data, and produce file_b.txt.

There are two ways to tell EAGLE what job you want to execute: You can load a job from a file (*.cam) or create a new job. Either way, the tool to use is the CAM Processor.

7.1.1 The CAM Processor

The CAM Processor is just a dialog box. To open it, go to File > CAM Processor in the board editor or File > New > CAM Job in the Control Panel's main menu. Figure 7.1 shows what the dialog looks like.

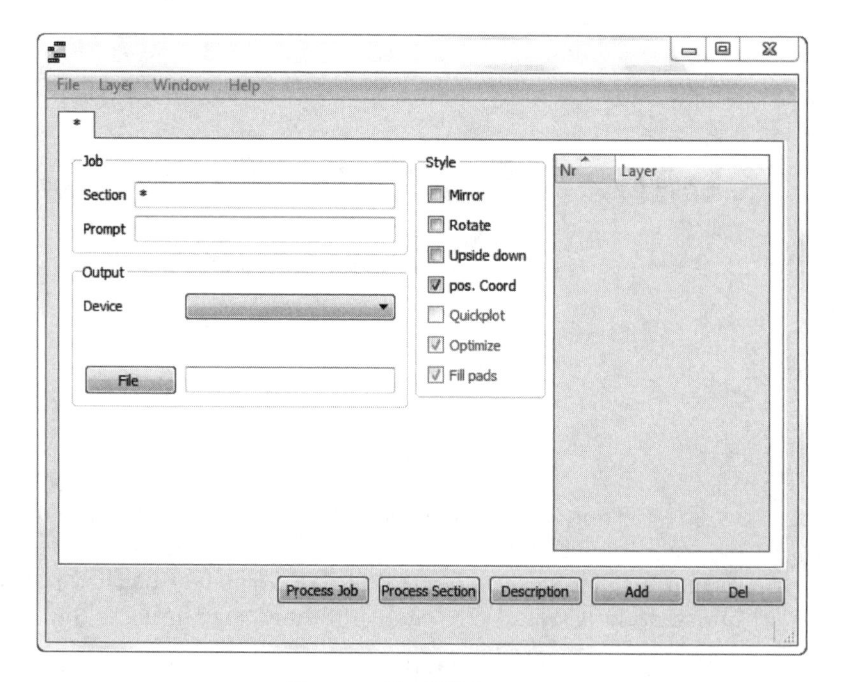

Figure 7.1: The CAM Processor

The process of using the CAM Processor consists of three steps:

1. Select the board file to be processed.
2. Select a job file or create a new job.
3. Execute the job.

To select the board file, go to File > Open > Board... on the dialog's main menu and choose the *.brd file in the file navigator. After this is selected, the CAM Processor displays the file's path near the bottom of the dialog.

To select a job file, go to File > Open > Job... on the dialog's main menu. The dialog opens the cam folder in EAGLE's top-level directory. This contains EAGLE's default job files, each identified by the *.cam suffix. When you select a file and press OK, the CAM Processor reads its tasks and updates the dialog.

7.1.2 The femtoduino.cam Job File

The Ch7 folder of this book's example archive contains a job file called femtoduino.cam, which defines seven tasks. The first six generate Gerber files and the seventh generates an Excellon file. Table 7.1 lists each of the tasks, the layers processed by the task, and the resulting output file.

Table 7.1

Processing Jobs in femtoduino.cam (%n Identifies the Board Name)

Name	Layers	Output File
Copper_Top	1, 17, 18, 20	%n.cmp
Copper_Bottom	16, 17, 18, 20	%n.sol
SilkScreen_Top	21, 25, 20	%n.plc
SilkScreen_Bottom	22, 26, 20	%n.pls
SolderMask_Top	29, 20	%n.stc
SolderMask_Bottom	30, 20	%n.sts
Drill	44, 45	%n.txt

You may want to look at Table 5.1 in Chapter 5, "Layout and Design Rules," to see what the numbered layers represent. The first six tasks process Layer 20 because it defines the boundary of the circuit board. The first two tasks process Layers 17 and 18, which define the board's pads and vias.

The Gerber format doesn't specify any particular file suffixes. The suffixes defined in the job are given by EAGLE's convention for naming Gerber files. Different design tools will generate Gerber files with different suffixes.

It's important to note that this job is only suitable for two-layer boards. For boards with more than two layers, the job must be modified to create a Gerber file for each additional layer. EAGLE convention uses the suffix *.ly*N* for additional Layers 2–9, where *N* is the number of the additional layer. For additional Layers 10–15, the suffix *.l*N* is used.

For example, suppose a board design has sixteen layers. The Gerber file for the top layer is *.cmp and the file for the bottom layer is *.sol. For the layer beneath the top layer, the suffix is *.ly2, and this progresses to *.ly9. For the rest of the inner layers, the suffixes range from *.l10 to *.l15.

NOTE For a multilayer board, each inner layer must process the vias layer (18).

7.1.3 Loading the Job File

The following steps show how to use the tasks in femtoduino.cam to generate files for the Femtoduino design:

1. If the board file isn't already selected, go to File > Open > Board... on the CAM Processor's main menu.
2. Find femtoduino.brd (or whatever you called the Femtoduino board design) and click Open.
3. Go to File > Open > Job... and navigate until you find the femtoduino.cam file. Click Open.

When the CAM Processor finishes reading the file, the dialog should look similar to that shown in Figure 7.2.

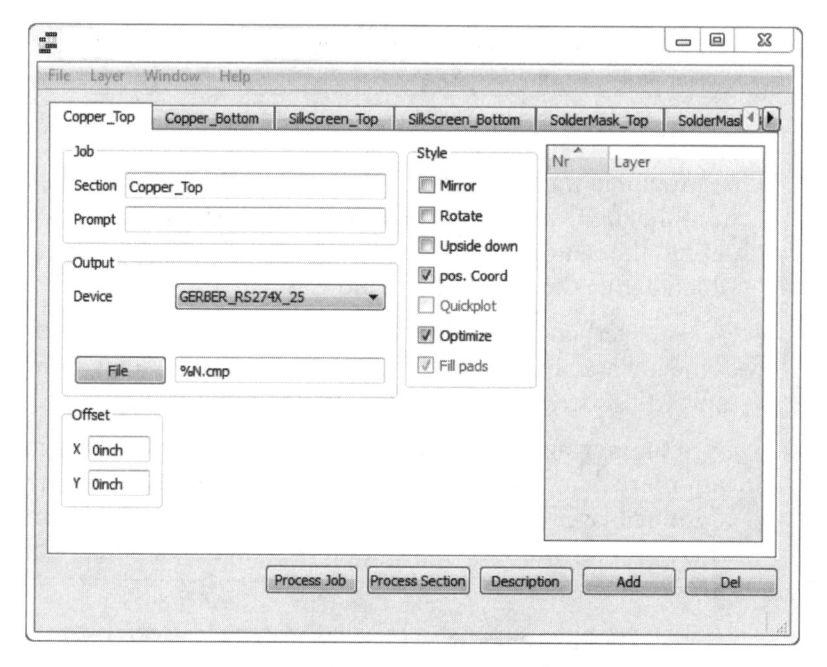

Figure 7.2: The CAM Processor with a Processing Job

The seven tabs near the top of the dialog correspond to the job's seven tasks. I recommend that you click these tabs to see how each task is configured.

The Device box is set to GERBER_RS274X_25 for the first six tasks and EXCELLON for the last. In essence, the Device box identifies the target format of the generated file, and if you scroll through the options, you'll see that the CAM Processor can generate files for many different types of printers and photoplotters.

The GERBER_RS274X_25 option produces the same type of Gerber file as generated with GERBER_RS274X. The only difference is that its floating-point values have an extra digit of precision. Appendix B, "The Gerber File Format," provides a great deal more information concerning the Gerber format.

7.1.4 Executing the Job

After you're satisfied that a job has been properly configured, you can process the job by clicking the Process Job button at the bottom of the dialog. If you execute the femtoduino.cam job for a two-layer board, the CAM Processor will generate a series of files corresponding to the job's tasks. If the design's name is femtoduino, you should find femtoduino.cmp, femtoduino.sol, femtoduino.plc, and so on.

In addition to the Gerber files and Excellon file, the CAM Processor generates two files that provide information about the design: one with a *.gpi suffix and another with a *.dri suffix. The *.gpi file contains data for the Gerber photoplotter that identifies the design's measurement units and tolerances along with which apertures are used (see Appendix B for information on apertures). If a fabrication facility requests a file defining an "aperture list," they're referring to the *.gpi file.

The *.dri file contains information about the drill settings used in the design. I'll discuss this file and the Excellon file in Section 7.3.

7.1.5 Creating New Jobs—Solder Paste Stencils

In the process of assembling a circuit board, it's common to cover SMT pads with solder paste. To ensure that the paste touches the pads only and not the surrounding board, assemblers frequently use stencils with cut-out rectangles for the pads.

Many fabrication facilities can create these stencils, but they require Gerber files that specifically identify where the cut-out shapes should be placed. In EAGLE, the solder paste regions on the top layer are defined in Layer 31 (tCream) and the solder paste regions on the bottom layer are defined in Layer 32 (bCream). Neither layer is processed by the femtoduino.cam job, so to generate Gerber files for these layers, a new job must be created.

The process of creating a job is simple. In the control panel, go to File > New > CAM Job. Set the name of the task, select the layers, and choose the device. For tasks involving Gerber files, the default settings in the CAM Processor should be sufficient. To add new tasks, click the Add button at the bottom of the dialog.

The following steps demonstrate how to configure the first task of a job that generates Gerber files for a board's solder paste layers.

1. Open the CAM Processor and go to File > Open > Board... to open the target board file.
2. In the text box marked Name, enter a name for the first task such as Stencil_Top.

3. In the combo box labeled Device, make sure GERBER_RS274X_25 is selected.
4. In the text box marked File, enter **%N.crc**. By EAGLE convention, this suffix denotes the solder paste layer on the top side.
5. In the box on the right, select Layers 31 and 20.

The other processing options can be left in their default configurations, which means the first task is complete. The steps for creating the second task are almost exactly similar:

1. Click the Add button to create a new task.
2. In the Name box, enter a name for the task such as Stencil_Bottom.
3. For the file suffix, enter **%N.crs**, which denotes solder paste on the bottom side.
4. In the box on the right, select Layers 32 and 20.

If both tasks have been created, the CAM Processor dialog will display two tabs. Figure 7.3 shows what this looks like.

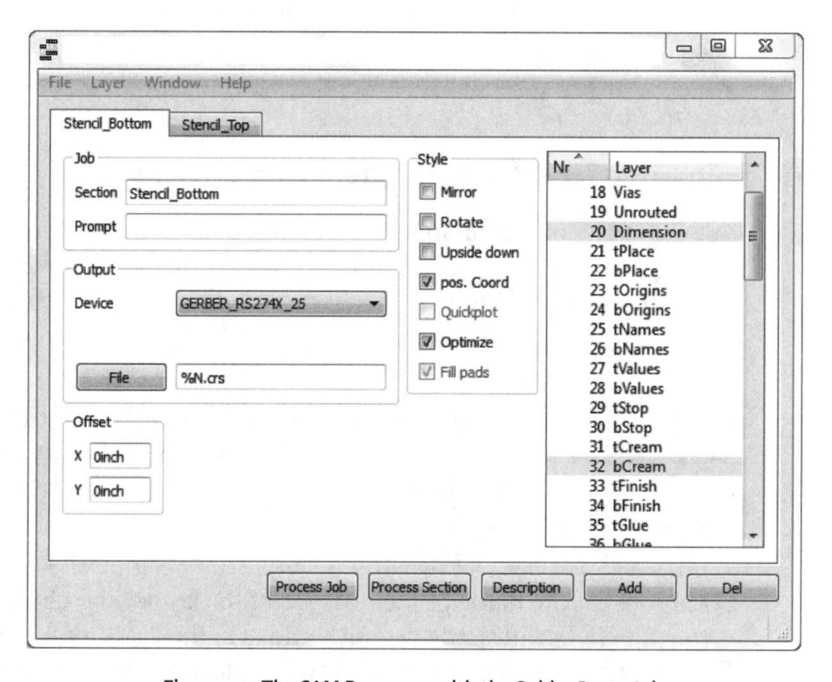

Figure 7.3: The CAM Processor with the Solder Paste Job

Press the Process Job button to generate the two Gerber files. To save the job, go to File > Save Job... and tell EAGLE where to store the new *.cam file.

7.2 Viewing Gerber Files

No matter what design tool you use, you should view your design files before you send them out for fabrication. This gives you one last chance to make sure your components are placed properly and the board's layers are aligned.

EAGLE doesn't provide any way to view Gerber files, but there is a free tool that you can download from the web. This is called gerbv, and at the time of this writing, its download site is http://sourceforge.net/projects/gerbv/files/gerbv.

When you arrive at this site, click the link corresponding to the most recent version (in my case, it's gerbv-2.6.1). Select the *.exe file if you're a Windows user or the *.tar.gz file if you're a Linux user. In both cases, the installation process is straightforward. When you start the application, the user interface should look like that shown in Figure 7.4.

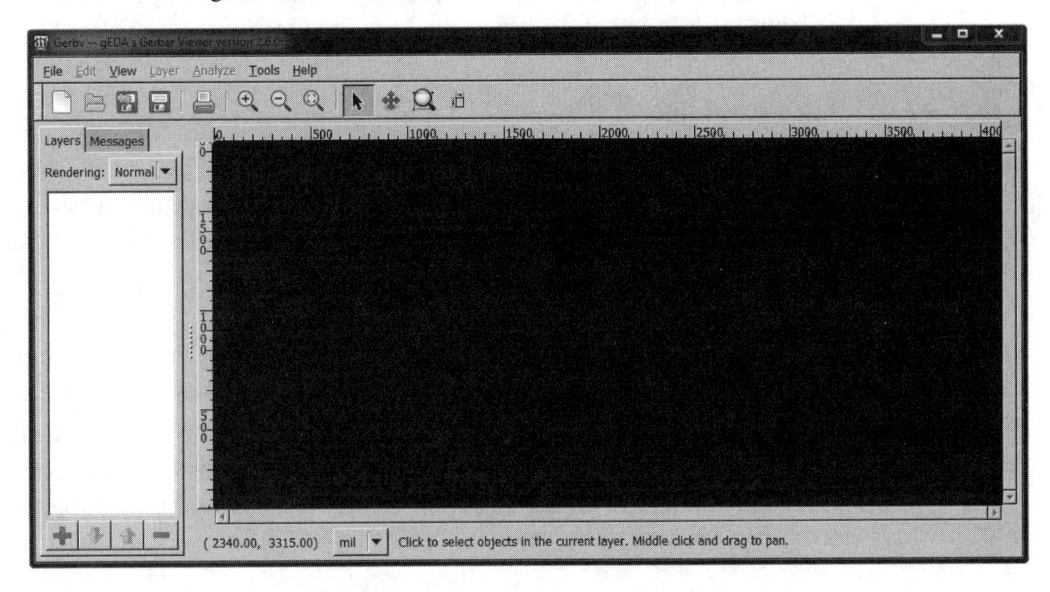

Figure 7.4: The gerbv User Interface on Windows 7

To view a Gerber file, go to the File menu, select Open layer(s)..., and navigate to a Gerber file. The file will be listed in the left box. When the board layout appears in the window, you can select shapes. Selected shapes are colored white, and by pressing the Ctrl key, you can select multiple shapes at once. When you right-click a selection, you have the option of viewing its properties or deleting it from the layer.

If you open multiple files using File > Open layer(s)..., gerbv will superimpose their images on top of one another in order of their position in the left box. To move a layer's position, right-click its entry and choose Move Up or Move Down.

To save an arrangement of layers, go to the File menu and choose Save project. gerbv will create a project file (*.gvp) that stores the current state.

7.3 Drill Files

Gerber files are so complex that it takes an entire appendix to explain their structure (see Appendix B for more on this topic). Drill files are easier to understand. If you execute the femtoduino.cam job, the CAM Processor will create two files with drill data:

- ***.txt**—Machine-readable Excellon file
- ***.dri**—Human-readable tool file

The first file is created explicitly by the femtoduino.cam job and its text is structured according to the Excellon format. The second file is generated automatically by EAGLE. It uses EAGLE's human-readable format and isn't defined or affected by the processing job. This section discusses both types of files and then explains a third type of drill file called the drill rack.

7.3.1 Excellon Files

During the 1980s, the most prominent company in the field of PCB drilling was the Excellon Automation Company. Its file format is still popular and every fab house I've encountered assumes that drill data will be provided in Excellon format.

Excellon files are frequently referred to as NC drill files because their content can be used to program numerically controlled (NC) machines. An NC machine translates text instructions into operations for a tool. In a drill file, the NC instructions tell the drill what bit to use, where to move the bit, and when to start drilling.

These drill instructions are given as a list of commands and each command occupies one line. An Excellon file groups its commands into two sections: the header and the program body. This discussion won't discuss all the available commands but focuses on those commonly encountered in EAGLE's Excellon files.

Header

Excellon files start with a header whose commands provide general information about the drilling tasks. Table 7.2 lists seven common header commands and their meanings. For a full list of commands, visit http://www.excellon.com/manuals/program.htm.

Table 7.2

Common Commands in Excellon Headers

Command	Meaning
M48	Beginning of header
M72	All dimensions given in inches
M71	Metric measuring mode
M60	Reference scaling enable
M61	Reference scaling disable
TxC*y*.*z*	Tool definition (*x*, *y*, and *z* represent one or more digits)
%	Start/end of the header section

The second-to-last entry is the most important to understand. Tool selection commands tell the machine which drills will be needed in the operation. This command consists of two parts:

1. T*x*—The drill's unique identifier. No two drills can have the same designation.
2. C*y*.*z*—The drill's diameter given as a floating-point value in the specified units.

For example, consider the following three commands at the start of a header:

```
M48
M72
T01C0.0650
```

The first command identifies the start of the header section. The second tells the machine that all dimensions are given in inches. The third command states that the drill designated as T01 should have a diameter of 0.0650 inches.

A common role served by an Excellon header is to associate drill diameters with identifiers such as T01. These unique identifiers will be used later in the file as part of the program body.

Program Body

An NC program consists of commands that tell a tool where to move and what to do when the intended position is reached. In an Excellon file, the program body consists of drill names followed by coordinates. As defined in the header, drill names are given as T*x*, where *x* is one or more numerical digits.

Coordinates are given by X*x*Y*y* commands, where *x* and *y* are absolute coordinates of the drill point. It's important to note that each command also tells the machine to drill a hole. Therefore, each X*x*Y*y* command produces a hole at the point (*x*, *y*).

As an example, the following Excellon file declares two drill names, T01 and T02, and creates three holes with the T01 drill and two with T02:

```
%
M48
M72
T01C0.0160
T02C0.0320
%
T01
X2767Y3970
X3337Y4570
X5117Y4190
T02
X3147Y12170
X2767Y13600
M30
```

These tool designations and coordinate locations make up the majority of the program body. The M30 command tells the machine that the end of the program has been reached.

7.3.2 EAGLE Tool Information Files

By default, EAGLE creates a *.dri file whenever an Excellon job is executed. Unlike Excellon files, which are intended to be read by machines, the text in *.dri files is intended for humans. It provides information about the drills used in the design and the generated Excellon file.

More specifically, these tool files provide four pieces of information:

- **File information**—The Excellon file's name, date, and directory location
- **Parameter settings**—CAM Processor settings including drill tolerance, coordinate offset, and which board layers are processed
- **Excellon settings**—Units of measurement, whether coordinate data is absolute or relative
- **Drills used**—The tool designation and diameter of each drill defined in the Excellon file

As an example, the following text is taken from the fourth section of the *.dri file that EAGLE generated when I executed the femtoduino.cam job:

```
Drills used:

 Code  Size          used
 T01   0.0160inch     47
 T02   0.0354inch      2
 T03   0.0400inch     36
 T04   0.0650inch      4
```

To the best of my knowledge, no PCB fabrication facility requires *.dri files. However, they make it possible to verify that the data in the Excellon file is correct without having to read Excellon commands.

7.3.3 Drill Rack

The Excellon file header defines the names and diameters of the drills needed for the design. However, some fab houses ask for this information to be provided in a separate file called a *drill rack*.

The CAM Processor doesn't generate drill racks on its own. Instead, EAGLE provides a ULP (User Language program) file called drillcfg.ulp. This program examines the current board design and generates a drill rack with the suffix *.drl. This program file can be found in EAGLE's top-level ulp directory.

To execute this program, open a design in the board editor and find the long text box above the editor area. This text box makes it possible to execute command scripts (discussed in Chapter 10, "Editor Commands") and ULPs (discussed in Chapters 11, "Introduction to the User Language [UL]," through 13, "Creating Dialogs and Menu Items"). To create the drill rack, type the following command in the box:

```
run drillcfg.ulp
```

Press Enter and the program will display a dialog box asking for the desired units. Make a selection and press OK. Next, the program will display a second dialog box listing the different drill sizes in the board design. Figure 7.5 shows what this dialog looks like.

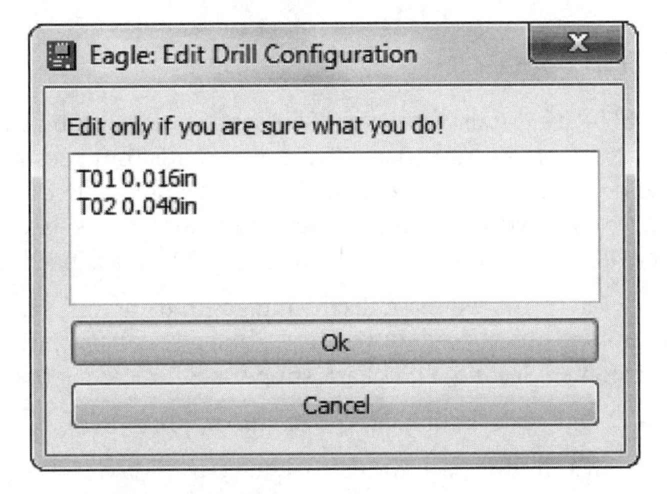

Figure 7.5: The Drill Configuration Dialog Box

This dialog associates the drill diameters needed for the board, such as 0.016in, with identifiers used in the Excellon file, such as T01 and T02. If you click OK, the program will ask where the drill rack file (*.drl) should be saved. After choosing the filename and directory, press Save, and EAGLE will write the drill rack data to the file.

The format of the drill rack file couldn't be simpler. The content of the drill file for the Femtoduino circuit (femtoduino.drl) is given as follows:

```
T01 0.016in
T02 0.040in
```

Chapters 11 through 13 provide more information about ULPs like drillcfg.ulp. With the information in these chapters, you'll be able to customize drillcfg.ulp in any way you like.

7.4 Submitting Design Files

I admire designers who construct their own circuit boards, but I'm not one of them. I lack the equipment, the coordination, and the patience, so I submit design files to companies that specialize in fabricating PCBs. The Internet makes this process easy.

This section presents five fabrication companies that accept Gerber/Excellon files over the Internet: OSH Park, Advanced Circuits, SunStone, Eurocircuits, and Seeed Studio. I don't have personal experience with all of them, but each has been sufficiently well-reviewed that I'd feel safe doing business with them.

7.4.1 OSH Park

OSH Park doesn't fabricate PCBs, but instead combines designs from multiple customers into large panel designs. It sends these designs out for fabrication and delivers the individual boards to customers. Because panel designs are so cost-effective, OSH Park is one of the most inexpensive fabrication options I've encountered. In addition, the ordering process is particularly simple for EAGLE users.

There are two main disadvantages to using this service. First, it takes time to combine board designs, so it may take weeks before the board is shipped to the customer. Second, OSH Park supports a limited set of fabrication features:

- Only 2-layer or 4-layer designs supported
- Minimum trace width/spacing: 6 mils
- Minimum clearance between traces and the board's edge: 15 mils
- Minimum drill size: 13 mils
- Minimum annular ring: 7 mils

OSH Park's web site is http://www.oshpark.com and the web interface couldn't be easier to use. The following instructions show how to order a board using an EAGLE board design file (*.brd):

1. Go to http://www.oshpark.com and click the large button entitled Get Started Now.
2. Click the file selection button and select the board design file (*.brd) on your computer.
3. As the file uploads, enter a name for the project, a description, and an email address.
4. When the file upload completes, a purple button named Continue will appear. Click this button, and on the next page, verify that the design's layers have been read successfully.
5. If you're satisfied with the board layers, click the Approve and Order button at the bottom of the page.
6. On the next page, enter the number of boards requested (must be a multiple of 3), a shipping address, and a shipping method. Then click the Order it! button.
7. On the next page, select a payment method and complete the order.

The pricing and wait time depends on the nature of the order. If you click the Pricing&Specs link at the top of the page, you'll see that OSH Park supports three types of orders:

- **2-layer order**—$5/square inch for three boards. Orders are placed 2–3 times per week and are received by OSH Park in about 12 days.
- **4-layer order**—$10/square inch for three boards. Orders are placed every 3–4 weeks and are received by OSH Park in about 14 days.
- **Medium run order**—$1/square inch for boards of 150 square inches minimum. Must be ordered in multiples of 10. 2-layer and 4-layer boards only. The designs must be emailed directly to support@oshpark.com.

OSH Park provides a design rules (DRU) file that makes it possible to check EAGLE board designs against their fabrication criteria. The current download link is http://www.oshpark.com/LaenPCBOrder.dru.

7.4.2 Advanced Circuits

Advanced Circuits, more commonly known as 4PCB, is a fabrication company whose main site is http://4pcb.com. Placing an order with 4PCB consists of five steps:

1. Create an account with 4PCB.
2. Use FreeDFM.com to validate the Gerber/Excellon files.
3. Generate a quote for your desired board.

4. Select a price from the price matrix.
5. Place the order.

The first step is the easiest. Click the Register Now link in the upper-right part and create a new account. Then log into your account and visit the Quote and Order History Page.

FreeDFM

In the upper-left part of the Quote and Order History Page, the first link is named Go to FreeDFM.com. This opens a page that allows 4PCB to check your design files before you create an order. This service is free, and though it's not required, I strongly recommend it.

To use FreeDFM, the design's Gerber/Excellon files need to be compressed into a zip file. FreeDFM (and 4PCB) recognizes all the suffixes in Table 7.1, so there's no need to rename any of the Gerber/Excellon files generated by EAGLE.

After combining the Gerber/Excellon files into a zip file, click the Choose File button and select the file on your computer. Then click Upload File to transfer the design to FreeDFM. A new page, called FreeDFM Quote Specifications, will appear and list the files in the zip file. Make sure each file corresponds to the desired aspect of the design.

NOTE You may have to specify that the *.txt file in the archive contains NC Drill information.

Quote Generation

The bottom of the FreeDFM Quote Specifications page requests information needed to determine prices for fabricating circuit boards. Figure 7.6 displays the top portion of this page.

Part #		Revision #		Layer Count	Select
X Dimension		Y Dimension		Array	☐
Array X Dim		Array Y Dim		Array Up	
Tab-Rout?	☐	Scoring?	☐		
Material Type	FR4				
Finish Thickness	0.062"	Finish Plating	Lead Free Solder	Gold Fingers	None
Soldermask Sides	Both Sides	Soldermask Color	Green	Copper Wt (Outer)	1 oz
Silkscreen Sides	Top Side	Silkscreen Color	White	Copper Wt (Inner)	None

Figure 7.6: The Quote Generation Form for Advanced Circuits

The form's default settings are sufficient for most hobbyist designs, but there are some points I'd like to make:

- Unless stated otherwise, the dimensions should be given in inches.
- If you want to fabricate a panel with multiple boards, check the Array box and identify the panel's dimensions in the Array X Dim and Array Y Dim boxes. To separate boards in the panel by tabs, check the TabRout box. To separate boards by grooves, check the Scoring box.
- If your design contains silk-screen files for the top and bottom sides, set the Silkscreen Sides field to Both Sides.
- Finish thickness refers to the final thickness of the board and finish plating refers to the material used to cover pads. If bare copper is selected, the pads will be left uncovered.
- The Quantities field is set to 5, 10, 50, and 150 by default. The quote will return prices for fabricating boards in these quantities. If you want only one board, be sure to change one of the four values to 1.
- If the board is part of an item on the United States Munitions List (USML), set the ITAR field to Yes. If not, set the field to No. This is a required field.

After the form is filled out, the last step is to press the Submit button at the bottom of the page. This delivers the design files to 4PCB, which will analyze the design and send an email containing the results. The analysis generally takes between 5 and 30 minutes.

Price Matrices

In addition to providing analysis results, the email response from 4PCB will have a link to a quote that provides prices for board fabrication. These prices are presented in a grid called a price matrix, where the rows represent different board quantities and the columns represent different delivery dates. Two price matrices are provided:

1. **Standard Spec matrix**—Lower prices, but the fabrication is limited to FR4 boards with lead-free plating, no microvias, white silk-screen, and green soldermask.
2. **Custom Spec matrix**—Higher prices, but the fabrication's parameters can be customized to a greater extent.

Figure 7.7 presents an example Custom Spec price matrix. Beneath the matrix, a note mentions that a Tooling charge will be added to the final price. This is necessary for the custom fabrication, and therefore is not present for Standard Spec prices.

Qty	Same Day	1-day	2-day	3-day	4-day	1-week	2-week	3-week	4-week
5	$272.18	$246.46	$143.59	$117.87	$102.44	$97.30	$87.01	$76.72	$66.44
10	$137.17	$124.20	$72.33	$59.37	$51.59	$48.99	$43.81	$38.62	$33.43
50	$29.16	$26.39	$15.33	$12.56	$10.90	$10.35	$9.24	$8.14	$7.03
150	$13.82	$12.49	$7.16	$5.83	$5.03	$4.76	$4.23	$3.70	$3.16
Tooling NRE = $199.00 (Tooling waived when re-ordered) * Testing = $156.66									

Figure 7.7: A Custom Spec Price Matrix

NOTE These prices are given on a per-board basis. That is, if the quantity is 5, the price of an order equals five times the stated price.

Each price is a hyperlink. If clicked, the link will open an order form. This is a traditional order form in many respects, but I'd like to call attention to the Electrical Test option. If checked, 4PCB will test the circuit board to ensure that points that are meant to be electrically connected are connected and that points that aren't meant to be electrically connected aren't connected.

After filling out the form, the Preview Order button opens a page that makes it possible to verify that the order information is correct. When the Process Order button is pressed, the order process is complete.

7.4.3 Sunstone

Sunstone Circuits serves many different types of PCB fabrication customers. To place an order, Sunstone's web site, http://www.sunstone.com, provides four options:

- **ValueProto**—Small quantities of two-layer PCBs
- **PCBExpress**—Limited-feature PCBs with 2 to 6 layers
- **Full Feature**—Expanded-feature PCBs with 2 to 14 layers
- **Custom Quote**—Fully customized PCBs

This discussion focuses on the first two options, ValueProto and PCBExpress.

NOTE ValueProto is only available for shipping in the USA. PCBExpress is available for international customers.

ValueProto

The ValueProto option makes it very easy to order simple circuit boards. By simple, I mean that the board design must meet the following criteria:

- One or two sides with green soldermask
- Optional white silk-screen on the top side, no silk-screen on the bottom

- Minimum trace and spacing of 6 mil
- All leads plated with tin
- Drill sizes limited to 23 options

After selecting ValueProto on Sunstone's main page, the ValueProto page asks for board dimensions, board quantity, and the desired ZIP code for shipping. It also asks for a zipped archive containing seven files:

1. **Top Silk-Screen**—Corresponds to *.plc in Table 7.1
2. **Top Solder Mask**—Corresponds to *.stc in Table 7.1
3. **Top Copper**—Corresponds to *.cmp in Table 7.1
4. **Bottom Copper**—Corresponds to *.sol in Table 7.1
5. **Bottom Solder Mask**—Corresponds to *.sts in Table 7.1
6. **Plated Holes**—Corresponds to *.txt in Table 7.1
7. **Tool Size Report**—Must be generated separately

It's important to see the difference between the six files generated by femtoduino.cam and the six files required by ValueProto: ValueProto doesn't support silk-screen on the bottom side, so it won't accept the *.pls file from Table 7.1.

But ValueProto does require a tool size report, which corresponds to the drill rack file discussed earlier in this chapter. This file (*.drl) can be generated from a board design by executing `run drillcfg.ulp` in the board editor. After the seven files are combined into a zip file, the design can be uploaded.

After ValueProto receives the design, it will ask you to check the files' content. Make sure the *.drl file is chosen as the tool size report. For the board outline, I recommend that you check the box labeled "Not specified, generate rectangular outline." Then click the Next Step button.

In the following page, read the board parameters and click Continue to Checkout. At this point, the ordering process is similar to any online order site. That is, after you provide shipping and billing information, the order will be complete.

PCBExpress

The process of ordering boards through Sunstone's PCBExpress option is markedly different than ordering boards through ValueProto. When the PCBExpress option is selected from Sunstone's main site, a form will request design parameters for the circuit board. Figure 7.8 shows what it looks like.

Q1/E1 (Next Day) 2 Layer w/o Mask | Rogers (Next Day) 2 Layer w/o Mask | Q2/E2 (2 Days) 2 Layer w/ Mask | Q4/E4 (3 Days) 4 Layer w/ Mask | Q6/E6 (4 Days) 6 Layer w/ Mask

2 Layer Prototype Board w/ silkscreen and soldermask. Orders of 50 boards or less will ship in 2 days

Need more boards?

Quantity: 2 Layers: 2 Build Time: 2 days Solder Mask: 2 Side Green

Surface Finish: Tin Lead Board Size: (in.) 0 X 0 Silkscreen: 1 Side (Top) White

Number Of Holes: 0 Slots/Cutouts: 0 Native File Upload: Silkscreen Clip: 1 Day Expedite:

Figure 7.8: PCBExpress Design Form

The tabs along the top make it possible to select how quickly the boards should be delivered and whether they should have soldermask. Below, the form fields are easy to understand, but I'd like to call attention to the Native File Upload checkbox. If this is checked, Sunstone will accept native board files from specific design applications, including EAGLE's *.brd file.

After the form is filled out, clicking the Order Boards button opens a dialog entitled Submit File Information. Clicking Submit Files Now opens a page that asks you to create an account. After the account is created and you've logged in, a new dialog asks for the design files.

To create a board based on an EAGLE *.brd file, go to the combo box labeled CAD System and select the Eagle EE option. Then click the Upload File button and select the *.brd file. After receiving the file, the dialog asks for layer mapping information. That is, it wants numbers for the following layers:

- **Top silk-screen**—Default layers are 21 (tPlace), 25 (tNames), and 27 (tValues)
- **Bottom silk-screen**—Default layers are 22 (bPlace), 26 (bNames), and 28 (bValues)
- **Outline Layer**—Default layer is 20 (Dimension)

After the layer values are set, clicking Done brings up a standard order form requesting shipping and mailing information. When this information is provided with the payment, the order is complete.

7.4.4 Eurocircuits

Eurocircuits provides many services related to PCB fabrication, and in addition to a design blog, it provides videos of its fabrication process. Its primary web page is http://www.eurocircuits.com, and to see its different services, click the Price Calculator button in the upper right. This brings up a page that presents six options:

- **PCB Proto**—Inexpensive PCB fabrication with limited quantities and features
- **STARDARD pool**—Pooled and non-pooled fabrication of PCBs with up to 16 layers
- **RF pool**—Pooled fabrication of PCBs with radio-frequency (RF) circuitry
- **IMS pool**—Pooled fabrication of PCBs with insulated metal substrate
- **Stencil**—Fabricates stencils that make it possible to apply solder paste to the board's pads
- **BINDI pool**—Pooled fabrication of two-sided PCBs using facilities in India

I'm only familiar with the first two, and the ordering process is similar in both cases. The following instructions show how to place an order for Eurocircuits' PCB Proto service or its STANDARD pool service:

1. From the Eurocircuits' home page (http://www.eurocircuits.com), click the Price Calculator button in the upper right.
2. Find the box entitled PCB Proto or the box titled STANDARD pool. Click the Price calculator button in the box.
3. In the new form, fill out the basic parameters of the board design. This includes the number of boards requested, the board dimensions, and properties of the soldermask and silk-screen. When finished, click the Add to Basket button.
4. If you're not already registered with Eurocircuits, create a new account and log in.
5. In the Complete action box, enter a name for the PCB and click the Choose File button. Select the EAGLE board file (*.brd) for the design. Click the Continue button.
6. The new page will display your shopping basket. The center of the page shows a pair of spinning arrows next to the word Processing. The arrows continue to spin as the PCB Visualizer analyzes the circuit design.
7. As the analysis proceeds, you may want to click the PDF under the Order column. This contains the full quote for the order and it will be valid for 30 days.
8. When the analysis is complete, click the PCB Visualizer link. This displays the different layers of the circuit board, including the copper layers, silk-screen layers, and soldermask layers.
9. After you verified that the design has been read without error, select the order by checking the box to the left. Then click the orange button at the top labeled Proceed to checkout.
10. The checkout form displays the delivery address, shipping address, and the details of the PCB order. In the lower right, the Submit button allows you to complete the order.

As you work with the Eurocircuits site, keep in mind that *buildup* has the same meaning as *stackup* in this book. Also, you may notice that you can order stencils for your design. Stencils cover the circuit board, allowing solder paste to be applied only to the pads.

7.4.5 Seeed Studio

Seeed Studio sells a vast array of electronic products, including components, kits, tools, and accessories. In addition, it provides low-cost PCB fabrication through its Propagate service. To see this, go to its main site at http://www.seeedstudio.com and click Propagate toward the top of the page.

As shown on the new page, Seeed Studio offers six services:

1. **Fusion**—Low-quantity PCB fabrication
2. **PCBA Prototype**—PCB fabrication and assembly
3. **PCB Stencil**—Fabricates stencils for application of solder paste
4. **Laser cutting**—Provides laser cutting for stencils
5. **Propagate**—Large-quantity PCB fabrication and assembly
6. **Custom cables**—Constructs cables based on customer requirements

This discussion focuses on the Fusion and PCBA Prototype services. The only difference between them is that PCBA Prototype performs two tasks: It fabricates the board and solders its components to the board's pads. This second step is called assembly, which provides the A in PCBA.

Fusion

If you click the box marked Fusion PCB, you'll be taken to a form that requests preliminary information about the circuit board. This includes the quantity requested, the number of layers, and the dimensions. Note that the PCB Color field is really asking for the color of the soldermask.

The last option on the page, surface finish, identifies the material that will cover the board's copper traces to prevent corrosion. The default is HASL, which stands for Hot Air Solder Leveling. A HASL board is dipped in hot solder, which is then cleaned off to leave a thin, level coating.

The other option is ENIG, which stands for Electroless Nickel Immersion Gold. An ENIG board is coated with nickel and a thin coating of gold. ENIG is more expensive than HASL, but the gold coating provides greater resistance to corrosion than solder. However, the gold finish may require additional heat for soldering.

After the initial form is filled out, the Next button opens a page that asks for Gerber files. For the most part, these are the same files discussed earlier in this chapter, but with different suffixes. For example, in EAGLE's convention, the top copper Gerber file is called *.cmp. Seeed Studio wants the same file, but with the *.gtl suffix.

The last Gerber file on Fusion's list is the board outline, which identifies the board's dimensions. It can be generated by running a CAM job in which only the Dimensions layer (20) is processed.

After the Gerber files are compressed into a zip file, the next step is to upload it. This can be accomplished by clicking the Choose File button, selecting the *.zip file, and clicking Next to upload the design.

The next page presents your fabrication order with a price. From this point forward, the remainder of the ordering process is similar to that of other online marketplaces.

PCBA Prototype

The process of using the PCBA Prototype service closely resembles that of using Fusion. That is, both services have the same initial form and require the same Gerber files. The final ordering pages are also the same.

The difference between them is that the PCBA Prototype service performs assembly (component soldering) in addition to board fabrication. To tell Seeed Studio how to assemble your board, you need to provide two files:

- **Bill of materials (BOM)**—A list of the board's parts in comma-separated value (CSV) format
- **Component placement drawing (CPD)**—Displays location and orientation of each component in the circuit

It's important to note that Seeed Studio supports assembly for only a limited set of components. These components form the Open Parts Library, or OPL. To make design easier, Seeed Studio provides an EAGLE library file that contains all the supported devices. This is called OPL EAGLE library.lbr, and to obtain it, click the OPL link on the assembly page. Then click the OPL EAGLE library link on the following page and select the most recent version of the library.

7.5 Conclusion

Designing a circuit board consists of four main steps: schematic design, layout, routing, and generating files. With EAGLE, the last step is the simplest. The CAM Processor is easy to use, and if you already have the right job, you can generate output files with a button click.

Before sending them out for fabrication, it's important to check these files to make sure the design has been formatted correctly. In general, output files come in two types: Gerber files contain the design's geometry and Excellon files contain drill data. This chapter has explained how to view Gerber files with gerbv and how to understand the Excellon format. Appendix B explains how commands are structured within a Gerber file.

The last part of this chapter has provided an overview of five fabrication services: OSH Park, Advanced Circuits (4PCB), Sunstone, Eurocircuits, and Seeed Studio. As you compare these services, keep in mind the trade-offs involved. You can have high-density boards with custom parameters, but you'll pay a greater price. Also, if you're willing to wait longer for the boards, you can pay a lower price by combining your design with others into larger panels.

Chapter 8

Creating Libraries and Components

As explained in the preceding chapters, the Femtoduino design relies on devices in the eagle-book.lbr library. EAGLE provides thousands more devices for your use, but even these might not be sufficient. Maybe you need to insert old or rare components into your design. Or maybe your company has a new device that needs to be made available through EAGLE. In either case, you'll need to create a new library with new components. This chapter explains how to do this.

The goal of this chapter is to create a new library called example-lib.lbr. By the end of this chapter, it will contain three devices:

- **SIMPLE-TQFP**—A 16-pin integrated circuit with one gate and a thin quad flat pack (TQFP) package
- **VACUUM-TH**—An irregularly shaped vacuum tube with five through-hole leads
- **TW9920**—An integrated circuit with 100 pins in a ball grid array (BGA) package

This chapter demonstrates how to create the library file and the three components, but it doesn't discuss the format of the library file. This is a new feature of EAGLE 6, and Appendix A, "EAGLE Library Files," presents the library's XML format in detail.

The section on the TW9920 device makes use of a combination of editor commands and User Language programs (ULPs). You don't have to understand these topics to design the component, but if you have questions, Chapters 10 through 13 discuss these topics in detail.

8.1 Creating the Library

Before I explain how to create symbols and packages, I'd like to address two preliminary concerns. First, I want to review the terms underlying EAGLE components. Then I'll explain how to create the example-lib library.

8.1.1 EAGLE Terminology

Chapter 3, "Designing a Simple Circuit," and Chapter 4, "Designing the Femtoduino Schematic," explained many of the terms used in EAGLE designs. Just in case you've forgotten, there are four crucial terms to know:

1. **Gate**—A single element that can be moved in a schematic. Most devices have only one gate but some have multiple gates.
2. **Symbol**—A gate's graphical representation. Each gate corresponds to a single symbol.
3. **Package**—A device's graphical representation in the board editor.
4. **Device**—A unique combination of one or more symbols and a package.

Suppose an integrated circuit contains an array of eight resistors. When the device is added to a schematic, each resistor can be moved and placed individually. These separate elements are the gates and each gate's appearance is defined by its symbol.

Packages and symbols are both graphical representations, but they serve different purposes. A symbol is drawn to make it easy for designers to make connections in a schematic. A package, in contrast, needs to resemble the physical device. That is, if the resistor array is manufactured as a surface-mount chip, the package drawn in EAGLE should have the same shape and dimensions as the chip.

In addition to the terms given previously, it's also important to understand the difference between a pin, pad, lead, and wire:

- **Pin**—An electrical connection in the schematic editor. Pins are defined inside a symbol definition.
- **Pad**—An electrical connection in the board editor. Pads are defined inside a package definition.
- **Lead**—The electrical connection point on the device. The device's leads must be soldered onto the board's pads.
- **Wire**—A line drawn as part of a symbol or package.

Note that wires are not electrical connections. I wish EAGLE would use *line* instead of *wire* to refer to the geometric lines of a drawing, but CadSoft never returns my calls.

8.1.2 Creating the Library

If you haven't already, launch the EAGLE application. In the Control Panel, go to File > New > Library. The window that appears is the library editor.

When you first open the editor, EAGLE assumes you want to edit a library file named untitled.lbr. To change this, go to File > Save As and choose a new name. In this chapter, the library's name is example-lib.lbr. Library files are saved to the lib folder inside EAGLE's installation directory.

There are fewer toolbar items in the library editor than in the schematic editor or board editor. You've already encountered most of the items on the horizontal toolbar, but three of them deserve special mention: Device, Package, and Symbol. The next section explains how these work together to create a simple EAGLE component.

8.2 Creating the SIMPLE-TQFP16

Integrated circuits are frequently represented by rectangular symbols in the schematic editor and rectangular packages in the board editor. These are easy to create, and in this section, we'll build a component for a 16-pin device whose package is a thin quad flat pack (TQFP). The first step is to design the symbol.

8.2.1 Creating the SIMPLE Symbol

In the library editor, the horizontal toolbar has an entry called Symbol that looks like a two-input AND gate. Click this and a dialog box will appear. Enter **SIMPLE** in the text box, click the OK button, and click Yes to create a new symbol.

The library editor will change its appearance, and the body of the window will have a grid. This is your editing space for laying out the shapes that will make up your symbol. The cross in the center identifies the origin (0, 0). It's a good idea to center the symbol on this point.

To the left of the grid, the vertical toolbar contains many items that you've already seen in the schematic editor, such as Move, Delete, and Name. At the bottom of the toolbar are eight items needed to draw shapes. Figure 8.1 shows what they look like and the shapes they create.

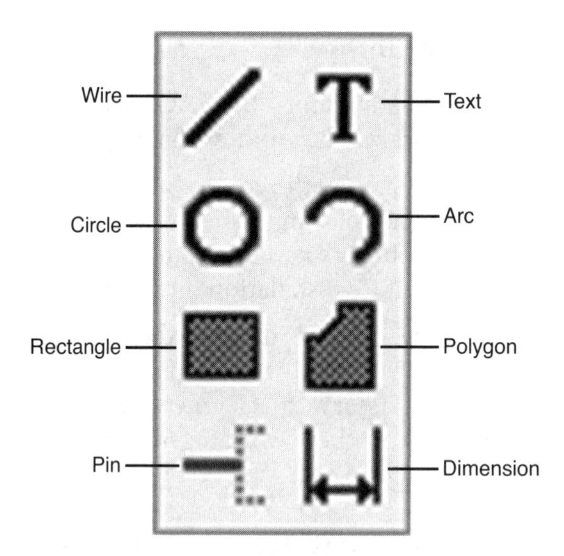

Figure 8.1: Toolbar Items for Drawing Symbols

The SIMPLE symbol consists of a rectangular body with eight pins on the left and eight pins on the right. When creating symbols for integrated circuits, I prefer to set the pins first and then draw the body. The following discussion shows how this works.

Adding Pins

Select the Pin tool at the bottom of the vertical toolbar, but don't add any pins to the design just yet. Instead, look at the new options in the horizontal toolbar. They make it possible to configure six characteristics:

- **Rotation**—Sets the pin's orientation. Default is R0 but other options include R90, R180, and R270.
- **Function**—Defines the pin's electrical characteristics. Default is active-high, but it may be set to active-low (dot), a clock signal (triangle), or both.
- **Length**—Identifies whether the pin is drawn with a small, medium, or long line. The pin may also be drawn as a point. Default is a medium line.
- **Visibility**—Specifies which labels, if any, should be set for the pin. By default, both the pin name and the pad name are given. This can be changed to pin, pad, or off.
- **Direction**—Sets the pin's direction. Default is io (input/output), but this can be set to in, out, pwr (power), and not connected (nc). Other options include pas (passive), sup (supply), hiz (high impedance), and oc (open-collector).

- **Swaplevel**—Indicates whether the pin can be swapped with other pins. For example, if Pins A, B, and C have a swaplevel of 1, they can be interchanged in the design. If Pins D and E have a swaplevel of 2, they can be interchanged with one another but not with Pins A, B, and C. The default swaplevel is 0, which means the pin can't be swapped.

With the Pin tool active, add eight pins to the left of the origin, each one grid mark below the last. With each new pin, EAGLE displays the name of the pin and the corresponding pad. When you finish, the result should look similar to the left side of Figure 8.2.

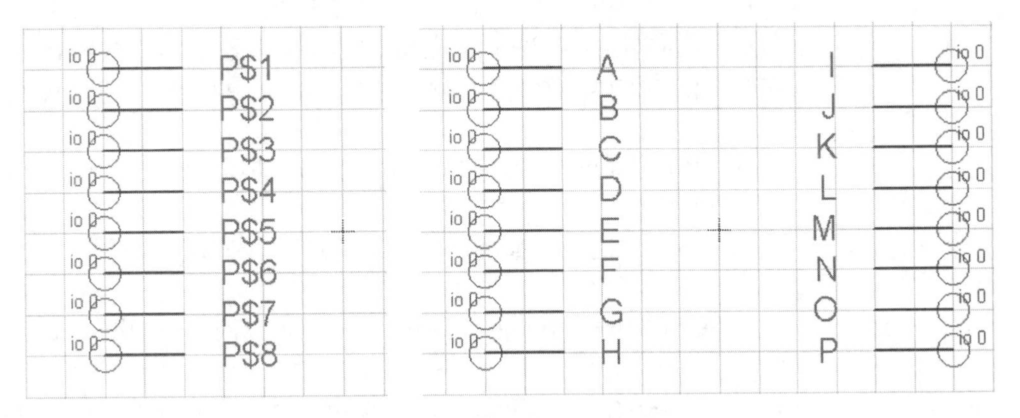

Figure 8.2: Pins of the SIMPLE Symbol (Before and After)

If you right-click before inserting a pin, it will rotate counterclockwise by 90°. Therefore, you'll need to right-click twice before inserting pins on the right side of the symbol. Create a column of eight pins opposite the column created earlier, from top to bottom.

By default, each pin displays the pin name and the pad name. To change a pin's name, click the Name entry in the vertical toolbar and select the pin in the design.

In the schematic, name the pins A through P, proceeding from the upper-left to the lower left and from the upper right to the lower right. The result should look similar to that shown on the right side of Figure 8.2.

Creating the Symbol Body

Now that the pins are set, you're ready to draw the rest of the symbol. For the rectangular body, you might think you'd use the Rectangle item. But rectangles drawn with the Rectangle item are solid in color. This color blocks the names of the symbol's pins.

Rectangular outlines are drawn with the Wire tool. With this tool active, each pair of mouse clicks draws a line. By clicking the corners, this tool can easily create a rectangle for the symbol's border. This is shown in Figure 8.3.

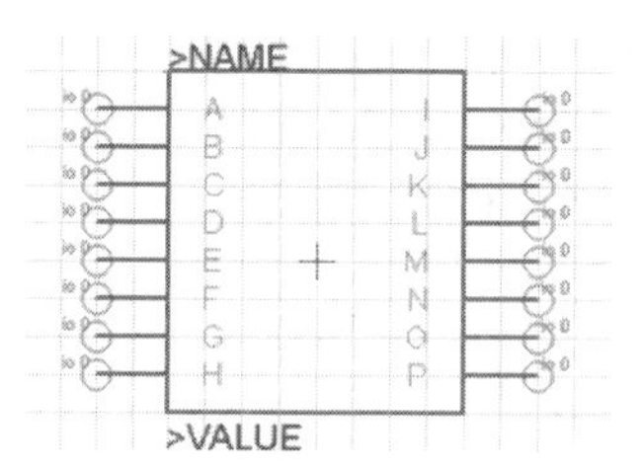

Figure 8.3: The Body of the SIMPLE Symbol

The following instructions explain how to complete the process of drawing the symbol:

1. Select the Wire item in the vertical toolbar and draw four lines in a rectangle, making sure the vertical lines touch the pins' internal edges. Click twice to stop drawing.
2. Select the Text item and set the text to >NAME. Place the text to the upper left of the symbol.
3. Select the Text item again and set the text to >VALUE. Place the text to the lower left of the symbol.
4. Click Ctrl-S or go to File > Save to save the package to the example-lib library.

The exact position of the text won't affect your design. By holding down the Alt key, you can position text using the alternate grid spacing.

8.2.2 Creating the TQFP16 Package

Now that the symbol is complete, the next step is to draw a package for the example device. A quad flat package (QFP) is a package for integrated circuits whose surface mount leads extend on all four sides. A common variant of the QFP is the thin quad flat package, or TQFP, whose profile isn't as thick. Most TQFP devices have between 32 and 256 leads, but our example SIMPLE-TQFP device has 16 leads with 4 on each side. Figure 8.4 shows what the actual device looks like.

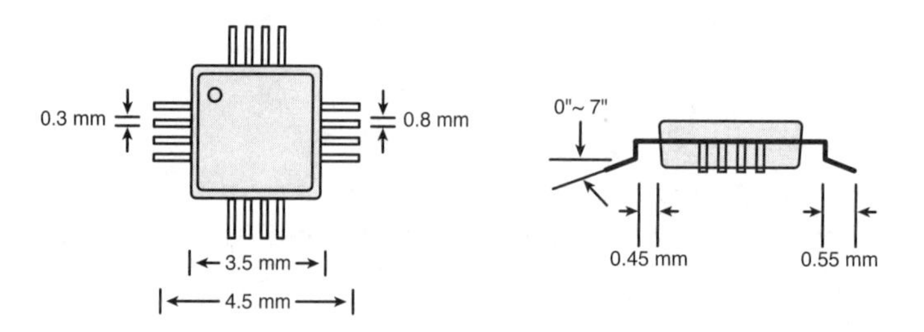

Figure 8.4: The 16-LeadThin Quad Flat Package

Figure 8.5 shows what the corresponding EAGLE package looks like.

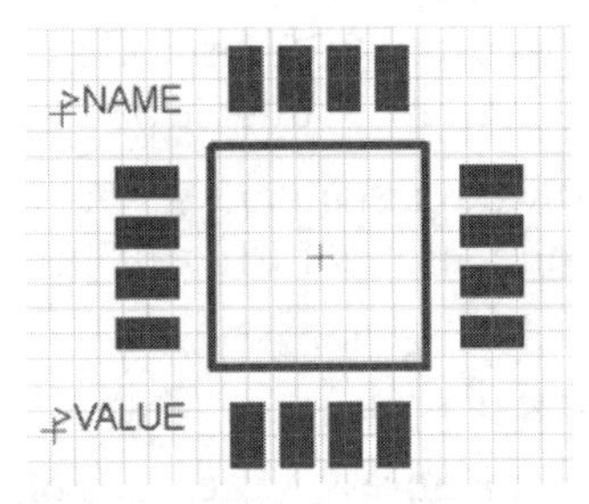

Figure 8.5: The SIMPLE-TQFP Package in EAGLE

To start creating this package, go to the horizontal toolbar in the library editor. Click the Package item to the left of the Symbol item. The Edit dialog will appear and list the different packages in the library. Enter TQFP16 in the text box and click OK.

Drawing the Package Body

The following instructions show how to draw the body of the TQFP16 package:

1. As shown in Figure 8.4, the dimensions of the package body are 3.5 x 3.5mm. To draw this in the editor, the grid spacing should be set to one-half the minimum dimension, or 1.75mm. Select the Grid tool, set the dimension to mm and the size to 1.75. Click OK and the grid marks will be spaced 1.75mm apart.
2. Select the Wire item and draw four 3.5mm wires around the origin.
3. Select the Circle item and draw a small circle in the upper right of the square. This will identify the location of Pad 1 of the device. If needed, press Alt to use the alternative grid spacing. The circle's size isn't important, but the radius can be set by selecting the Info tool and clicking the circle.
4. Select the Text item and enter >NAME. In the horizontal toolbar, set the text size to 0.4064. Place the text to the upper left of the package body.
5. Select the Text item again and enter >VALUE. Place the text to the lower left of the package body.

When the package is placed in a board design, the >NAME field is replaced by the name of the device instance.

Drawing SMD Pads

When designing a package, it's important to know the exact size and position of the component's leads. Otherwise, the leads won't come in contact with the board's pads and the device won't connect to the circuit properly.

In deciding on the size of the pads, there are two constraints: the pads must touch the device's leads and they must not come in contact with one another. As shown in Figure 8.4, the leads are 0.35mm wide and their centers are spaced 0.8mm away from one another. Therefore, a suitable pad width should be around 0.5mm.

EAGLE provides a standard set of SMD sizes, and the one whose width is closest to 0.5mm has dimensions 1.016 x 0.508.

The following directions show how to add these pads (called SMDs) to the SIMPLE package design:

1. As shown in Figure 8.4, the leads of the device are spaced 0.8mm apart, so the grid spacing should be set to 0.4mm. To configure this, select the Grid item in the toolbar, set the dimension to mm, and set size to 0.4. You may also want to set the alternative grid spacing to 0.04mm.
2. Select the Smd item in the vertical toolbar. In the horizontal toolbar, change the pad's dimension to 1.016 x 0.508. Place 16 pads around the square body, keeping approximately 0.6mm (a square and a half) between the edges of the pads and the lines that form the body.
3. Select the Name item and assign names to the SMD pads. Starting from the top left pad, assign numbers from 1 through 16 in a counterclockwise fashion.
4. Click Ctrl-S or go to File > Save to save the package to the example-lib library.

8.2.3 Creating the SIMPLE-TQFP Device

At this point, you've created a symbol with 16 pins and a package with 16 pads. Now it's time to create a device that will combine the symbol and package and associate the symbol's pins with the package's pads. The following steps show how this is done:

1. In the horizontal toolbar, the Device item can be found to the immediate left of the Package item. Click this, and in the Edit dialog's text box, enter **SIMPLE**. Click OK and then click Yes to create the new device.
2. Click the Add item in the vertical toolbar. In the dialog, choose the symbol called SIMPLE and click OK. Place the symbol somewhere near the center of the editor area in the upper left.
3. There are three buttons in the lower-right corner of the editor: New, Connect, and Prefix. Click the New button and a dialog will appear titled "Create new package variant for SIMPLE." In the upper right of the dialog, a box contains the names of the library's packages. Select TQFP16 but don't click OK just yet.
4. Below the box containing package names, enter **-TQFP** as the variant name. This will be appended to the device name to identify the device with the TQFP16 package. Click OK.
5. Click the Connect button. This brings up a dialog that creates associations between pins and pads. Figure 8.6 shows what it looks like.

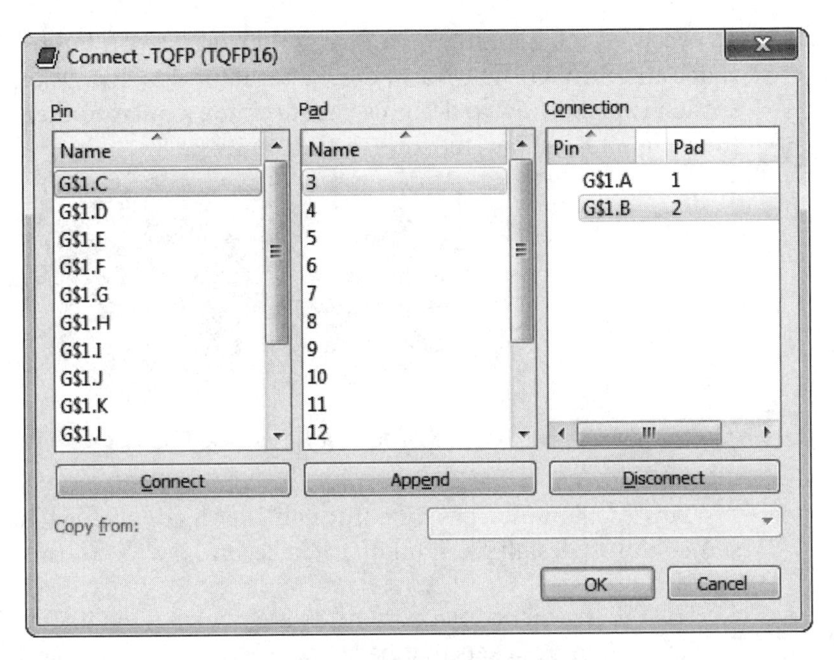

Figure 8.6: Creating Pin/Pad Connections

6. To associate a pin with a pad, select a pin name in the first box, such as G$1.A, and select the corresponding pad in the second box, such as 1. Then click the Connect button in the dialog to form the connection.
7. Continue connecting pins and pads until the first two boxes are empty. Click OK.

At this point, the new component is finished. If you would like to view its definition in XML, open the example-lib.lbr file in EAGLE's lbr directory and search for the SIMPLE symbol, the TQFP16 package, and the SIMPLE-TQFP device.

NOTE Multiple symbols can be associated with a package so long as the total number of pins equals the number of pads. These symbols form gates in the schematic editor.

8.3 Creating the VACUUM-TH

If you're working with surface-mount ICs, the SIMPLE-TQFP discussion should be sufficient for creating new devices. But if a device has an irregular shape or needs through-hole leads, the creation process changes. To demonstrate this, this section explains how to create an EAGLE component that represents a vacuum tube.

Vacuum tubes aren't as popular as they used to be, but because of their irregular shape and through-hole leads, they're ideal for this discussion. In particular, this section explains how to design a vacuum tube similar to the venerable 12AX7 vacuum tube. Figure 8.7 shows the tube resting on its side.

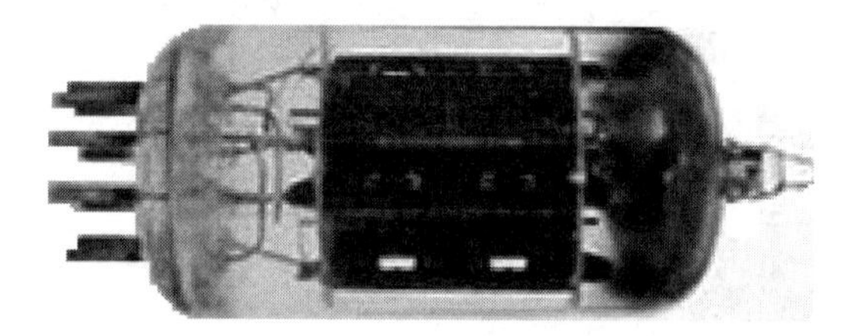

Figure 8.7: The 12AX7 Vacuum Tube

This vacuum tube has nine through-hole leads arranged in a circle. This section shows how to design the symbol, package, and device for this component.

NOTE In practice, vacuum tubes are not soldered to circuit boards. They're inserted into sockets that have been soldered to the board.

8.3.1 Creating the VACUUM Symbol

The component's symbol will be circular to match the actual device. It will contain nine smaller circles representing the leads, and each lead will be connected to a pin. Figure 8.8 shows what the symbol looks like.

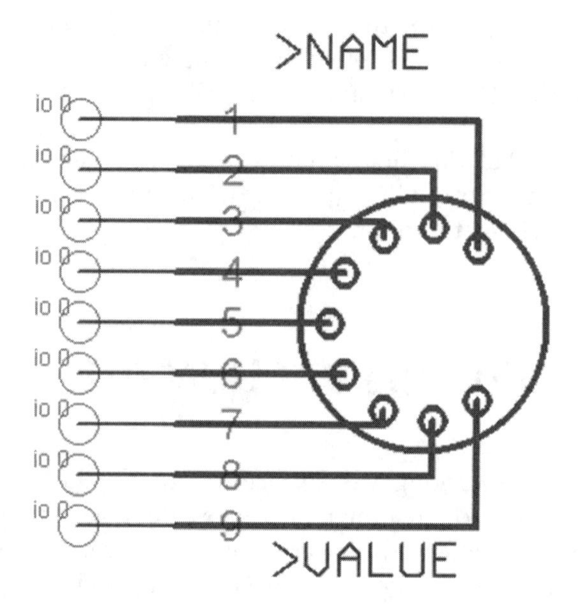

Figure 8.8: The VACUUM Symbol

The following directions show how to create this symbol in EAGLE.

1. Click the Symbol toolbar item in the library editor. In the Edit dialog, enter VACUUM in the text box and click OK. Click Yes to create a new symbol.
2. Make sure the grid is set so that the primary grid spacing is 0.1 inches and the alternate grid spacing is 0.01 inches.
3. Select the Circle tool and add a circle to the design. Using the Info tool, set its center to (0, 0) and its radius to 0.25 inches. This will serve as the outermost shape of the symbol.
4. Add another circle to the design. Set its radius to 0.025 inches and its position to (–0.19, 0.0). Create eight copies of this circle and position their centers at (–0.16, 0.1), (–0.16, –0.1), (–0.08, 0.17), (–0.08, –0.17), (0.02, 0.19), (0.02, –0.19), (0.11, 0.15), and (0.11, –0.15).
5. Activate the Pin tool and add nine pins to the left of the drawing in the manner shown in Figure 8.8. Make sure the pins are positioned symmetrically around the x-axis.
6. Activate the Name tool and assign the names 1 through 9 to the nine pins, proceeding from top to bottom.
7. Activate the Wire tool and draw lines between the centers of the nine circles to the nine pins. Hold the Alt key down to use the alternate grid.
8. Select the Text item and set the text to >NAME. Place the text above the topmost wire.
9. Select the Text item again and set the text to >VALUE. Place the text just beneath the lowest wire.
10. Click Ctrl-S or go to File > Save to save the symbol to the example-lib library.

In a real-world symbol, the pins' directions would be set appropriately. For the moment, this isn't a concern.

8.3.2 Creating the TH9 Package

The 12AX7 has a complex shape and it would take a great deal of work to draw a package that resembles it. In the interests of keeping matters simple, the package in this discussion is similar to the symbol from the preceding discussion: a circle containing circles. But when drawing the package, proper positioning is crucial and the inner circles are replaced by through-hole pads. Figure 8.9 shows what the package looks like.

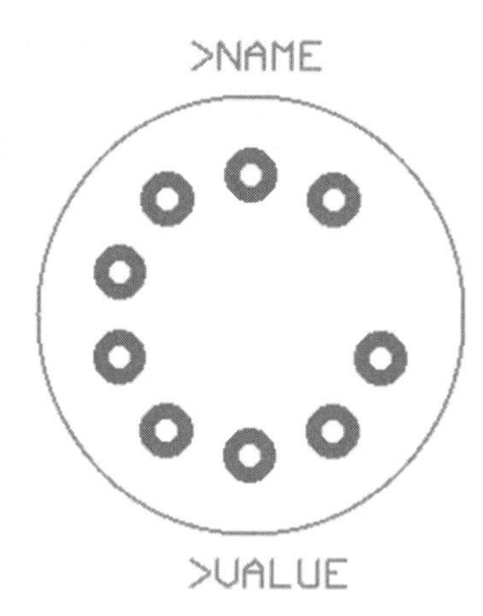

Figure 8.9: The TH Package for the Vacuum Tube Device

The following directions show how to construct this package in the EAGLE editor.

1. Click the Package toolbar item in the library editor. In the Edit dialog, enter **TH9** in the text box and click the OK button. Click Yes to create a new package.
2. Add a circle to the design. Set its center to (0, 0) and its radius to 0.36 inches. This will serve as the package's outline, so make sure its layer is set to tPlace (21).
3. Activate the Pad tool and add a through-hole pad to the design. Set its shape to a circle, its diameter to 0.086, its drill seting to approximately 0.0433, and its position to (0.14, 0.19). Create eight copies of this circle and position their centers at (0.0, 0.23), (–0.14, 0.19), (–0.22, 0.07), (–0.22, –0.07), (–0.14, –0.19), (0.0, –0.23), (0.14, –0.19), and (0.22, –0.07).
4. Find the pad centered at (0.14, 0.19) and set its name to 1. In counterclockwise order, number the rest of the pads 2 through 9.
5. Select the Text item and set the text to >NAME. Place the text above the package.
6. Select the Text item again and set the text to >VALUE. Place the text just beneath the package.
7. Click Ctrl-S or go to File > Save to save the package to the example-lib library.

8.3.3 Creating the VACUUM-TH Device

The VACUUM-TH device associates the VACUUM symbol and the TH9 package. The process of creating the device is similar to the process presented in Subsection 8.2.3. The following steps show how this is done:

1. Click the Device item in the horizontal toolbar. Enter **VACUUM** in the dialog's text box and click OK. Click Yes to create the new device.
2. Click the Add item in the vertical toolbar or press Ctrl-A. In the dialog, choose the symbol named VACUUM and click OK. Place the symbol somewhere near the center of the editor area in the upper left.

3. There are three buttons in the lower-right corner of the editor: New, Connect, and Prefix. Click the New button and a dialog will appear titled "Create new package variant for VACUUM." In the upper right of the dialog, a box contains the names of the library's packages. Select TH9, but don't click the OK button yet.
4. Below the box containing package names, enter **-TH** as the variant name. This will be appended to the device name to identify the device with the TH9 package. Click OK.
5. Click the Connect button. The dialog will present a list of pins named G$1.1 through G$1.9 and a list of pads named 1 through 9. Click the Connect button, and continue clicking until each pin is connected to its corresponding pad. Pin G$1.1 should be connected to Pad 1, Pin G$1.2 should be connected to Pad 2, and so on.
6. Click OK to complete the device. Click Ctrl-S or go to File > Save to save it to the example-lib library.

These steps are almost exactly similar to those needed to create the SIMPLE-TQFP device. Despite differences in technology and packaging, the process of creating devices doesn't change significantly.

8.4 Creating the TW9920

As discussed in Chapter 2, "An Overview of Circuit Boards and EAGLE Design," BGA stands for ball grid array. Instead of providing leads around the perimeter, the underside of a BGA contains tiny metal balls. These balls can be positioned together very closely, so BGA packages are a popular choice for modern high-density devices. Common BGA chips have hundreds of leads and Intel's latest processors are packaged as 479-pin FCBGAs (flip-chip ball grid arrays).

EAGLE doesn't provide direct support for BGA components, and given the number of leads, creating the symbol and package is a long and tedious effort. To automate these tasks, I developed two ULPs (User Language programs): make_symbol.ulp and make_bga_package.ulp. Both can be found in the ulp directory of this book's file archive.

This section shows how to create a new component called a TW9920. This is a multistandard video decoder/encoder chip from Intersil that has 100 pins in a ball grid array. I'll explain how to use make_symbol to construct the symbol and make_bga_package to create the BGA package.

> **NOTE** Command scripts aren't discussed until Chapter 10, "Editor Commands," and ULPs are introduced in Chapter 11, "Introduction to the User Language (UL)." But for this section, you don't need to understand either topic. All you need to know is that when a ULP file (*.ulp) is placed in EAGLE's ulp directory, it can be executed by executing `run ulp_name` in the text box above the editor area.

8.4.1 Making the TW9920 Symbol

The make_symbol.ulp program launches a dialog that creates a symbol based on the given pin configuration. Figure 8.10 shows what this dialog looks like for a symbol (TOUGH16) with 16 pins.

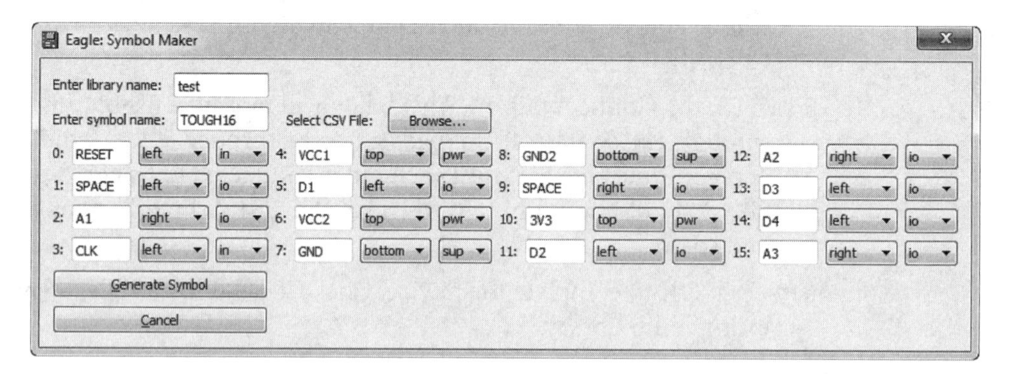

Figure 8.10: The make_symbol Dialog

The dialog requests a library name and a symbol name. Then it provides a numbered series of entries that accept information about the symbol's pins. Each entry accepts a pin's name, its side (left, right, top, and bottom) and its configuration (io, in, out, oc, pas, pwr, sup, hiz, or nc).

After the pin data is entered, the Generate Symbol button opens EAGLE's symbol editor and creates the symbol. The program examines each entry in sequence and creates a pin on the appropriate side. Figure 8.11 presents the symbol corresponding to the configuration data in Figure 8.10.

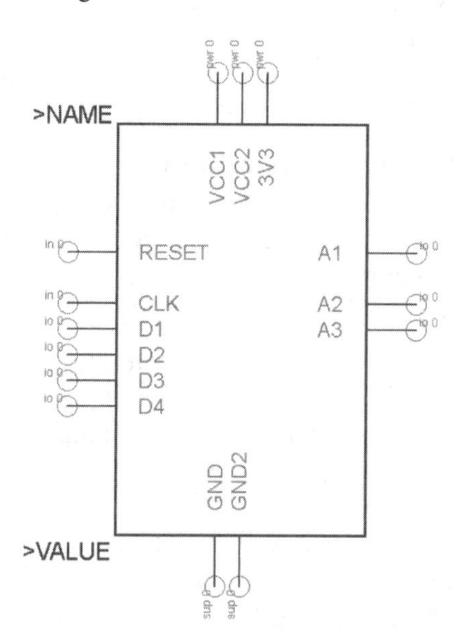

Figure 8.11: The Automatically Generated Symbol

The dialog has 16 entries but the symbol has 14 pins. This is because 2 of the dialog entries are named SPACE. When the program encounters a SPACE entry, it inserts a space on the appropriate side instead of a pin. This is why the generated symbol has one space below the RESET pin and another below the A1 pin.

The make_symbol program can create symbols with many more than 16 pin entries. The number of entries is determined by how the program is executed. The dialog in Figure 8.10 is created with four rows and four columns. This is created by typing the following command in an editor's text box and pressing Enter.

```
run make_symbol 4 4
```

Similarly, the following command creates a dialog with 160 entries (20 rows and 8 columns.

```
run make_symbol 20 8
```

Suppose a symbol's number of pins can't be evenly divided into rows and columns, such as a symbol with 131 pins. In this case, configure the dialog with a number of entries greater than the number of pins (such as 20 rows and 7 columns) and leave the final unwanted pins blank.

Instead of entering configuration data directly in the dialog, users can store the data to a file and load the file in the dialog. The file format is formatted with comma-separated values (CSV), in which each line consists of three fields: the pin's name, the pin's side, and the pin's configuration.

For example, the Ch8 directory in this book's file archive contains a file called TW9920.dat. Its first three lines are given as follows:

```
MUX0,left,in
MUX1,left,in
MUX2,left,in
```

The following directions show how to use this text file to create the TW9920 symbol for the example-lib library.

1. Find the make_symbol.ulp and make_bga_package.ulp files in the Ch8 directory of this book's file archive. Copy both files to EAGLE's top-level ulp directory.
2. Open an editor and find the text box above the editor area. Enter `run make_symbol 20 5` and press Enter.
3. The program will open the Symbol Maker dialog. Enter **example-lib** as the library name and **TW9920** as the name of the symbol.
4. Click the Browse... button and find the Ch8 directory in this book's file archive. Select the TW9920.dat file and press Open.
5. The ULP will read the data file and display the pin data in the dialog. Click the Generate Symbol button at the bottom of the dialog.
6. The ULP will open the symbol editor and draw the new TW9920 symbol as part of the example-lib library. To save this, go to File > Save in the menu.

NOTE For Boundary Scan Description Language (BSDL) files, EAGLE has a ULP called make-symbol-device-package-bsdl.ulp. This converts a BSDL description into a symbol, package, and device, and it can be found in EAGLE's top-level ulp directory. The BSDL format is described in the IEEE 1149.1 standard.

8.4.2 Creating the VFBGA100L-8X8 Package

BGA packaging makes it possible to fit more leads in a given area than other surface mount techniques. In general, the following assumptions can be made regarding these packages:

- The spherical leads (balls) are attached to the bottom of the rectangular (usually square) package.
- Every pair of adjacent leads are separated by the same distance. This distance is called the pitch.
- The leads are arranged in rows and columns, but there may be empty spaces inside the grid.
- The leads are named according to their row and column location. Columns are identified by letter and rows are identified by number starting from 1. Therefore, the lead in the second row and second column is B2.

Figure 8.12 displays a common arrangement of leads in a ball grid array. This contains two groups of leads: those that make up the outer rectangle and those that make up the inner rectangle.

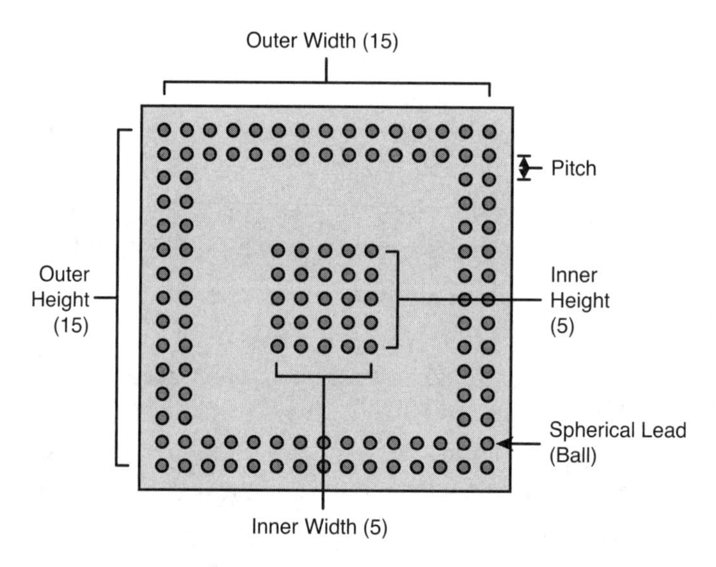

Figure 8.12: Leads in a Common Ball Grid Array (BGA) Package

Many BGA packages arrange their leads in rectangles, but others arrange their leads in irregular shapes. The make_bga_package progam can generate BGA packages in only three categories:

- BGAs with leads arranged in a solid rectangle
- BGAs with leads arranged in a rectangle containing rectangular space
- BGAs with leads arranged in an outer rectangle containing an inner rectangle (as shown in Figure 8.12)

Like the make_symbol.ulp program file discussed earlier, make_bga_package.ulp is located in the Ch8 folder of this book's file archive. The program can be executed in two ways: by going to File > Run ULP... in an editor's menu or by placing the ULP in EAGLE's top-level ulp directory and executing run make_bga_package in a text box.

When the ULP executes, it creates a dialog that accepts configuration information needed to generate a BGA package. Figure 8.13 shows what the dialog looks like.

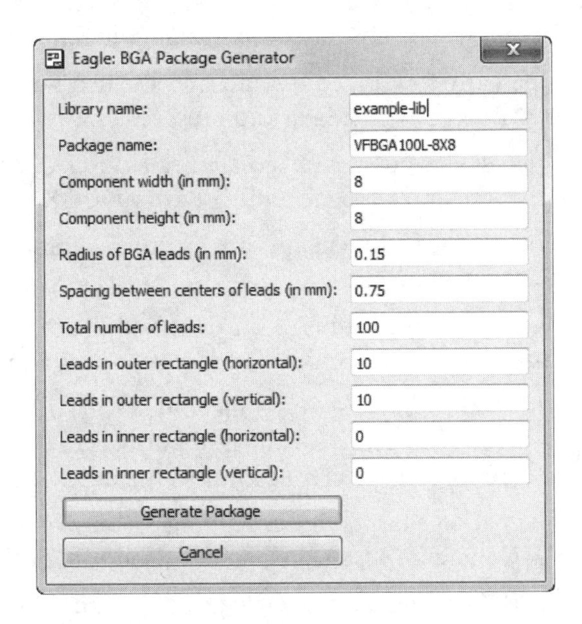

Figure 8.13: The Dialog Created by make_bga_package

In addition to the library name and package name, the dialog accepts nine parameters that define a BGA package:

- **Component width**—The component's horizontal dimension
- **Component height**—The component's vertical dimension
- **Total number of leads**—The number of spherical leads
- **Lead radius**—Radius of the BGA's spherical leads

- **Pitch**—Space between centers of the BGA's leads
- **Outer rectangle width**—Leads in the outer rectangle's horizontal dimension
- **Outer rectangle height**—Leads in the outer rectangle's vertical dimension
- **Inner rectangle width**—Leads in the inner rectangle's horizontal dimension
- **Inner rectangle height**—Leads in the inner rectangle's vertical dimension

It's important to note that the dialog expects the package's dimensions to be given in millimeters instead of inches. This is because the datasheets of large-scale devices frequently provide dimensions in millimeters.

The goal of this section is to create a package that corresponds to the TW9920 symbol created earlier. The name of this package is the VFBGA100L-8X8, where VFBGA stands for Very-thin Fine-pitch Ball Grid Array. The following directions explain how to use make_bga_package to create the VFBGA100L-8X8:

1. If it's not already there, copy the Ch8/make_bga_package.ulp file from this book's file archive to EAGLE's top-level ulp directory.
2. Open any EAGLE editor and find the text box above the editor area. Enter `run make_bga_package` and press Enter.
3. The program will open a dialog called BGA Package Generator. Enter example-lib as the library name and VFBGA100L-8X8 as the name of the package.
4. The package's dimensions are 8x8mm. Enter 8 for the component width and the component height.
5. Each lead has a radius of 0.15mm and the centers of adjacent leads are 0.75mm apart. Enter 0.15 for the lead radius and 0.75 for the lead spacing.
6. The package's leads are arranged in a solid 10x10 square. Enter 100 for the total number of leads, 10 for the number of leads in the outer rectangle (horizontal), and 10 for the number of leads in the outer rectangle (vertical).
7. The package has no inner rectangle, so leave the last two values at 0. Then click the Generate Package button at the bottom of the dialog.
8. The ULP will open the package editor and draw the new VFBGA100L-8X8 package as part of the example-lib library. To save this, go to File > Save in the menu.

Figure 8.14 shows what the resulting package looks like. This square package has a line across the upper-left corner. This identifies the lead in the upper-left corner as being A1. The leads in the first row proceed from A1 to A18 and the leads in the first column proceed from A1 to R1.

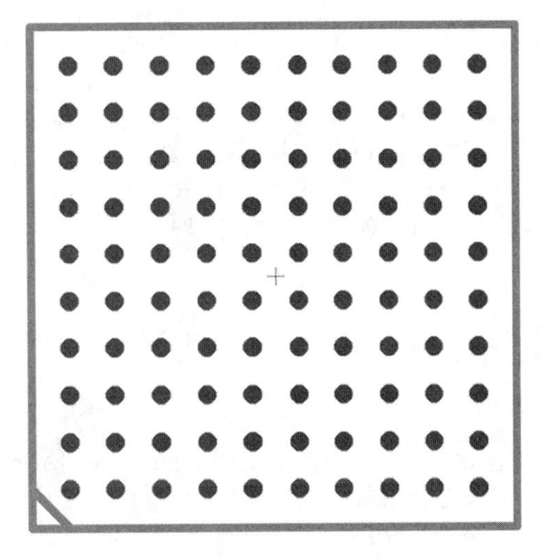

Figure 8.14: The VFBGA100L-8X8 Package

As a second example, Figure 8.15 depicts a 16x13mm package with 248 pads. In this case, the outer width is 20, the outer height is 16, the inner width is 4, and the inner height is 6.

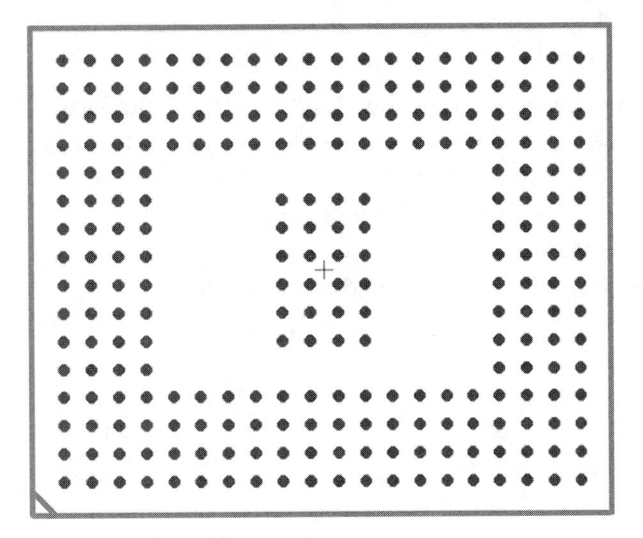

Figure 8.15: A 248-Pad BGA Package

8.4.3 Creating the TW9920 Device

When it comes to creating devices from symbols and packages, I can't write a ULP that improves on EAGLE's device editor. Therefore, no matter how complex the device, I recommend that you use this editor in the manner described in preceding sections.

For the TW9920, this means matching the pins from the TW9920 symbol to the pads in the VFBGA100L-8X8 package. Making 100 pin-pad associations is a tedious process, but in this case, there's an easier way. The TW9920 is a real component provided in the intersil-techwell library. Therefore, all you need to do is open the library file and copy the XML markup that defines the component's device. Appendix A explains the format of EAGLE library files.

8.5 Conclusion

Despite the breadth of EAGLE's library, I've found it necessary to create many custom components. Thankfully, EAGLE makes this process straightforward with three editors: a symbol editor, a package editor, and a device editor.

This chapter has presented three examples of creating new components. The first is a surface-mount device with 16 pins. The symbol and the package are simple rectangles that can be drawn with the Wire tool. Similarly, the pads of the package are rectangular and extend from the package's perimeter.

The second example component was a vacuum tube. In this case, the symbol and package are circular, and are drawn with the Circle tool. This device is through-hole, so its pads are created with the Pad tool instead of the Smd tool.

The last example involved creating a large-scale BGA device. The symbol and package can be drawn manually, but given the number of leads, it's better to automate the process. This chapter has presented two programs: The make_symbol program generates a symbol and make_bga_package generates a package with BGA leads.

The two programs described in this chapter were written using a combination of EAGLE's editor commands and the User Language (UL). Chapter 10 introduces the fascinating topic of editor commands and Chapters 11 through 13 present the UL in detail.

Chapter 9

Simulating Circuits with LTspice

One of the most recent and most interesting improvements to EAGLE is the ability to exchange designs with LTspice. This makes it possible for EAGLE users to export their designs to LTspice for simulations. It also allows LTspice users to export their schematics to EAGLE to create board designs and generate output files.

LTspice can't simulate circuits with complex digital devices like microcontrollers. But if you want to know how a switching-power supply or amplification circuit responds to different input frequencies, LTspice can help. The tool provides access to a wide array of devices, including RC filters, voltage converters, power regulators, and every type of transistor under the sun. If that isn't sufficient, users can create their own components and configure properties like gain, tolerance, and power rating.

The goal of this chapter is to explain how LTspice works and how to export LTspice schematics to EAGLE. A full presentation of LTspice's capabilities is beyond the scope of this book. Instead, this chapter explains how to design schematics and simulate their responses to input. But first, it's important to understand what LTspice is all about.

9.1 Introducing LTspice

I first encountered SPICE during an advanced electronics course in college. I needed to understand circuits with multiple transistors and simply studying the equations wasn't helping. It took me some time to learn the syntax behind SPICE commands, but once I did, I was able to simulate complex circuits and understand how they worked. SPICE was a tremendous help.

SPICE isn't as popular now as it was then, but its file format and command syntax are still used in academia and industry. This section explains the relationship between SPICE and LTspice and how to obtain the LTspice application.

9.1.1 SPICE and LTspice

The first version of the Simulation Program with Integrated Circuit Emphasis, or SPICE, was released in 1973 by researchers from the University of California at Berkeley. It wasn't the first circuit simulation application, but it was the first freely available tool that enabled general-purpose simulation. Instead of focusing on specific types of circuits, the second version of SPICE made it possible to analyze AC circuits, DC circuits, noise, and transient response.

Another feature that made SPICE attractive was its powerful language for circuit specification. For example, the following text defines a circuit with three 100-kQ resistors in series with a 5-volt voltage source.

```
Example netlist
v1 1 0 dc 5
r1 1 2 100k
r2 2 3 100k
r3 3 0 100k
.end
```

In addition to defining circuits, SPICE can also define new component models, including transistors and diodes. This allows engineers to simulate new semiconductor components before fabricating them. It also allows circuit designers to determine how a change in a device's characteristics will affect a circuit's behavior.

Now more than 40 years old, SPICE has improved dramatically since its initial release. It has acquired PC-compatibility, multiple graphical interfaces, and a library of thousands and thousands of analog components. Berkeley no longer provides updates, but many companies have released custom versions of SPICE, including PSpice from MicroSim and HSpice from Meta Systems. Linear Technology has released LTspice, which is both freely available and regularly maintained.

For designers of analog circuits, it's time-consuming to draw a schematic twice: once for design tools like EAGLE and once for simulators like LTspice. To simplify design and simulation, CadSoft augmented EAGLE with the ability to import and export schematics between EAGLE and LTspice.

Later in this chapter, I'll explain how to exchange schematics between EAGLE and LTspice. But first, it's important to install LTspice and see how the application works.

9.1.2 Obtaining LTspice

At present, Linear Technology provides LTspice for Windows and Mac OS only, but work is planned for Linux integration. On Windows, installing the executable is a three-step process.

1. Go to Linear's main page for circuit design software: http://www.linear.com/designtools/software.
2. Select the link entitled Download LTspice and save the installer to your computer.
3. Execute the installer, click Accept to accept the license agreement, and click Install Now to begin installation.

9.1.3 Simulation Example—Inverting Amplifier

Chapter 3, "Designing a Simple Circuit," presented the theory behind inverting amplifier circuits and showed how to design a simple amplifier in EAGLE. As a reminder, Figure 9.1 shows what the circuit looks like.

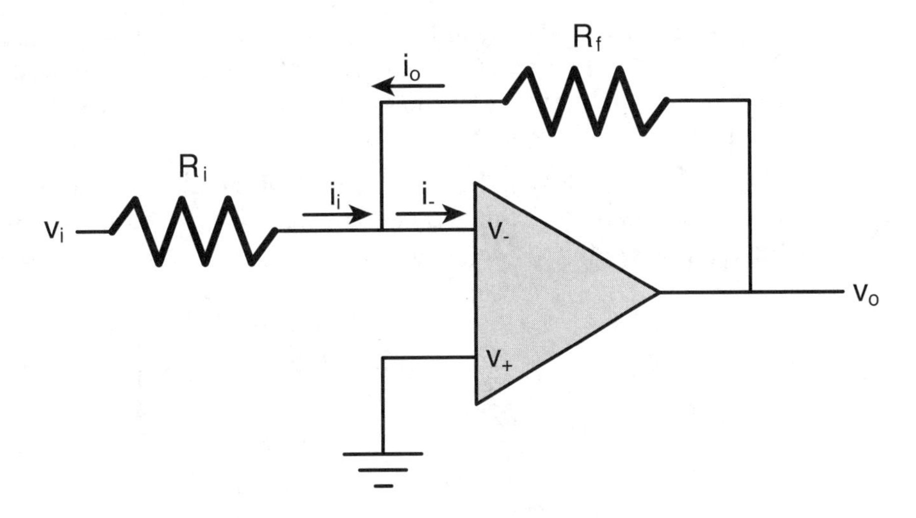

Figure 9.1: The Inverting Amplifier in EAGLE

This chapter shows how to design and simulate the same circuit in LTspice. Then we'll export the schematic for editing inside EAGLE.

9.2 Designing a Schematic

When installation is complete, I recommend that you launch the LTspiceIV executable. As you'll see, the user interface is simple, containing a title bar, a main menu, a toolbar, and a large, dark gray editor area.

To start using the tool, you need to create a schematic design. To do this, go to File > New Schematic or click the leftmost item in the toolbar. The editor area's color will change from dark gray to light gray, and the title bar will identify the current schematic with a *.asc suffix. Figure 9.2 shows what the application looks like.

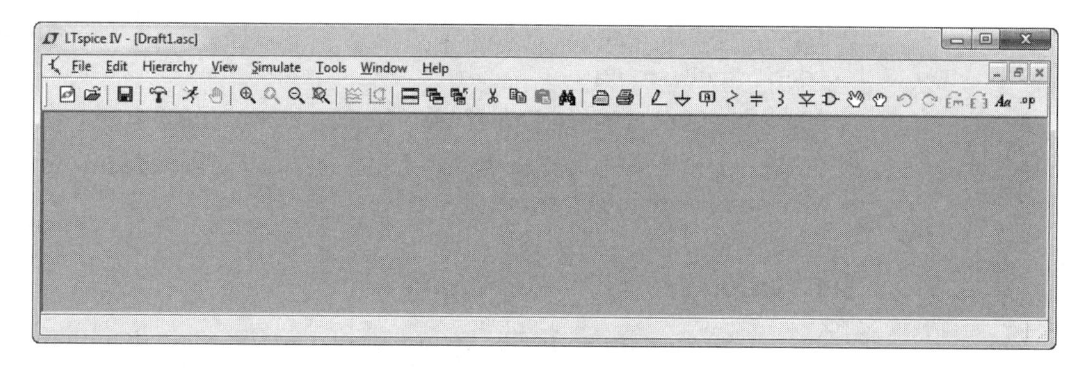

Figure 9.2: The LTspice Schematic Editor

Designing a schematic in LTspice is similar to designing a schematic in EAGLE. The process has four steps: add components, connect their pins, configure the components' properties, and add labels. This section walks through each of these tasks in LTspice.

Similar to the EAGLE toolbars, the LTspice toolbar contains an array of items whose functions aren't immediately apparent. This section focuses on a set of items on the right side of the toolbar. Figure 9.3 shows what this looks like.

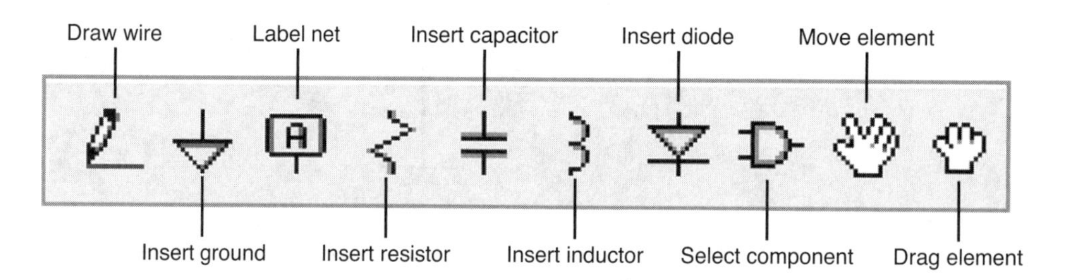

Figure 9.3: Toolbar Items for Schematic Design

The ultimate goal is to create a schematic for the inverting amplifier described in Chapter 3. You may want to familiarize yourself with this circuit before proceeding.

9.2.1 Adding Components

Simple components like resistors and diodes can be inserted into a schematic using items from the toolbar. The inverting amplifier circuit requires six simple components: two resistors and four ground elements. Therefore, the first step in drawing the schematic is given as follows:

1. Using the toolbar, add two resistors and four ground connections to the schematic. Their positions aren't important yet.

As with EAGLE, when you perform an operation with an LTspice tool, the application assumes you want to repeat it. To deactivate the current tool, right-click.

For the rest of the circuit's components, we'll need the Select Component Symbol dialog. This can be accessed in one of two ways: by selecting Edit > Component in the main menu or by selecting the Component item in the toolbar. This item is shaped like an AND gate, and in Figure 9.3, it's the third item from the right.

The Select Component Symbol dialog provides an assortment of advanced components to select from. Figure 9.4 shows what it looks like.

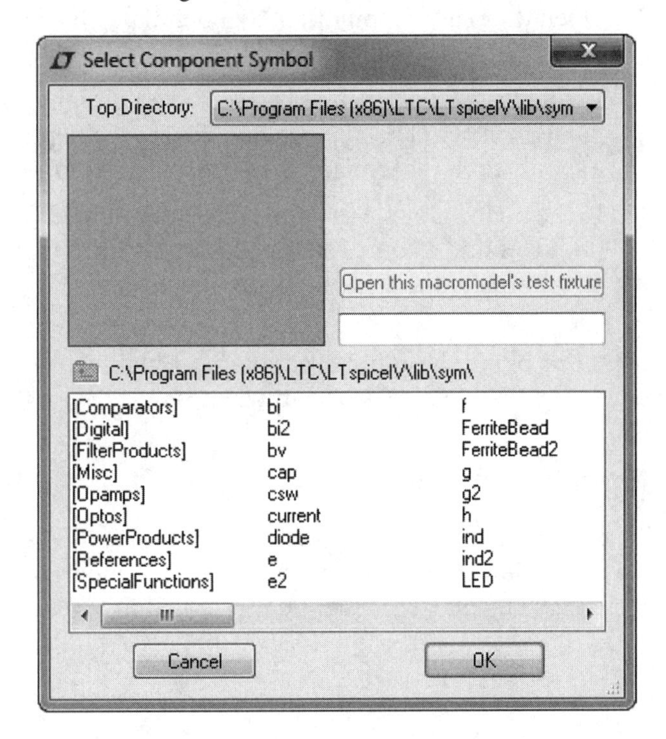

Figure 9.4: The Select Component Symbol Dialog

The bottom portion of this dialog contains a list that can be divided into two parts:

- Elements in the leftmost column surrounded in square brackets, such as [PowerProducts], identify directories containing advanced circuit components. Most of these are offerings of Linear Technology. For example, the [Optos] directory contains optoelectronic devices.
- Elements to the right of the directories are not surrounded by square brackets. These are generic circuit components, and include transistors (pnp, npn, nmos), diodes (zener, schottky), and power supplies (voltage, current).

NOTE SPICE and LTspice refer to circuit elements as models and macromodels. This chapter employs the term component.

When you select a circuit element, the upper-left portion of the dialog displays its representation in a schematic. The text in the upper right provides a brief description. If the device's name is preceded by LT, the button titled "Open this macromodel's text fixture" may become enabled. Clicking this button generates a schematic that displays how the component can be used in an example circuit.

To continue creating the inverting amplifier schematic, four components are needed: three voltage supplies and an operational amplifier. The following directions show how to insert them in the circuit.

2. Open the Select Component Symbol dialog by going to Edit > Component or selecting the toolbar item with the icon of an AND gate.
3. In the dialog's component list, scroll to the right until you find a component named voltage. This represents a general voltage supply. Double-click this to add a supply to the schematic. Add three voltage supplies in total.
4. In the dialog, double-click the directory named [opamps]. Scroll to the far right until you find a component named UniversalOpamp2. Double-click this to add an op-amp to the schematic.

At this point, the editor area should contain ten components: four grounds, three voltage supplies, two resistors, and an op-amp.

9.2.2 Moving Components

The components' exact positions aren't important, but an orderly arrangement makes it easier to view the circuit and create connections. For now, the goal is to position the 10 circuit elements in the manner presented in Figure 9.5.

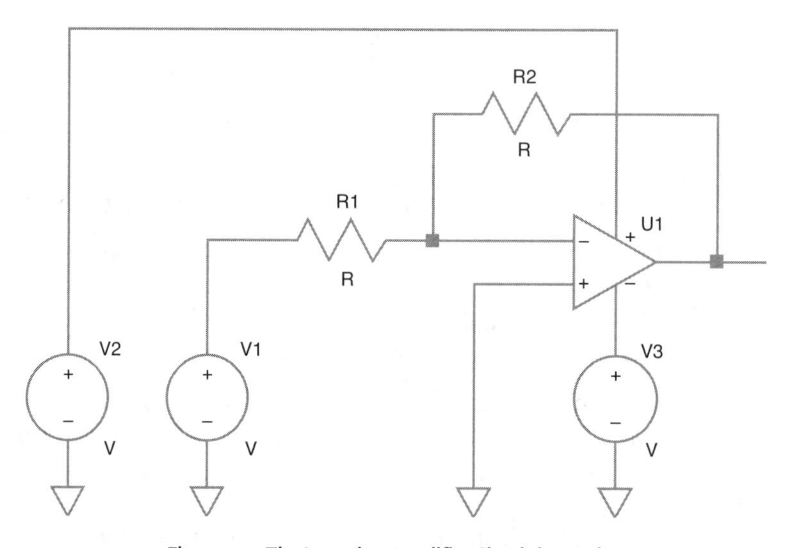

Figure 9.5: The Inverting Amplifier Circuit in LTspice

Toward the right of the toolbar is an item with a picture of an open hand. This is called Move, and if you select it (or go to Edit > Move), you can reposition the components within the design.

5. Rotate the two resistors from vertical to horizontal, and move all the components in the manner shown in Figure 9.5.

Unlike EAGLE, right-clicking in the editor doesn't rotate the selected component. Rotation is accomplished by activating the Rotate tool, which is fourth from the right in the LTspice toolbar.

9.2.3 Creating Connections

Once the design components are in position, the next step is to create connections between them. This is performed by the Wire tool, whose toolbar item is to the immediate left of the Add Ground item. The Wire tool can also be activated by pressing F3 or by going to Edit > Draw Wire in the main menu.

This leads to the sixth step in designing a schematic for the inverting amplifier:

6. Draw connections between the circuit's components as shown in Figure 9.5.

Each component has one or more small squares that represent its connection points, or terminals. When the Wire tool is active, clicking a terminal starts drawing a connection. Clicking a second terminal ends the connection. Any clicks between the starting and ending terminal form bendpoints in the wire.

LTspice doesn't have a Delete tool, but you can use the Cut tool instead. This is activated by selecting the toolbar item with the image of scissors or by pressing the Delete key.

9.2.4 Component Names and Values

EAGLE provides a Name tool for setting a component's name, such as R1, and a Value tool for setting its value, such as 100k. In LTspice, these operations are easy but not as obvious.

- To set a component's name, right-click the component's name and set the new name in the dialog.
- To set a component's value, right-click the component's body and set its value in the dialog.

For example, the leftmost voltage supply provides positive voltage to the op-amp, so it's name should be changed to V+. To make this change, right-click the supply's name and the dialog in Figure 9.6 appears.

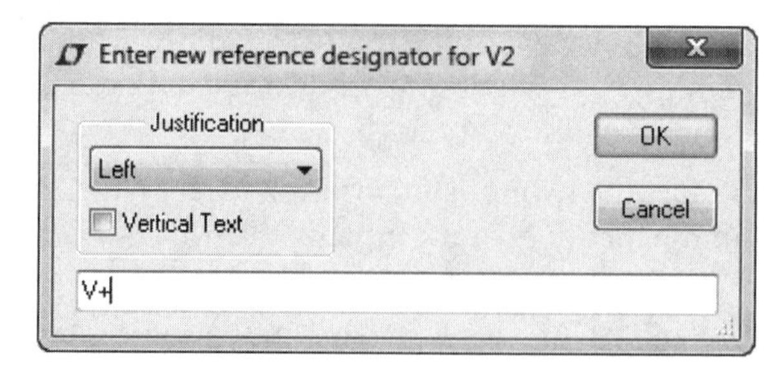

Figure 9.6: Changing a Component's Name

The text alignment options are Left, Right, Center, Top, and Bottom. In addition, the checkbox makes it possible to print the text vertically.

Setting the value is just as easy. To set this supply's voltage to 10V, right-click the supply. Figure 9.7 shows what the dialog looks like.

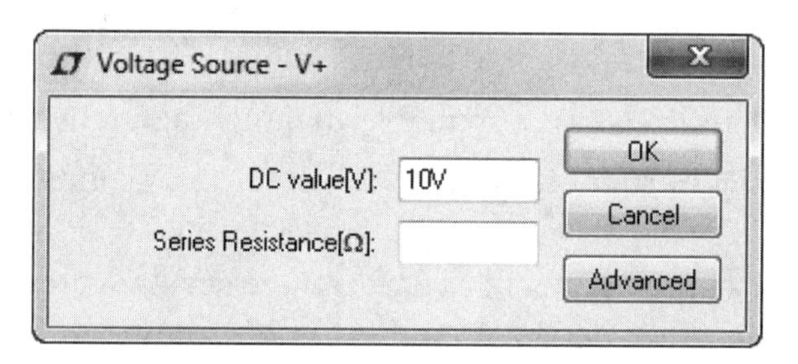

Figure 9.7: Changing a Component's Value

To simulate the inverting amplifier, the following name/value changes need to be made in the schematic:

7. The input resistor (R1 in Figure 9.5) connects a voltage supply to the op-amp's negative terminal. Change the resistor's name to Ri and its value to 1k. Change the other resistor's name to Rf and its value to 5k.
8. The name of the leftmost voltage supply should be set to V+ and its value should be set to 10V. The name of the rightmost voltage supply should be set to V– and its value should be set to –10V.
9. The central voltage supply provides the input signal, so its name should be changed to Vin. Its voltage should be set to a sine wave with an amplitude of 3V and a frequency of 120 Hz. To set this value, right-click the supply and click the Advanced button in the dialog.
10. In the new dialog, select the SINE function on the left. Toward the bottom of the dialog, set the frequency to 120 and the amplitude to 3. All the other fields can be left blank. (Do not set the AC Amplitude on the right).

At this point, the schematic is almost finished. The only step that remains is to set the name of the output net.

9.2.5 Named Nets

Simulating a circuit involves measuring voltage and/or current at different points in a design. For the sake of clarity, it helps to assign names to these points. In LTspice, the process of assigning a name to a net consists of three steps:

- Activate Label Net, which is located on the toolbar between the Ground and Resistor.
- In the dialog, assign a name and choose a port type (None, Input, Output, or Bidirectional).
- Select the net to receive the name.

In the case of the inverting amplifier, the output leaving the op-amp must be measured during the simulation. The following steps show how to set the net's name.

11. Activate the Label Net tool. In the Net Name dialog, enter **Vout** in the text box. Set the port type to Output.
12. Click the net leaving the output terminal of the operational amplifier. An output label should appear with the name Vout.

Now the inverting amplifier schematic is complete and the circuit is ready for simulation. Figure 9.8 shows the finished design.

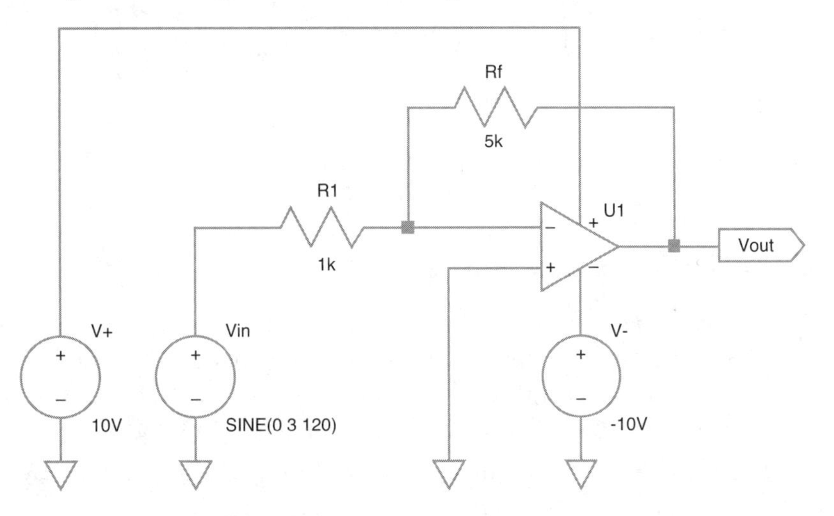

Figure 9.8: The Inverting Amplifier Design in LTspice

Before simulating, it's a good idea to examine the circuit and estimate how it should behave. The first part of Chapter 3 derived an equation that links the output voltage to the input voltage: $v_{out} = -(R_f/R_i)v_{in}$. In this case, Rf/Ri = 5 and the input voltage has an amplitude of 3V. Therefore, we would expect the op-amp to produce a sinusoidal wave with an amplitude of 15V.

But an operational amplifier can't produce voltage beyond that provided by its supply rails, which are set to +/– 10V. Therefore, we should expect to see a sinusoidal output in which the maximum output is clipped to 10V and the minimum is clipped to –10V. The next section verifies that this is the case.

9.3 Simulating the Circuit

Simulating a circuit in LTspice consists of three simple steps:

- Set simulation parameters.
- Execute the simulation.
- Configure how the simulator displays the results.

Performing these tasks requires a different set of toolbar items than those used to design the schematic. Figure 9.9 presents each of them.

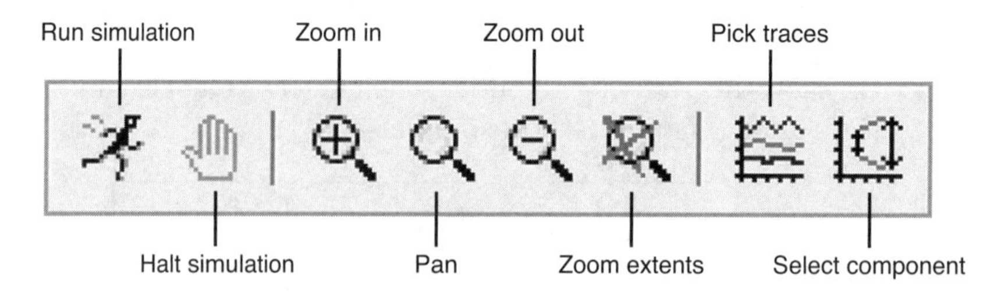

Figure 9.9: Toolbar Items for Circuit Simulation

9.3.1 Setting Simulation Parameters

If you open the Simulate entry in the main menu, the bottom option is called Edit Simulation Cmd. Selecting this option opens the Edit Simulation Command dialog, which provides options for controlling how the simulation should be performed. This dialog divides the simulation's parameters into six groups:

- **Transient**—Settings related to timing and initial voltage and current
- **AC Analysis**—Changing the AC frequency over the course of the simulation
- **DC Analaysis**—Changing the DC voltage over the course of the simulation
- **Noise**—Adding noise to the circuit

- **DC Transfer**—Computing the transfer function (relationship of output signal to the input signal)
- **DC op pnt**—Computing the DC operating point, where the capacitors are treated as open-circuits and inductors are treated as closed-circuits

The inverting amplifier is so simple that there's no need for advanced analysis. But it's a good idea to limit the time of the simulation. This not only reduces the amount of computation involved, but also makes it easier to view the results.

13. Open the Edit Simulation Command dialog by going to Simulate > Edit Simulation Cmd on the main menu.
14. In the Transient tab, set the Stop Time value to 0.05. Click the OK button and click in the schematic design to place the configuration data.

In addition to setting the time, LTspice also lets you choose which values to test during the course of the simulation. This is made possible by the Pick Visible Traces item in the toolbar. This bears the image of a graph and is located to the right of the zoom options.

When Pick Visible Traces is clicked, the Select Visible Waveforms dialog appears. This lists the different currents and voltages that can be tested. Figure 9.10 shows what it looks like.

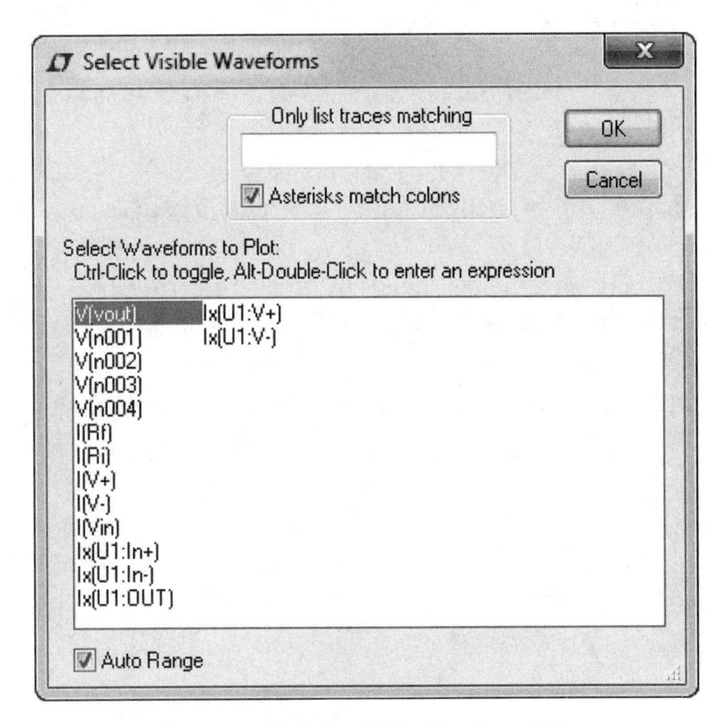

Figure 9.10: The Select Visible Waveforms Dialog

Like EAGLE, LTspice assigns its own names to nets, so it can be hard to see what each option represents. But the first option, V(vout), represents the voltage of the net named Vout. For the moment, this is the only net we'll simulate.

15. Launch the simulation by going to Simulate > Run in the main menu.
16. Click the Pick Visible Traces item in the toolbar. In the Select Visible Waveforms dialog, select the V(vout) option for testing and click OK.

9.3.2 Running the Simulation

When the simulation starts, LTspice divides the editor area into two horizontal halves. The bottom half contains the schematic and the upper half contains a graph of the test signal over the course of the simulation. Figure 9.11 shows what this graph looks like on my Windows 7 system.

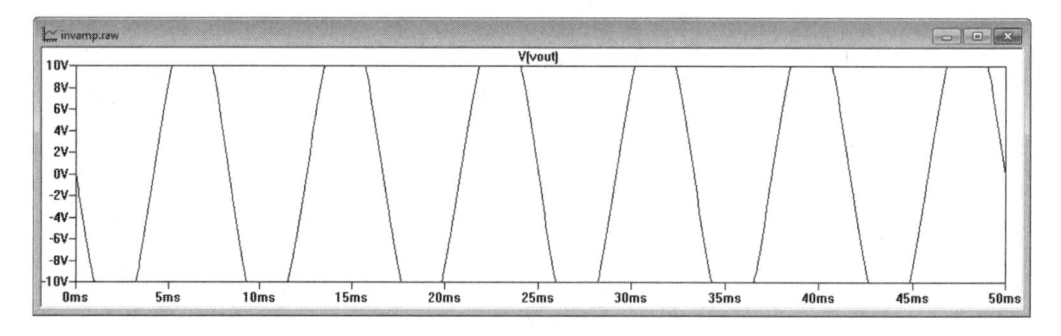

Figure 9.11: Simulation Result for the Inverting Amplifier (Voltage at Vout)

As shown, the voltage at Vout is a negative sine wave clipped at +/– 10V. To verify that this an inverting amplifier, it would help to see both the input and output voltages on the graph.

You can add new signals using the Select Visible Waveforms dialog described earlier, but there's an easier method: Select nets in the original schematic. The following steps show how this is done.

17. Click anywhere in the lower window to bring focus to the schematic editor.
18. In the circuit design, click the net just above the voltage supply named Vin. This adds the net's voltage to the simulation.

Figure 9.12 shows the simulation graph after the Vin net is selected. Now it's clear that the amplifier's output is inverted, or 180° out of phase with the input.

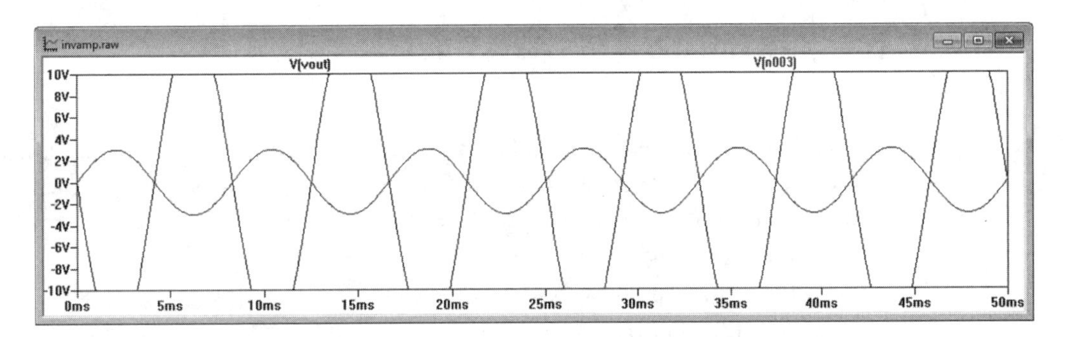

Figure 9.12: The Full Simulation Result (Vout and Vin)

If you hover your mouse over the Vin voltage supply, the pointer will change shape. This new shape represents an ammeter for measuring current. LTspice assumes you want to measure the current because the voltage inside the supply is undefined.

In addition to choosing new signals to simulate, you can also change the characteristics of the circuit's elements. For example, if you right-click Vin and reduce its amplitude from 3V to 2V, the resulting output amplitude of 10V won't be clipped by the positive and negative rails.

9.3.3 Configuring Simulation Appearance

LTspice provides many ways to change the appearance of the simulation results. These tools can be accessed through Plot Settings in the main menu or by right-clicking in the simulation window and selecting an option from the context menu.

- **Grid**—Draws horizontal and vertical lines that intersect the tick marks on the axes.
- **Manual limits**—Set the boundary values of the horizontal and vertical axis and define the distance between ticks.
- **Add Trace/Delete Trace**—Add signals to the simulation or remove them from the simulation.
- **Add Plot Pane**—Adds a new graph to the simulation window. The new graph will serve as the target for future simulation operations.
- **Mark Data Points**—Draws a dot for each data point.

The Color Palette Editor makes it possible to change the color of each aspect of the simulation, from the background to the axes to the graph of the signal values. This dialog can be accessed by going to Tools > Color Preferences on the main menu. I use this dialog to change the backgrounds of the editor and simulation graph to white.

9.4 Exchanging Designs with EAGLE

As of version 6.4, EAGLE makes it possible to import LTspice schematics into EAGLE and export EAGLE schematics to LTspice. Both operations are performed using a toolbar item in EAGLE's schematic editor. This item is located to the far right of the toolbar and is labeled LTCspice.

This toolbar item has an arrow, and if you click the arrow, you'll see a context menu with four entries:

- **Export**—Exports current schematic to LTspice
- **Export Setup**—Finds locations of the EAGLE and LTspice executables
- **Export Group**—Exports the selected group of components to LTspice
- **Import**—Reads an LTspice schematic file (*.asc) into the schematic editor

This section explains each of these options and starts with a discussion of importing LTspice schematics into EAGLE.

NOTE The first time you select any of the import/export operations, EAGLE will determine where LTspice is installed and present a dialog entitled EAGLE: Auto Setup Info. If the settings are correct, click OK. Then you'll be able to import/export designs normally.

9.4.1 Importing LTspice Schematics into EAGLE

EAGLE makes it easy to import LTspice designs into the schematic editor. If you click the Import option in the LTCspice context menu, the Eagle: LT-Spice Import dialog will appear. This dialog has two text boxes whose values need to be set:

- **Spice schematic**—the path of an LTspice schematic file (*.asc)
- **Eagle schematic**—the path of the EAGLE schematic file (*.sch) into which the LTspice schematic should be imported

When the LTspice schematic file is selected, EAGLE will assume you want to create a similarly named EAGLE schematic in the same directory. If you press OK, EAGLE will create a new schematic based on the LTspice schematic.

EAGLE recognizes many LTspice elements, but not all of them. In particular, it doesn't create signals based on LTspice's named nets.

9.4.2 Exporting EAGLE Schematics to LTspice

The first three options in the LTCspice context menu are Export, Export Setup, and Export Group. The second option, Export Setup, doesn't perform any export operations, but checks for the installations of EAGLE and LTspice. The Eagle: Auto Setup Info dialog presents the paths of the required applications. If any file can't be found, the design exchange between EAGLE and LTspice won't work properly.

The Export option converts the current schematic to an LTspice design, saves it to a file (*.asc), and opens LTspice to edit this file. The Export Group option is similar, but instead of exporting the entire schematic, it exports only the selected group of components.

Many components in EAGLE's library have no equivalents in LTspice. If this happens, EAGLE will display a warning about a "Different quantity of SpiceOrder and PINs in symbol XYZ." In general, this means the schematic can't be exported.

To ensure that designs can be simulated with LTspice, EAGLE provides a number of libraries with compatible components. These are located in the ltspice folder in the lbr directory. Table 9.1 lists each of them and the types of components it provides.

Table 9.1

LTspice-Compatible Libraries

Compatible Library	Components
capmeter.lbr	Impedance meter
Comparators.lbr	Signal comparators
Digital.lbr	Simple logic gates, buffers, and flip-flops
FilterProducts.lbr	Amplifiers and filters
lt-spice-simulation.lbr	Voltage/current sources
lt-supply.lbr	+V, –V, 0, COM, IN, OUT
Misc.lbr	Capacitors, resistors, miscellaneous
Opamps.lbr	Operational amplifiers
Optos.lbr	Optoelectronic devices
PAsystem.lbr	Resistors, capacitors, diodes
PowerProducts.lbr	Battery chargers, regulators, converters, controllers
References.lbr	Voltage references
rload.lbr	Rload resistor
SpecialFunctions.lbr	Voltage monitors, regulators, overvoltage protection
sym.lbr	Diodes, transistors, inductors

To show how these components can be used, EAGLE provides an example project featuring the LT3518 DC/DC converter. The can be found in the Control Panel by going to Projects > examples > ltspice.

The schematic design in the example project is 3518-eagle.sch. If you open this file in the schematic editor, you can export it to LTspice by selecting the arrow near the LTCspice toolbar item and selecting Export in the context menu.

9.5 Conclusion

When designing circuits, it's vital to find mistakes as early as possible. The longer it takes to find an error, the more expensive it is to fix it. For this reason, being able to simulate a circuit in the schematic stage is a great help.

LTspice builds on a long tradition of SPICE-related circuit simulators and provides many capabilities. The tasks of designing and simulating circuits are both straightforward, and in most cases, all you need are items from the toolbar. LTspice supports hundreds of circuit elements, including passive components, power supplies, and many types of power related elements. In addition, the simulation can be configured to analyze many different aspects of the circuit.

For the purposes of this book, the most important feature of LTspice is the ability to exchange schematic designs with EAGLE. LTspice can't simulate the majority of the components available in EAGLE's libraries, but some libraries contain components specifically created for LTspice simulation. As future versions of EAGLE are released, it's safe to hope that the compatibility between the two tools will improve and that we'll be able to simulate new types of designs.

Chapter 10

Editor Commands

At this point, you've learned a great deal about designing circuits with EAGLE, from defining a schematic to routing connections to generating artwork files. All the interaction with EAGLE has been accomplished with point-and-click operations, and this is fine for simple designs. But when designs have hundreds or thousands of components, mouse clicks become tedious and error-prone. For this reason, EAGLE makes it possible to automate operations using editor commands.

The syntax of these commands is easy to grasp. For example, the command `MOVE U3 (2.5 3.5)` moves the origin of component U3 to the position (2.5, 3.5). But it takes time to become familiar with all the new commands and their arguments. It's time well spent. Understanding EAGLE's commands has dramatically improved my productivity as a designer. By combining them with the User Language programs (ULPs) introduced in the next chapter, I've succeeded in automating tasks that would normally take weeks if I'd used traditional points and clicks.

10.1 Introducing Editor Commands

This chapter doesn't present all EAGLE's editor commands but focuses on those that create and configure circuit board designs. But before you learn about the individual commands, it's important to understand the basic features of the language. This section explains editor command syntax, command execution, wildcards, and the process of defining points.

10.1.1 EAGLE Command Syntax

Every editor command has the following form:

```
<cmd_name> <cmd_property> <cmd_property> <cmd_property> ...
```

`cmd_name` identifies the command's name and each `cmd_property` constrains how the operation should be performed. For example, the following command makes the editor's grid visible:

```
grid on
```

In a similar manner, this command sets the grid to use units of millimeters:

```
grid mm
```

This command chains both commands together:

```
grid on mm
```

This chained command performs two operations in sequence. It makes the grid visible first and sets the grid's units to millimeters second. Most of EAGLE's commands, including `grid`, can be executed in a similar manner, but keep in mind that order is important.

10.1.2 Executing Commands

Every editor in EAGLE has a text box above the design area. If you type a command and press Enter, the command executes. Figure 10.1 shows what the text box looks like in the schematic editor.

Figure 10.1: Entering a Command in an Editor's Text Box

To the right of the text box, you can see an arrow pointing downward. This opens a drop box that provides a list of previously entered commands. Similarly, if you press the down arrow on your keyboard, the most recent command appears in the box.

Rather than enter commands over and over again, you can store them in a file. This file is referred to as a script, and EAGLE has two requirements for scripts:

- The file name must have the suffix *.scr.
- Each command must end with a semicolon.

For example, if you want to create a script that uses separate commands to turn on the grid and set its units to millimeters, you could save the following lines to simple.scr:

```
grid on;
grid mm;
```

To execute a script from the user interface, go to File > Execute Script… in any of the EAGLE editors. In the dialog, navigate to the script file and click Open. EAGLE executes each of the script's commands in order. It's important to note that some commands are only applicable for certain editors. For example, the `AUTO` command launches the autorouter, and if you try to execute it in the schematic editor, an error results.

Scripts can also be executed with the `script` command. If you open an editor and execute `script simple.scr`, EAGLE executes the commands in the simple.scr file. You can also enter the script name without the suffix.

By default, EAGLE searches for scripts in its top-level scr folder. If a script has been saved to this directory, it can be executed without the full path. Otherwise, the script's full path must be provided.

> **NOTE** Commands operate on the design in the editor executing the script. There is a way to code scripts so that different commands execute for different editors. The last topic in this chapter discusses this in detail.

10.1.3 Wildcards and Shortened Forms

Many commands accept names, such as library names, component names, or package names. You can tell a command to operate on multiple items by using wildcards. A wildcard is a special character that takes the place of one or more regular characters.

EAGLE provides three wildcards: `?`, `*`, and `[]`. For the most part, they behave similarly in editor commands as they do in traditional programming languages.

- `?` can replace any single number or character, so `WI?E` can be used in place of `WIRE`, `WINE`, `WISE`, `WI2E`, and so on.
- `*` can replace any group of characters or numbers, including a group with no characters or numbers at all. Therefore, `WI*E` can be used in place of `WIE`, `WIRE`, `WIGGLE`, `WINDPIPE`, and so on.
- `[]` is similar to `?`, but can be used to match any single character inside the brackets. Therefore, `WI[RSL]E` can be used in place of `WIRE`, `WISE`, and `WILE`, but not `WIFE` because `F` isn't contained in the brackets.

In addition to providing wildcards, EAGLE allows shortened forms of commands. So long as there are enough characters to distinguish the command, EAGLE can execute it.

For example, if `rot` is entered, EAGLE understands that the intended command is `rotate`. But if `ro` is entered, EAGLE doesn't know what to do because the command could be `rotate` or `route`.

10.1.4 Defining Points

Many commands presented in this chapter require locations in the editor. I'll refer to these locations as *points*. For example, positioning a package in the board editor requires a single point—the intended position of the package's origin. Drawing a net in the schematic editor requires two points—the locations to be connected by the net.

There are two ways to define a point in an editor command:

- An x-y coordinate pair surrounded by parentheses, such as `(0.5 1.5)`. Note that the coordinates are not separated by a comma.
- A mouse click in the editor area.

As an example, consider the `move` command, which accepts the name of the component to be moved and its intended location. The following command moves the origin of component U3 to the point (–0.2, 3.6).

```
move U3 (-0.2 3.6)
```

The coordinates can be made unit-specific by following each coordinate with `mm`, `mic`, `mil`, or `in` without a space. This is shown by the following command.

```
move U3 (-0.2mm 3.6mic)
```

If `move` is executed without x-y coordinates, EAGLE assumes that the position is intended to be set with the mouse.

10.2 Schematic Editor Commands

The goal of this section is to show how operations in the schematic editor can be performed using editor commands. Table 10.1 lists 12 commands and provides a description of each.

Table 10.1

Editor Commands for Schematic Designs

Command	Description
`use`	Makes a library active or inactive
`add`	Selects a component from a library to be placed in the design
`name`	Changes the component's name
`value`	Changes the component's value
`attribute`	Adds/modifies a component's attribute

smash	Enables a component's name/value to be moved separately
move	Repositions a component in the design
rotate	Changes a component's angular orientation
net	Draws a net between two pins
bus	Draws a bus between pins
label	Assigns a label to a net or bus
frame	Creates a frame around the schematic

10.2.1 Use

In general, the first step in designing a schematic involves selecting components from a library. If you go to the EAGLE Control Panel and open the Libraries option in the main window, you'll see that many of the libraries have large green dots. A large dot implies that the library is marked. EAGLE only allows you to search for components in marked libraries.

The `use` command makes it possible to mark and unmark libraries. If `use` is followed by a library name, the library will be marked. The following command marks a library whose file is example.lbr.

```
use example.lbr
```

The suffix (*.lbr) can be omitted. Also, if `use` is followed by `-*` and a library name, the library will be unmarked. The following command unmarks the example library.

```
use -* example
```

10.2.2 Add

The `add` command selects a component to be placed in the schematic. This component could be a device in the schematic editor or a package in the board editor.

The only required properties of `add` are the name of the desired component and the point where it should be placed in the editor. The following command adds a `gnd` component to the editor at point (2.5, –1.5).

```
add gnd (2.5 -1.5)
```

When this is executed, EAGLE searches through all the marked libraries and presents a dialog that lists available `gnd` components. If the component's name is followed by `@` and a library name, EAGLE searches for the component in the named library. This provides the important advantage of not requiring user interaction.

For example, the following command tells EAGLE to select the `gnd` component from the `supply1` library.

```
add gnd@supply1 (2.5 -1.5)
```

In addition to setting the library, the `add` command can assign a name for the new component. To specify the name, follow the component's name and library by the component's intended name. For example, the following command adds the `VUSB` element from `eagle-book.lbr`, gives it the name `VTOP`, and places it at (0.5, 0.25).

```
add VUSB@eagle-book VTOP (0.5 0.25)
```

`add` also accepts an initial angular orientation for the component. This is set with `R` followed by an angle given in degrees. In the schematic editor, the angle must be either 0, 90, 180, or 270.

As an example, the following command adds `VUSB` with an initial angle of 90 degrees:

```
add VUSB@eagle-book R90
```

In the board editor, this rotation can take additional properties. The angle can take any value from 0 to 359.9, with resolution down to 0.1 degrees. If the `R` is preceded by `S`, the component's text won't remain aligned to the drawing's bottom or right sides. If `R` is preceded by `M`, the component will be mirrored in the y-axis.

10.2.3 Name

The `name` command can be used in one of two ways:

- If a component has been selected in the editor, the command `name` *`new_name`* sets the component's name to *`new_name`*.
- If a component's name is *`old_name`*, the command `name` *`old_name new_name`* changes the component's name to *`new_name`*.

I don't use the `name` command often. I usually set a component's name when I add it to a schematic with `add`.

10.2.4 Value

Many components accept a parameter that defines a value. A resistor accepts a value for its resistance and a capacitor accepts a capacitance. This can be assigned with the `value` command. This can be invoked with or without a component name.

If no component name is given, the value is assigned to every selected component in the design. If there are no selected components, the value will be assigned to each new component as it's added to the circuit. For example, the following command gives each new component the value of 10uF if it accepts a value:

```
value 10uF
```

If a component name is given before the value, the value will be assigned to the named component. The following command assigns the resistor R1 a value of 10k:

```
value R1 10k
```

The `value -*` command clears all the values in the design. Also, values can be assigned to multiple components by adding further name-value pairs. The following command assigns R3 the value of 100k and C3 the value of 20uF:

```
value R3 100k C3 20uF
```

10.2.5 Attribute

As explained in Chapter 4, "Designing the Femtoduino Schematic," an attribute represents a name/value pair associated with a component or design. To see a component's attributes, enter `attribute` without arguments and select the component in the editor.

`attribute` can also be used to create new attributes or modify existing ones. In this case, the command is followed by a part name, an attribute name, and a value. This is given as follows:

```
attribute part_name attribute_name 'attribute_value'
```

As shown, the value string must be contained inside single quotes. For example, the following command assigns an attribute to U2 whose name is TOL and whose tolerance is 0.05.

```
attribute U2 TOL '0.05'
```

By default, EAGLE displays attribute values in the editor. After this command is executed, EAGLE expects a mouse click to define where the value should be placed. Until the mouse is clicked, the attribute isn't assigned. This behavior can be configured with the `change` command, which will be discussed in a later section.

If an attribute for the component already exists with the same name, the attribute's value will be modified if the attribute isn't constant. If the attribute name is followed by `DELETE`, the attribute will be deleted. The following command removes the TOL attribute from U2.

```
attribute U2 TOL DELETE
```

The described method for setting attributes is valid for the schematic editor and the board editor. But when interfacing libraries, `attribute` can be used in different ways.

10.2.6 Smash

The `smash` command is as easy to understand as the tool. It can be used in one of two ways:

- If one or more components have been selected in the editor, `smash` separates the component's name and value text so that they can be moved separately.
- If a component's name is *`name`*, `smash` *`name`* separates the named component's name and value text so that they can be moved separately.

To the best of my knowledge, there is no command that unsmashes a component. Instead, with the Smash tool active, hold the Shift button and click the component in the editor.

10.2.7 Move

After a component has been placed in the schematic, its position (or more precisely, the position of its origin) can be set with the `move` command. This is usually called with two properties: the name of the component and the coordinates of its intended location. The following command moves the component V1 to (3, 3) in the editor's grid.

```
move V1 (3 3)
```

These coordinates identify a grid position by default, but you can set different units for x and y: `mm` (millimeters), `mic` (microns), `mil` (thousandths of an inch), and `in` (inches). The following command moves V1 to 5mm in the x-direction and 0.5 inches in the y-direction.

```
move V1 (5mm 0.5in)
```

In addition to moving components, `move` can reposition a smashed component's name text, value text, or attribute text. To identify the aspect to be moved, follow the part's name with `>NAME`, `>VALUE`, or `>`*`ATTR_NAME`*. The following command moves the name text of R3 to the position (4.8, 0.3).

```
move R3>NAME (4.8 0.3)
```

10.2.8 Rotate

`rotate` is similar to `move`, but the angle precedes the part's name instead of following it. The rules for setting the angle are given as follows:

- The angle must be given in degrees and preceded by R.
- In the schematic editor, the angle must be 0, 90, 180, or 270. In the board editor, the angle can be any value between 0.0 and 359.9.

- By default, the angle is added to the component's current angle. If the angle is preceded by `=`, the angular measure will be made the final angle.
- If M precedes R, the component will be mirrored in the y-axis.
- If S precedes R (or M) in the board editor, the text will not be required to be vertically or horizontally readable.

The following command adds 90° to the angle of U1 and sets the angle of U2 to 270°.

```
rotate R90 U1 =R270 U2
```

The following command mirrors R3 in the y-axis and adds an angle of 180°.

```
rotate MR180 R3
```

As with `move`, the component's name/value/attribute text can be rotated independently if the component has been smashed. For example, the following command mirrors R3's value text and sets its orientation to 180°.

```
rotate =MR180 R3>VALUE
```

10.2.9 Net

The net command makes it possible to draw connections in the schematic editor. Its only required arguments are the two points to be connected. For example, the following command draws a net from (0, 0) to (1, 1).

```
net (0 0) (1 1)
```

It's important to understand that `net` accepts points (grid locations) instead of pin names. This means that, to connect two pins, you need to know where the pins' connection points are located. This information can't be obtained using editor commands, but it can be obtained using the User Language, which will be introduced in Chapter 11, "Introduction to the User Language (UL)."

If `net` is followed by a name, the name will be assigned to the net's signal. Remember that EAGLE treats all nets with the same signal name as being electrically connected. The following command assigns the name `RESET` to the net between (0.75, 1.25) and (2.5, 3.5).

```
net RESET (0.75 1.25) (2.5 3.5)
```

If the name is preceded by `!`, EAGLE will place a line over the net's name, indicating that it's active-low. A net's properties can be further configured in three ways:

- If `net` is followed by `net_wire_width` and a value, EAGLE creates the net with the desired width.

- If `net` is followed by `auto_end_net_off`, EAGLE doesn't assume the net ends at another net, pin, or bus wire.
- If `net` is followed by `auto_junction_off`, EAGLE doesn't assume a junction should be created where the net ends at another net, pin, or bus wire.

10.2.10 Bus

The `bus` command is similar to the `net` command. In both cases, the only required arguments are the points to be connected. One difference between them is that a bus name can have indices specifying the nets that make up the bus. These indices must be enclosed by square brackets and separated by two periods.

As an example, the following command creates a bus called `MEMBUS`. Its width is set to 16 and its connection points are (0.5, 1.5) and (2.5, 1.5).

```
bus MEMBUS[0..15] (0.5 1.5) (2.5 1.5)
```

In addition to indices, a bus's wires can be specified by following the bus name with wire names separated by commas. For example, the following command creates a bus called `CLKBUS` that contains nets `ACLK` and `BCLK`, and a seven-signal bus called `D`:

```
bus CLKBUS:ACLK,BCLK,D[0..8] (0.5 1.5) (2.5 1.5)
```

If the `bus` command is followed by `bus_wire_width` and a value, EAGLE draws the bus with the specified width. The default width is 30 mil.

10.2.11 Label

After a net or bus has been added to the design, the `label` command can be used to display its name. As with `net` and `bus`, this command requires two points. `label` is used in the following manner:

1. Enter `label` in the schematic editor's command box and press Enter.
2. Click a net or bus in the design or follow `label` with a point on a net or bus.
3. Click to set the position of the label or follow `label` with the desired location.

It's important to note that `label` doesn't accept the text that should be printed. The label of a net or bus is automatically assigned the name of its signal.

10.2.12 Frame

After a schematic's components are positioned and connected, the `frame` command can be used to draw a frame around the design. This accepts four arguments:

1. The number of columns
2. The number of rows

3. A point specifying one corner of the frame
4. A point specifying the opposite corner of the frame

For example, the following command tells EAGLE to create a frame with eight columns and four rows. Its opposite corners are (–3.0, –1.5) and (3.0, 1.5):

```
frame 8 4 (-3.0 -1.5) (3.0 1.5)
```

By default, EAGLE labels the columns and rows on all four sides of the frame (numbers on top and bottom, letters on left and right). To change this behavior, specify which of the sides (`top`, `bottom`, `left`, and `right`) should have labels. The following command creates a frame with labels on every side except the bottom.

```
frame 8 4 top left right (-3.0 -1.5) (3.0 1.5)
```

10.3 Commands for Board Designs

Like the schematic editor, the board editor provides a command box for entering commands. Designers can enter many of the commands discussed in the preceding section, but nine are particularly useful for board designs: `grid`, `layer`, `mirror`, `signal`, `ratsnest`, `route`, `ripup`, `via`, and `auto`. Table 10.2 lists these along with their functions and properties.

Table 10.2

Editor Commands for Board Designs

Command	Description
`grid`	Configures the properties of the editor's grid
`layer`	Sets the active layer
`display`	Selects which layer or layers should be displayed in the editor
`mirror`	Flips a part's layer from top (1) to bottom (16) or from bottom to top
`signal`	Creates connections between pads
`ratsnest`	Minimizes the length of the design's airwires
`route`	Creates traces from airwires
`ripup`	Opposite of route—returns traces to airwires
`via`	Creates a via
`auto`	Launches the autorouter

10.3.1 Grid

Before laying out devices in the board editor, it's important to configure the editor's grid. This is accomplished with the `grid` command, which can set a number of properties, including these three:

- **Visibility**—`on` or `off`
- **Units**—`mic`, `mm`, `mil`, or `inch`
- **Display**—`dots` or `lines`

In addition, the grid spacing can be set by adding a number to the command. If two numbers are given, the second number determines the grid factor—the distance between displayed dots/lines.

For example, the following command tells EAGLE to show the grid as a series of dots where adjacent dots are spaced 0.5mm apart and every fifth dot will be displayed:

```
grid on dots mm 0.5 5
```

To configure the editor's alternate grid, follow `grid` with `alt`. The following command sets the alternate grid to use a spacing of 0.05mm.

```
grid alt mm 0.05
```

One interesting feature of `grid` is that names can be assigned to grids and then configured by name. To set a grid's name, follow `grid` with an equals sign and the name. The following command tells that the name `fine_grid` should be associated with a dotted grid whose spacing is set to 0.1 inches:

```
grid = fine_grid dots inch 0.1
```

To set a name for the current grid, use the command `grid = name @`. To delete a name, use `grid = name`.

10.3.2 Layer

The `layer` command accepts a number that sets the current layer in the editor. New circuit elements and electrical connections will be placed in this layer.

As discussed in Chapter 5, "Layout and Design Rules," Layers 1 through 16 are routing layers, with 1 being the top layer and 16 being the bottom. The following command makes the bottom layer the current layer:

```
layer 16
```

Any name after the layer number serves as the layer's identifier. EAGLE's predefined layers range from 1 through 100, so the following command defines a new Layer 105 and gives it the name `NewLayer`:

```
layer 105 NewLayer
```

If the layer number is preceded by a minus sign, the layer will be deleted if it is empty. If the layer isn't empty, EAGLE will display an error message.

10.3.3 Display

The `layer` and `display` commands are confusingly similar. `layer` tells the editor which layer should be active. That is, when new elements are inserted in the design, they're placed in the layer made active by `layer`.

The `display` command tells the board editor which layer or layers to make visible. In general, this is followed by one or more numbers corresponding to layers. In Chapter 5, Table 5.1 lists a number of important layers and their contents.

The following command displays five layers containing information related to the top surface of the board.

```
layer 1 21 23 25 27
```

In addition to numbers, `layer` can be invoked with any of the following strings:

- `ALL`—All layers should be displayed.
- `NONE`—No layers should be displayed.
- `TOP`—Displays the top layer (Layer 1).
- `BOTTOM`—Displays the bottom layer (Layer 16).

10.3.4 Mirror

The `mirror` command moves a circuit element from the top layer to the bottom layer or from the bottom to the top. This is important for components in the board editor, which EAGLE positions on the top layer by default.

To mirror a component, follow the `mirror` command with its name. The following command mirrors the layer of U16.

```
mirror U16
```

If no name is given, EAGLE applies the mirror operation to the next element selected in the editor.

10.3.5 Signal

`signal` creates unrouted connections (airwires) between pads in a board design. Like the `net` command, the airwire can be defined with two points that correspond to the pads' locations. In addition, `signal` can specify a name for the airwire's signal.

The following command creates an airwire named `GND` from (2.0, 2.0) to (4.0, 4.0).

```
signal GND (2.0 2.0) (4.0 4.0)
```

Unlike the `net` command, `signal` can also accept the *names* of the connection points. For `signal`, each pad name must be preceded by the name of the pad's circuit element. The following command creates a connection named `DATA` between Pad A of U3 and Pad B of U4.

```
signal DATA U3 A U4 B
```

The `signal` command isn't limited to connecting two points. If `signal` is called with more than two points/pads, the airwire connects each of the given positions. This is shown by the following command.

```
signal DATA U3 A U4 B U5 C U6 D
```

10.3.6 Ratsnest

Before routing a design in the board editor, designers may find it helpful to minimize the lengths of its airwires. This is accomplished by the `ratsnest` command.

By default, `ratsnest` affects every airwire on the board. But this command also accepts names of individual signals. The following command optimizes the airwires carrying signals `sig1`, `sig2`, and `sig3`:

```
ratsnest sig1, sig2, sig3
```

To prevent airwires from being optimized, precede the list of signal names with an exclamation point. The following command performs `ratsnest` on every airwire except those carrying signals `sig4` and `sig5`:

```
ratsnest ! sig4 sig5
```

In the process of optimizing airwires, `ratsnest` computes the areas of polygons carrying signals. The board editor normally depicts polygons as outlines, but `ratsnest` draws the region of each polygon in full. This can be time-consuming for large-scale designs, so EAGLE allows you to control this behavior with the `set` command and the `polygon_ratsnest` property, which can be set to `on` or `off`.

For example, the following commands turn off polygon computation and optimize the board's airwires.

```
set polygon_ratsnest off
ratsnest
```

10.3.7 Route

The `route` command creates traces from airwires. Its usage requires two points. The first specifies the airwire to be routed. The end of the airwire closest to the point will be chosen as the starting point of the trace. The second point specifies its ending point.

If `route` is followed by a numeric value, that value will set the width of the trace. In addition, the `set` command can configure aspects of the routing such as the wire bend type (`set wire_bend #`) and the snap length (`set snap_length #`). The `set` command is discussed in a later section.

10.3.8 Ripup

`ripup` is the opposite of `route`. That is, it unroutes traces and converts them to airwires. If executed without parameters, this command unroutes every visible routed wire. But its behavior can be constrained in the following ways:

- `ripup sig1 sig2`—Unroutes signals `sig1` and `sig2`
- `ripup ! sig1 sig2`—Unroutes every signal except `sig1` and `sig2`
- `ripup @`—Converts all polygons into outlines
- `ripup @ poly1 poly2`—Converts polygons `poly1` and `poly2` into outlines
- `ripup @ ! poly1 poly2`—Converts all polygons into outlines except `poly1` and `poly2`

Keep in mind that `ripup` affects traces and polygons in visible layers. Features in hidden layers remain routed. To change layer visibility, use the `display` command followed by the layer number(s) or name(s).

10.3.9 Via

The `via` command creates a conductive hole (via) connecting signals on different layers. At minimum, its usage consists of the command name followed by a point.

`via` can be configured by following the command name with one of the following properties:

- **Signal name**—EAGLE will connect the via to the named signal.
- **Diameter**—Specifies the via's diameter.
- **Shape**—The via's shape can be set as `square`, `round`, or `octagon`. In inner layers, vias are always round.
- **Layers**—Connecting layers can be set with *x-y*, where *x* and *y* are layer numbers.
- **Solder mask**—As explained in Chapter 6, "Routing," if a via's diameter is less than the Masks/Limit setting in the Design Rule Check, it will be covered with solder mask. If the `via` command is followed by `stop`, it will not be covered with solder mask, regardless of its diameter.

As an example, the following command creates a via connected to the VCC signal that connects Layers 1 and 16. It's located at (150mil, 1340mil) and its diameter is 16 mil.

```
via 'VCC' 16mil 1-16 (150mil 1340mil)
```

Note that the signal's name must be surrounded in single quotes. By default, when a signal name is set for a via, that name will be applied to all other vias.

10.3.10 Auto

After a board's components are put into position, the autorouter can convert its airwires into traces. This is launched with the `auto` command. If executed without parameters or a semicolon, `auto` brings up the Autorouter Setup dialog discussed in Chapter 6. If executed with a semicolon, `auto` immediately starts the autorouting process with default settings.

The `auto` command accepts three properties that configure its operation:

- **Signal names**—If `auto` is followed by the names of one or more signals, only the nets carrying those signals will be routed. If the signal names are preceded by `!`, all the signals except those carrying the named signals will be routed.
- **Followme**—If `followme` is given as one of the parameters, `auto` will launch EAGLE's Follow-me router. Chapter 6 explains how this works.
- **File access**—If `auto` is followed by `LOAD` and the name of a *.ctl file, the autorouter will use the settings given in the named file. If auto is followed by `SAVE`, the autorouter will save its settings to the file.

As an example, the following command launches the autorouter for every signal except `data1` and `data2` and saves its operating parameters to config.ctl.

```
auto ! data1 data2 save 'config.ctl'
```

After the autorouter starts, its routing can be halted with the `stop` command. Then `ratsnest` can be used to unroute the connections.

10.4 Commands for Library Interface

The main reason I like EAGLE commands is they allow me to automate the creation of libraries and library components. Table 10.3 lists the commands that make this possible.

Table 10.3

Editor Commands for Interfacing Libraries

Command	Description
`open`	Opens an existing library or creates a new one
`edit`	Opens a design file or library component for editing
`write`	Saves library changes to the library file
`wire`	Draws a line
`arc`	Draws a circular arc
`rect`	Draws a filled rectangle
`circle`	Draws a filled or unfilled circle
`pin`	Creates a pin for a symbol
`pad`	Creates a through-hole pad for a package
`smd`	Creates an SMD pad for a package
`prefix`	Associates the currently edited component with a name prefix
`package`	Changes the package associated with the currently edited device
`technology`	Changes the technology associated with the currently edited device
`connect`	Create associations between a device's pins and pads

10.4.1 Open

The open command tells EAGLE to open a library file or create a new one. It has only one usage: `open` *`libfile_name`*, where *`libfile_name`* is the name of the library file. The *.lbr suffix isn't required, but if the library isn't in EAGLE's top-level lbr folder, the file's path is required. The frames.lbr library is in the lbr folder, so it can be opened with the following command.

```
open frames
```

It's important to see the difference between `open` and the `use` command discussed earlier. `use` marks a library as active for the purpose of adding components to a schematic. In contrast, `open` tells EAGLE to open an editor to edit a library.

10.4.2 Edit

Like `open`, `edit` accepts only one argument: the target to be edited. This can be a design file (circuit.sch or circuit.brd) or an individual sheet of a schematic (.s1, .s2, and so on). This discussion focuses on editing library components.

As discussed in Chapter 8, "Creating Libraries and Components," a library component has three instantiations: a symbol, a package, and a device. All three can be edited with `edit`, but the component name must have the appropriate suffix:

- **name.sym**—Edits a symbol for the name component
- **name.pac**—Edits a package for the name component
- **name.dev**—Edits a device for the name component

If the component hasn't been created, EAGLE will display a dialog asking to create the new component. If Yes is clicked, the new component will be created and EAGLE will open the appropriate editor. This behavior can be turned off with the following command.

```
set confirm yes
```

10.4.3 Write

After a design or library has been modified, the `write` command will save the modifications to the library file. Without arguments, this writes data to the file that's currently open in the editor. If `write` is followed by a file name, the data will be written to the named file instead.

If `write` is followed by a filename starting with `@`, EAGLE will write the design or library data to the named file and then open this file in the editor. The following command writes the current result to new_circuit and loads this in the editor.

```
write @new_circuit
```

Note that EAGLE always tries to keep the schematic and board design synchronized. Therefore, if a schematic design is saved to a new file, a new board file will be created as well.

10.4.4 Wire

When designing a new symbol or package, the first step usually involves drawing a border representing the shape. For a symbol, the most common shape is an unfilled rectangle, which is commonly used to represent integrated circuits. This can be created easily with the `wire` command.

At its simplest, `wire` can be followed by points representing the starting and ending points of the line. The following command creates a line from (–1, 0) to (0, 1).

```
wire (-1 0) (0 1)
```

`wire` isn't limited to two points. The following command draws three lines: one from (–1, 0) to (0, 1), one from (0, 1) to (1, 1), and one from (1, 1) to (1, 0).

```
wire (-1 0) (0 1) (1 1) (1 0)
```

10.4.5 Arc

`arc` draws a circular arc in the design. Although `wire` requires two points, `arc` requires three points:

- The first point sets a point on the arc.
- The second point defines the circle containing the arc.
- The third point sets the line through the arc's circle containing the endpoint.

Figure 10.2 shows how these points define an arc. They can be specified using coordinates such as `(x y)` or with mouse clicks.

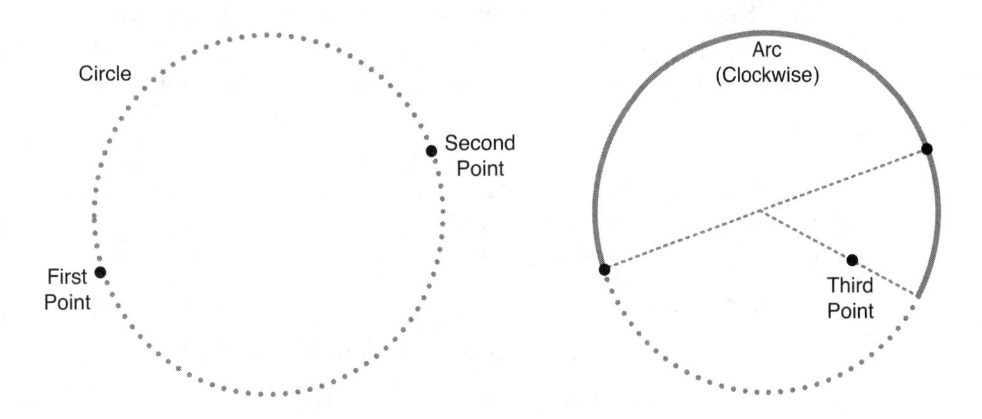

Figure 10.2: Defining an Arc with Three Points

By default, the arc is drawn in a clockwise orientation. To draw a counterclockwise arc, precede the coordinates with `ccw`. The following command draws a counterclockwise arc starting from (–1, –0.5).

```
arc ccw (-1 -0.5) (1 0.5) (-1 -1)
```

In addition to the orientation, two other aspects of the arc can be configured. The width of the arc's line can be set by following `arc` with a floating-point value. Also, the arc's endpoint can be set as rounded with the `round` argument or set to a flat edge with `flat`.

10.4.6 Rect

Like `wire`, the `rect` command requires two points (coordinates or mouse clicks) to define its shape. But in this case, the shape is a filled rectangle. `rect` can't draw an unfilled rectangle. Drawing an unfilled rectangle requires four calls to `wire`.

By default, the rectangle is oriented so that its sides are parallel to the x-y axes. If `rect` is followed by an angle, the rectangle will be oriented at that angle. The angle's value is set in the same way as the argument of `rotate`.

10.4.7 Circle

The circle command draws a circle using two points: the center and a point on the circle. As an example, the following command draws a circle that has a center as the origin (0, 0) and passes through the point (2, 2).

```
circle (0 0) (2 2)
```

If circle is followed by a real value, that will be used as the width of the line forming the circle. If this width is set to 0, the circle will be filled.

10.4.8 Pin

The `pin` command creates an electrical connection for a symbol. At minimum, this command requires a name (surrounded by single quotes) and a position in the symbol design. The following command creates a pin named test and places it at the origin.

```
pin 'test' (0 0)
```

In the symbol editor, the pin is drawn as a line with two ends: one intended for the net connection and one drawn with a circle. The second end is placed at the location given by the `pin` command.

By default, each pin created in the symbol editor has the following properties:

- The pin is bidirectional. This is identified by the io designation.
- The pin is drawn 0.2 inches long. In EAGLE, this is a middle-sized pin.
- Both the pin and pad names are displayed in the editor.
- The connection point is to the right. This means the pin's angular orientation is R0.
- The pin has a swaplevel of 0, which means it can't be swapped for any other pin.
- The pin isn't inverted or designated as a clock signal.

Each of these properties can be individually configured. For the pin's direction, the pin command and name must be followed by the appropriate direction name. The first column of Table 10.4 lists the direction names available.

Table 10.4

Pin Directions

Direction Name	Description
`io`	Input/output (bidirectional)
`in`	Input only
`out`	Output only
`oc`	Open-collector or open-drain
`pas`	Passive component

pwr	Power input pin
sup	General supply pin
hiz	High-impedance output (3-state)
nc	Not connected

When specifying a pin's direction, it's important to remember that its name is surrounded by single quotes and its direction isn't. The following command creates a pin named `GND3` at the origin and configures it as a general supply (`sup`) pin.

```
pin 'GND3' sup (0 0)
```

A pin's length can't be set numerically. To configure length, `pin` must be followed by one of the four length designations listed in Table 10.5.

Table 10.5

Pin Lengths

Name	Pin Length
point	No length—the pin is simply a point
short	0.1 inches
middle	0.2 inches
long	0.3 inches

As an example, the following command creates a pin named D4 at point (1, 1) and sets its length to 0.3 inches.

```
pin 'D4' long (1 1)
```

Like the direction, a pin's visibility is identified by a string that follows the pin's name. Table 10.6 lists the possible values. The following command creates a power supply pin whose pin and pad names won't be displayed.

```
pin 'VCC' pwr off (-1 0)
```

Table 10.6

Pin Name Visibility

Name	Pin Length
both	Both the pin and pad names are displayed
pin	Only the pin name is displayed
pad	Only the pad name is displayed
off	No name is displayed for the pin

A pin's orientation is specified using the same arguments as rotate: R0, R90, R180, and R270. Its swaplevel, which determines which pins it can be swapped with, is given by a number greater than 0. As an example, the following command creates a pin that will be oriented 180° (with the connection point to the left) and given a swaplevel of 2.

```
pin 'N5' R180 2 (0 -2)
```

In addition to these five configuration parameters, a pin's function can also be set by the `pin` command. This changes how the pin is displayed in the editor. Table 10.7 lists the pin function options.

Table 10.7

Pin Functions

Name	Description
`dot`	Displays the pin with an inverter symbol
`clk`	Displays the pin with a clock input symbol
`dotclk`	Displays the pin with an inverted clock input symbol

It's important to distinguish a pin's function (the entries listed in Table 10.7) from its direction (the entries listed in Table 10.4). The function changes how the pin is displayed and the direction tells EAGLE what types of signals can be connected to the pin.

10.4.9 Pad

Just as pins represent connection points in the schematic editor, pads represent connection points in the board editor. The `pad` command creates a specific type of pad: a through-hole pad. This is a drill hole intended to contain a lead of a through-hole component.

Like the `pin` command, `pad` can be followed by a number of configuration parameters. The most important of these is the hole's diameter, which is given by following `pad` with a real (floating-point) value. Table 10.8 lists the other parameters of the `pad` command and the different values that can be set.

Table 10.8

Pad Command Parameters

Parameter	Possible Value	Description
Name	`'name'`	The pad's assigned name

Shape	`round`
	`square`
	`octagon`
	`long`
	`offset`
		Defines the pad's shape on the outer layers

Orientation	`R0 through R359.9`	Sets the angular orientation of the pad
Location	`coordinates or` `mouse-click`	Set's the pad's location
Appearance	`first nostop` `nothermals`	Designate the pad as the first Don't generate solder-stop mask Don't generate thermal pads

The following command creates a round pad named `RESET` that is located at the origin with a diameter of 0.25. As shown, the name must be surrounded by single quotes and the other configuration strings must not be.

```
pad 0.25 round 'RESET' (0 0)
```

By default, `pad` sets the pad's shape to be square and its orientation to 0°. If the name isn't given, EAGLE will assign it a generic name, such as a number. By default, EAGLE assumes it can generate thermals and solder-stop mask features as needed.

10.4.10 SMD

An SMD represents a pad that connects to a lead of a surface-mount device. SMDs are generally rectangular, but the `smd` command makes it possible to control the shape in a number of ways:

- **Dimensions**—The `smd` command accepts the width (x-dimension) and height (y-dimension) as regular floating-point values.
- **Roundness**—The pad's roundness is given as an integer value between 0 (fully rectangular) and 100 (fully round corners). This value must be preceded by a ‘ -’ to distinguish it from the pad's dimensions.
- **Orientation**—The pad's angular orientation can be set with an angle between R0 and R359.9.
- **Name**—A specific name must be given between single quotes.
- **Position**—The pad's location can be set with a mouse-click or by x-y coordinates given in rounded brackets.

The following command creates a 0.6 x 0.4 SMD pad named CLK. Its orientation is set to 90° and its corners are fully rounded.

```
smd 0.4 0.6 'CLK' -100 (0 0)
```

As with the `pad` command, three additional properties can be set. `nothermals` tells EAGLE not to generate thermal pads. `nostop` tells EAGLE not to include the pad in the solder-stop mask, and `nocream` tells EAGLE not to generate solder-cream mask.

10.4.11 Prefix

The prefix command is easy to understand. In the schematic editor, component names frequently begin with a common designation. For example, integrated circuits are called U1, U2, U3, and so on, whereas resistors have names like R1 and R2.

EAGLE refers to the introductory letter(s) as a prefix. The `prefix` command associates the currently edited symbol with one or more letters that will be assigned to the symbol's names as the component is added to a design. The following command specifies that symbol names should be prefixed with B-. When the symbol is added to the design, the names will be B0, B1, B2, and so on.

```
prefix B
```

10.4.12 Package

As discussed in Chapter 8 and Appendix A, "EAGLE Library Files," a deviceset identifies a family of related devices. These devices differ in two ways: package and technology. Package refers to the device's physical shape. Device A and Device B may belong to the same deviceset, but if Device A has through-hole leads and Device B has surface-mount leads, we say that the devices differ by package.

To fully define a device, it must be associated with a package. When this association is made, the device's name is changed to reflect the package. For example, if Device ABC is associated with package TQFP144, its full name may become ABC_TQ. In this case, "_TQ" is called the package variant. This is usually appended to the end of the name.

The `package` command associates a package with the device and assigns a package variant. If called without parameters, EAGLE will open a dialog for choosing the package. Otherwise, `package` is called with two parameters: the name of the package and the name of the package variant. As an example, the following command associates the current device with the DIP32 package and assigns the _PQ variant.

```
package DIP32 _PQ
```

The following command accomplishes the same result but selects DIP32 from the dipchip library.

```
package DIP32@dipchip _PQ
```

`package` can also be used to rename package variants or delete them. The important character to know is '-' and it can be used in two ways:

- `package -old_var new_var`—Renames the `old_var` package variant to `new_var`
- `package -var`—Removes the package variant named `var`

10.4.13 Technology

In EAGLE parlance, a device's technology identifies aspects of its function, not its physical form. Suppose Device A and Device B belong to the same deviceset and have the same package. If Device A has a larger cache or higher clock speed, we say that the devices differ by technology. From what I've seen, technology is a catch-all term that refers to differences between physically identical devices in the same deviceset.

To define possible technologies of a device, follow the `technology` command with one or more technology designations. These will be added to the device's list of technologies. For example, the following command adds `C`, `CT`, and `R` to the technologies of the device being edited:

```
technology C CT R
```

Suppose the name of the current deviceset is A1*B1. In this case, the example command produces three new devices: A1CB1, A1CTB1, and A1RB1. If there is no * in the name of the deviceset, devices are named by appending the technology to the deviceset name (preceding the package name).

To remove a technology from the device's list of technologies, precede the technology name with -. This command removes DF from the list of technologies associated with the current device.

```
technology -DF
```

To remove all the technologies associated with a device, follow the `technology` command with `-*`.

10.4.14 Connect

After a device's symbol and package have been defined, the last step in defining a device involves matching pins to pads. This is the purpose of the `connect` command. If this is called without parameters, EAGLE will display a dialog for making pin-pad associations.

To associate pins to pads in a script, connect must be called in the following way:

```
connect gate.pin pad
```

In this command, *gate*, *pin*, and *pad* are the names of the gate, pin, and pad, respectively. As an example, the following command associates Pin CLK of Gate A with Pad 2 of its package.

```
connect A.CLK 2
```

The gate must be specified if the device has multiple gates. If the device has a single gate, the gate name can be omitted.

A single `connect` command can make multiple pin-pad associations. The following command makes the same association as before but also associates Pin RESET to Pad 5 and Pin D4 to Pad 9.

```
connect A.CLK 2 A.RESET 5 A.D4 9
```

If a pin must be associated with multiple pads, the syntax of the `connect` command is as follows:

```
connect gate.pin 'pad1 pad2 pad3 pad4 ...'
```

In this case, the list of pad names must be surrounded by single quotes. The pad names must also be separated by one or more spaces.

Now suppose a pin can be associated with multiple pads, but only one association is needed. The usage is similar to the command previously given, but now `connect` must be followed by the `ANY` keyword. The following command indicates that Pin PWR can be connected to VCC1, VCC2, or VCC3, but only one connection is needed.

```
connect ANY PWR 'VCC1 VCC2 VCC3'
```

Here, the gate name has been omitted because the example device is assumed to consist of only one gate.

10.5 The Assign, Change, and Set Commands

Up to this point, the commands presented in this chapter have focused on interacting with editors, designs, and components. This section discusses three commands that operate on a higher level. That is, they modify the state of EAGLE's execution environment.

These three commands are `assign`, `change`, and `set`. Their purposes are given as follows:

- `assign`—Associates keystrokes (hotkeys) with commands
- `change`—Alters a design parameter, such as the dimensions of an SMD pad or the height of text
- `set`—Alters an aspect of EAGLE's operation, such as the minimum grid size

10.5.1 Assign

The `assign` command associates a keystroke with an EAGLE command. This command is used frequently in eagle.scr, the default script file provided by EAGLE. This contains many hotkey assignments, and the following use of `assign` associates Ctrl-Z with the `UNDO` command.

```
assign C+Z UNDO;
```

As shown, the `assign` command accepts two parameters: a keystroke and a corresponding command. The keystroke can be an alphanumeric key (A–Z, 0–9), a function key (F1–F12), or the backspace key.

The keystroke can be specified further using modifier keys such as Alt, Ctrl, or Shift. On Mac OS systems, the Command key is also available. When assigning keystrokes with these keys, the modified keystroke is given by `X+K`, where X is the abbreviation of the modifier key(s) and K is the base key. The abbreviations are given as follows:

- **Ctrl**—Abbreviated as `C`
- **Shift**—Abbreviated as `S`
- **Command**—Abbreviated as `M` (only on Mac OS)
- **Alt**—Not abbreviated

With these rules in mind, it should be clear that `C+T` represents Ctrl-T and `A+F5` represents Alt-F5. In addition, Ctrl and Shift can be combined with `CS`, so `CS+Q` represents Ctrl-Shift-Q. Further modifiers can be combined with additional + signs. That is, `C+A+F5` represents Ctrl-Alt-F5 and `A+S+B` represents Alt-Shift-B.

To associate a keystroke with multiple commands, place the commands in a script file (*.scr) and associate the keystroke with `script`. For example, the following code associates Ctrl-9 with a script file called test.scr.

```
assign C+9 'script test.scr';
```

As shown, if the command portion of `assign` consists of multiple words, it must be placed in single quotes.

10.5.2 Change

The `change` command alters an EAGLE design property, such as pin length or text alignment. Its usage is given as follows:

```
change NAME VALUE
```

Here, *`NAME`* is an identifier representing the property to be changed and *`VALUE`* identifies the desired change. Table 10.9 lists the property identifiers available and the corresponding values that can be set.

Table 10.9

Property Identifiers for the Change Command

Identifier	Possible Value	Description
LAYER	layer ID (name or #)	Makes the specified layer active
TEXT	text height	Updates text
SIZE	text thickness	Sets the height of text
RATIO	text line distance	Sets the thickness of text
FONT	VECTOR, FIXED, PROPORTIONAL	Sets the text's line distance
ALIGN	BOTTOM, LEFT, CENTER, TOP, RIGHT	Sets the text alignment
WIDTH	wire width	Sets the width of wires
STYLE	Continuous, LongDash, ShortDash, DashDot	Changes how lines are depicted in the editor
CAP	ROUND, FLAT	Sets how an arc's edge is drawn
SHAPE	SQUARE, ROUND, OFFSET, OCTAGON, LONG	Sets the shape of through-hole pads
DIAMETER	Pad/via/hole diameter	Sets the diameter of pads, vias, and holes
DRILL	Drill size	Sets the drill size
VIA	start_layer-end_layer	Creates a via between two layers
SMD	width height	Sets the dimensions of SMT pads
ROUNDNESS	0-100	Sets the roundess of SMT pads, from 0 (not rounded) to 100 (fully rounded)
DIRECTION	NC, IN, OUT, IO, OC, HIZ, SUP, PAS, PWR	Configures a pin's direction
FUNCTION	NONE, DOT, CLK, DOTCLK	Identifies the pin's special function (if any)
LENGTH	POINT, SHORT, MIDDLE, LONG	Sets the pin's length
VISIBLE	BOTH, PAD, PIN, OFF	Specifies what aspects of the pin should be displayed
SWAPLEVEL	integer >= 0	Identifies what class of pins/gates can be swapped (0 - pin can't be swapped)
THERMALS	ON, OFF	Sets whether thermal pads can be generated
ORPHANS	ON, OFF	Sets whether orphans (isolated regions of a copper pour) should be allowed
ISOLATE	distance	The spacing between a polygon and the edge of a board
RANK	integer >= 0	Sets whether a polygon takes priority over an overlapping polygon

ADDLEVEL	NEXT, MUST, ALWAYS, CAN, REQUEST	Specifies if/how new gates are presented to the user in a schematic design
DISPLAY	OFF, VALUE, NAME, BOTH	Identifies how attributes are displayed
COLUMNS	positive integer	Sets the number of columns in the frame
ROWS	positive integer	Sets the number of rows in the frame
BORDER	NONE, BOTTOM, RIGHT, TOP, LEFT, ALL	Sets which borders of the frame should be displayed
XREF	ON, OFF	Identifies whether cross-reference labels are allowed
DUNIT	MIC, MM, MIL, INCH	Sets the unit to be used for dimensions

In many cases, `change` can be used in place of another command. For example, the following command creates an SMD pad with dimensions 0.6 x 0.4:

```
smd 0.6 0.4
```

The `change smd` command performs a similar operation, but makes 0.6 x 0.4 the default size of all SMD pads.

```
change smd 0.6 0.4
```

10.5.3 Set

The `set` command is similar to `change`, but in addition to setting design properties, it can also modify aspects of EAGLE's behavior. Like `change`, it accepts a property identifier and a value that specifies how the property should be changed.

The number of properties that can be configured with `set` is enormous. For a full listing, go to Help > General in the main menu, open up Editor Commands, and select the SET entry. Table 10.10 lists 13 properties that can be configured with `set`.

Table 10.10

Property Identifiers for the Set Command

Identifier	Possible Value	Description
USED_LAYERS	layer names/ numbers	Identifies which layers should be displayed in the Display dialog
WIRE_BEND	0-9	Sets the default wire bend for manual routing
CONFIRM	YES, NO	Automatically answers dialog boxes with the given response
POLYGON_ RATSNEST	ON, OFF	Controls whether RATSNEST causes polygons to be recalculated
OPTIMIZING	ON, OFF	Sets whether traces that form a single line should be connected into a single trace
VECTOR_FONT	ON, OFF	Controls the use of vector fonts

CHECK_CONNECTS	ON, OFF	Specifies whether ADD checks for matching pins-pads in components
AUTO_END_NET	ON, OFF	Sets whether nets and buses are automatically terminated when placed at another net or bus wire or a pin
AUTO_JUNCTION	ON, OFF	Controls whether a junction is automatically created when one net ends at another
NET_WIRE_WIDTH	integer	Sets the displayed width of nets in the schematic editor
BUS_WIRE_WIDTH	integer	Sets the displayed width of buses in the schematic editor
MIN_TEXT_SIZE	integer	Sets the minimum text size
MIN_GRID_SIZE	integer	Sets the minimum number of pixels between grid points

Chapter 6 discussed the different wire bend styles available for manual routing. This is normally set in EAGLE's horizontal toolbar when the Route tool is active. This can also be configured by calling `set` with `WIRE_BEND`. In this case, the `set` command accepts a value between 0 and 9. Table 10.11 explains what these values represent.

Table 10.11

Wire Bend Values

WIRE_BEND Value	Wire Bend Style
0	Straight, then bends at 90°
1	Straight, then bends at 45°
2	No bends—wire proceeds from start to finish
3	Bends at 45°, then straight
4	Bends at 90°, then straight
5	Arc, then straight
6	Straight, then arc
7	Arc, then straight, then arc
8	Follow-me router—route computed from initial pad
9	Follow-me router—route computed from both pads

As an example, the following command tells EAGLE to make Style 4 the active wire bend style:

```
set wire_bend 4
```

The third property in the table, `CONFIRM`, is very useful to know when you're automating tasks. In many instances, EAGLE will display a dialog asking for permission. The dialog can't be closed from a script, so the existence of the dialog makes it difficult to automate some operations.

For example, when the `EDIT` command is used to create a new component, EAGLE produces a dialog asking for permission. But if the `CONFIRM` property is set to `yes`, the dialog won't appear and a script can continue executing normally. The following code shows how this is used:

```
set confirm yes
```

10.6 Configuration Scripts

In general, command scripts are executed by the user in an EAGLE editor. But in some cases, you may want a script to execute automatically when a specific editor launches. This makes it possible to configure the editor's hotkeys or set new default options. This kind of script is called a configuration script.

If you look in EAGLE's top-level scr directory, you'll find eagle.scr, a configuration script that is automatically called when EAGLE starts and when each new editor launches. This provides different configuration settings for each of EAGLE's editors (schematic, board, library, device, symbol, or package).

To enable an editor-specific configuration, the content of eagle.scr is divided into sections delimited by special section names. Each of these section names (`BRD`, `SCH`, `LBR`, `DEV`, `SYM`, and `PAC`) occupies a line of its own and is followed by a colon. The overall structure of eagle.scr is given as follows:

```
BRD:
  ... Configuration commands for the board editor
SCH:
  ... Configuration commands for the schematic editor
LBR:
  ... Configuration commands for the library editor
DEV:
  ... Configuration commands for the device editor
SYM:
  ... Configuration commands for the symbol editor
PAC:
  ... Configuration commands for the package editor
```

As an example, the board editor section configures hotkeys for route and ripup while the schematic editor section sets hotkeys for net and smash. In the portion of the file preceding the board editor section, eagle.scr contains commands that apply to all editors in general.

This configuration process can be customized by modifying eagle.scr. In general, there are two ways to accomplish this. First, commands can be directly inserted into the appropriate section. For example, any command intended to configure the library editor should be placed in the section delimited by LBR.

The second method involves the `script` command, which executes a script file. If this command is placed in one of the eagle.scr sections, the script will be executed whenever the corresponding editor opens.

10.7 Conclusion

In my opinion, the ability to execute editor commands is one of EAGLE's greatest strengths. Used properly, a script can significantly reduce the time needed to perform repetitive tasks. Further, performing operations with scripts reduces the potential for error associated with points and clicks.

Editor commands can be executed in one of two ways. First, commands can be executed individually in the text box provided by each EAGLE editor. Second, the script command executes a batch of commands stored in a script file. There are two rules regarding script files: The file's suffix must be *.scr and each command must end with a semicolon.

Editor command syntax is easy to understand. In each case, the command's name is followed by one or more parameters that constrain its operation. The difficulty stems from the vast number of commands available. I still confuse `use` with `open` and I still need to check `via`'s usage every time I execute it. The key is practice.

These commands make it possible to configure editors, modify designs, and even create new libraries and components. But none of them provide information. That is, you can change a design's name but you can't find out what it is. You can add a new component to a design but you can't determine what components are already present.

Thankfully, EAGLE provides another language called the User Language or UL. This provides functions that obtain information about designs and components. UL also makes it possible to perform complex processing tasks using statements similar to those found in the C programming language. The next chapter provides an introduction to UL, and as you'll see, combining UL with editor commands makes it possible to automate just about any circuit design task you can think of.

Chapter 11

Introduction to the User Language (UL)

EAGLE's command language is great for automating actions, but it lacks many features of high-level programming languages such as loops, variables, and file access. To make up for these shortcomings, EAGLE provides the User Language, which this book shortens to UL. A program coded in UL is a User Language program, or a ULP.

ULPs serve two main roles: writing design data to files and automating tasks related to circuit design and library generation. These roles will be explored throughout Chapter 12, "Examining Designs with the User Language," and Chapter 13, "Creating Dialogs and Menu Items." For this chapter, the goal is simply to introduce UL and its features.

UL's syntax closely resembles that of the popular C language, but UL is not C or a library of C. This is important to keep in mind. There's no way to interface a ULP with C-coded applications or access C libraries. Traditional C/C++ compilers, such as gcc and Visual Studio, can't compile or link ULPs.

Now that I've explained what UL isn't, the rest of this chapter is devoted to explaining what UL is. More precisely, this chapter explains four aspects of the User Language: data types, builtin features, control structures, and the `exit` statement.

11.1 Overview of UL

I'd like to start this discussion by presenting a simple ULP. The code in Listing 11.1 accesses the active schematic design and prints its name to a file called test.txt. This illustrates two capabilities of UL programs: the ability to access designs and the ability to write data to files.

Listing 11.1: basic.ulp

```
string get_name() {
  schematic(S) {
    return S.name;
  }
}

output("test.txt") {
  printf("The schematic's name is: %s.\n", get_name());
}
```

This section introduces UL by comparing and contrasting it with C. But first, I'll explain how to execute ULPs inside EAGLE.

11.1.1 Executing ULPs

A ULP is a text file that contains code in the UL language. There are at least three ways to execute a ULP.

1. Open an EAGLE editor and go to File > Run... on the main menu. This opens a dialog that lets you select the ULP (*.ulp) in your file system. After you find the file, click Open to execute the program.
2. The `run` command of the EAGLE control language (discussed in the last chapter) executes a ULP and provides optional arguments within square brackets.
3. If a ULP file is part of an EAGLE project, it can be executed through the Control Panel. In the Control Panel, open the project's folder and right-click the *.ulp file. On the context menu, select one of the last three options: Run in Schematic, Run in Board, or Run in Library. Note that the corresponding editor must be open for the option to be available.

Returning to Listing 11.1, if you try to execute basic.ulp in any editor except the schematic editor, you'll get an error message. That's because the ULP accesses the current schematic and reads its name. If there is no current schematic, or the editor isn't the schematic editor, the ULP won't execute properly.

11.1.2 ULPs and C Programs

Presenting the C language is beyond the scope of this book, but you can find excellent documentation elsewhere, such as *C Primer Plus* by Stephen Prata. I also strongly recommend the tutorial-based web site http://www.cprogramming.com/tutorial.html.

When I first encountered a ULP, I was pleased to see C-like functions but surprised by the lack of a `main` function. In regular C programs, the `main` function tells the compiler which code should be executed first. But ULPs don't include `main` because the UL processor reads from the top of the file to the bottom, executing commands in order.

With this in mind, most ULP statements can be divided into three categories:

- **Variable declarations**—Declarations of global variables that can be accessed throughout the program
- **User-defined functions**—Definitions of C-like functions with user-specified names
- **Executable statements**—Statements that call on UL functions and user-defined functions

When EAGLE executes a ULP, it starts by executing the first executable statement it encounters. It proceeds executing these statements until it reaches the end of the program. User-defined functions execute only when they're called by an executable statement.

Looking back at Listing 11.1, it may appear that the ULP contains two user-defined functions, and you might wonder where the executable statement is. As it turns out, EAGLE recognizes `output()` as an executable statement. More precisely, `output()` defines a block in which all output routines, such as `printf`, will be directed to a file called test.txt. In constrast, EAGLE doesn't recognize the `get_name` function, so it doesn't execute it until it's called inside the `output` block.

Table 11.1 lists a number of similarities and differences between C and UL. This isn't a complete list, but I hope it will make UL more comprehensible.

Table 11.1

Similarities and Differences Between UL and C

Similarities	Differences
Subroutines defined using functions	ULPs don't have a `main` function.
Output displayed with `printf`	ULPs print output to a file, never the console.
Similar data types (`int`, `char`, `void`)	UL uses `string` instead of `char[]` and `real` instead of `double`. UL has special data types that start with `UL_`.
Casting between types	UL uses special statements to read elements of EAGLE designs.
Array syntax and initialization	No import statements or linking, can't access header files or libraries.

11.2 Simple Data Types and Functions

ULPs support many of the same basic types as C, including `void`, 8-bit `char`, and 32-bit `int`. For floating-point values, ULPs use `reals`. These types behave like their C counterparts and the same types of mathematical operators are available (`+`, `-`, `*`, `/`, `%`, `++`, `--`, `+=`, `-=`, `*=`, `/=`, and `%=`).

In addition to the basic mathematical operators, ULP supports math functions whose names and purposes resemble those declared in C's math.h header file. Table 11.2 lists the math functions provided by UL.

Table 11.2

Math Functions

Math Function	Purpose
`char/int/real abs(char/int/real)`	Returns the absolute value
`char/int/real max(char/int/real, char/int/real)`	Returns the maximum of two values
`char/int/real min(char/int/real, char/int/real)`	Returns the minimum of two values
`real ceil(real)`	Returns the smallest integer larger than or equal to the value
`real floor(real)`	Returns the largest integer less than or equal to the value
`real frac(real)`	Returns the value's fractional part
`real round(real)`	Returns the integer nearest x
`real trunc(real)`	Returns the value's integer part
`real sin(real), real asin(real)`	Computes the sine and inverse sine
`real cos(real), real acos(real)`	Computes the cosine and inverse cosine
`real tan(real), real atan(real)`	Computes the tangent and inverse tangent
`real sqrt(real)`	Returns the square root
`real pow(real, real)`	Returns the first value raised to the power of the second value
`real exp(real)`	Returns the exponential function
`real log(real)`	Returns the natural logarithm
`real log10(real)`	Returns the base-10 logarithm

11.2.1 Arrays

Arrays in ULPs look and behave like regular C arrays. For example, the following statement declares and initializes an array of five `int`s:

```
int example_array[] = {1, 2, 3, 4, 5};
```

The UL provides a `sort` function that sorts one or more arrays in ascending order. It accepts two values: the number of elements in the array and the array itself. The following code sorts the elements in `example_array`:

```
sort(5, example_array);
```

11.2.2 Strings

In addition to arrays, ULP supports a `string` type that's essentially an array of `char`s. The following line creates a new `string`:

```
string ex = "Example";
```

Like `char` arrays in C, ULP `string`s are zero-terminated, which means the processor adds the character `'\0'` to the end of the array. This is equivalent to the `int` value 0.

Table 11.3 lists the UL functions associated with `string`s. As with the math functions, most of them have the same names and roles as the functions in the C standard library.

Table 11.3

String Functions

String Function	Purpose
`printf(string text, ...)`	Prints formatted text to selected output
`sprintf(string dst, string text, ...)`	Prints formatted text to destination string
`int strchr(string st, char ch[, int pos])`	Returns the first location of `ch` within `st` starting at `pos`
`string strjoin(string array[], char sep)`	Concatenates the `string`s in `array` delimited by `sep`
`int strlen(string st)`	Returns the length of `st`
`string strlwr(string s)`	Converts uppercase to lowercase
`int strrchr(string st, char ch[, int pos])`	Returns the last location of `ch` within `st` starting at `pos`
`int strrstr(string st1, string st2[, int pos])`	Returns the last offset of the first character of `st2` in `st1` starting at `pos`
`int strsplit(string array[], string st, char sep)`	Splits `st` delimited by `sep` and places the result in `array`
`int strstr(string st1, string st2[, int pos)`	Returns the first offset of the first character of `st2` in `st1` starting at `pos`
`string strsub(string st, int start[, int len)`	Returns the substring of `st` starting at `start` and containing `len` characters
`real strtod(string st)`	Converts `st` to a real
`int strtol(string st)`	Converts `st` to an int
`string strupr(string st)`	Converts lowercase to uppercase
`int strxstr(string st, string reg [, int pos[, int len]])`	Returns the first offset of `st` that matches the regular expression `reg` starting from `pos`

These functions have the same names as their counterparts in C/C++ and perform the same operations. The only important difference involves `printf`. This accepts the same format characters (`%c`, `%d`, `%u`, `%s`, `%f`, `+`, and so on) as the C version, but UL has no concept of a console or standard output. In a ULP, `printf` always prints text to a file.

11.3 Builtins

Up to this point, UL's datatypes and functions have closely resembled those of C. This section discusses additional features that are specific to UL. They're called *builtins* and they come in four types:

- **Constants**—Global identifiers that provide information about the EAGLE environment
- **Variables**—Variables that provide access to the ULP's parameters
- **Functions**—Have the same usage and purpose as C functions
- **Statements**—Define blocks in which UL objects can be accessed

This section presents an overview of each of these builtin types. Afterward, I'll present an example ULP that demonstrates how builtins can be used in code.

11.3.1 Builtin Constants

Many ULPs direct their output to files, so it's important that they have access to information related to the user's computing environment, including EAGLE's directories and files. This access is provided through UL's builtin constants.

These constants come in two types. Constants of the first type have a single value and provide information about EAGLE directories and math limits. Constants of the second type are string arrays that identify files in EAGLE's top-level directories.

Single-Valued Constants

Table 11.4 lists the single-valued constants provided by the UL.

Table 11.4

Single-Valued Constants

Constant	Datatype	Description
`EAGLE_VERSION`	`int`	EAGLE's version number
`EAGLE_RELEASE`	`int`	EAGLE's release number
`EAGLE_SIGNATURE`	`string`	A string containing EAGLE's name, version, and copyright

EAGLE_DIR	string	Directory containing the EAGLE executable
EAGLE_PATH	string	Full path of the EAGLE executable
EAGLE_HOME	string	The user's home directory
OS_SIGNATURE	string	The user's operating system
REAL_EPSILON	real	Minimum positive real value such that 1.0 + REAL_EPSILON doesn't equal 1.0
REAL_MAX	real	Maximum possible real value
REAL_MIN	real	Minimum possible real value
INT_MAX	int	Largest possible int value
INT_MIN	int	Smallest possible int value
PI	real	The geometric PI constant (3.14159)
usage	string	Contains text from the #usage directive

The following code sets x equal to one less than the maximum possible integer:

```
int x = INT_MAX - 1
```

The last constant, `usage`, requires explanation. Professional ULPs have a message that tells the user what the ULP accomplishes and how it should be executed. This message is set by the `usage` directive, which consists of `#usage` and a string to be presented to the user. When the user selects the ULP in the Control Panel, the message is printed in the right window. The following code shows how `#usage` works:

```
#usage "This ULP serves no useful purpose."
       "<author>Author: Matt Scarpino fake@email.com</author><br>"
```

The `#usage` directive is usually inserted near the beginning of a ULP. The `usage` constant makes it possible to access this string throughout the program.

Array Constants

A ULP may need to know the directory path EAGLE uses to read different types of files, such as scripts, library files, or design rule files. UL makes this possible by providing constants that correspond to arrays of strings, where each string is a directory in EAGLE's path. Table 11.5 lists each of the arrays that can be accessed.

Table 11.5

Array Constants

Array Constant	Description
path_lbr[]	Returns the paths associated with EAGLE library files
path_dru[]	Returns the paths associated with EAGLE design rule files
path_ulp[]	Returns the paths associated with EAGLE UL program files
path_scr[]	Returns the paths associated with EAGLE scripts

`path_cam[]`	Returns the paths associated with EAGLE CAM files
`path_epf[]`	Returns the paths associated with EAGLE project files
`used_libraries[]`	A list of active libraries for the design

The last entry in this table doesn't return a list of directory names. Instead, the array contains the full path of each active library file (*.lbr) for the design. The following code uses the `output()` executable statement to print the first library path to path.txt.

```
output("path.txt") {
  printf("The first library path is %s.\n", used_libraries[0]);
}
```

11.3.2 Builtin Variables

If a ULP is executed using the `RUN` command, command-line arguments can be added to configure its operation. These arguments can be accessed with the following builtin variables.

- `argc`—An `int` identifying the number of arguments
- `argv`—An array of `string`s containing the arguments

These variables have the same names and purposes as those used in C code. The following statement prints the number of arguments used to invoke the ULP.

```
output("numargs.txt") {
  printf("The number of command-line arguments is %d.\n", argc);
}
```

The name of the ULP is always the first argument. Therefore, if `RUN` is executed without additional arguments, `argc` will equal 1 and `argv[0]` will equal `"RUN"`.

11.3.3 Builtin Functions

ULPs can't include header files or external libraries, so it's a good thing UL provides so many builtin functions. In preceding discussions, I presented functions related to math and strings, but many more are available. This discussion explains UL's functions related to file handling and unit conversion. These functions will play an important role in the code presented in Chapters 12 and 13.

File Handling Functions

As discussed earlier, one of the most important uses of UL involves creating files and writing data to them. For example, many fabrication facilities require a centroid file that lists the name, position, and orienation of each part in the design. It's easier to create this automatically with a ULP than to write the information by hand.

For this reason, every UL developer should have a thorough understanding of UL's file handling capabilities. Table 11.6 lists nine functions that relate to files and file handling.

Table 11.6

File Handling Functions

Function Name	Description
`fileerror()`	Returns zero if preceding I/O operations succeeded, returns a nonzero value if any I/O operations failed
`fileglob(string array[], string pattern)`	Searches the top-level directory for files whose names match the pattern—results placed in the array
`filedir(string filename)`	Returns the name of the file's directory
`fileext(string filename)`	Returns the file's extension
`filename(string filename)`	Returns the file's name, including the extension
`fileread(destination, string filename)`	Reads data from the file, stores in destination, which can take multiple data types
`filesetext(string filename, string new_extension)`	Sets the extension of the given file to new_extension
`filesize(string filename)`	Returns the size of the file
`filetime(string filename)`	Returns the file's timestamp

Most of these functions are easy to understand, but `fileglob` and `fileread` deserve further explanation. `fileglob` searches through EAGLE's installation directory for files whose names match the given pattern. This is *not* recursive, so by default, the search includes only the top-level directories, such as lbr, dru, and cam.

To search through a subdirectory, its name must be included in the pattern. As an example, the following code prints the first filename in EAGLE's top-level cam directory.

```
output("output.txt") {
  string cfiles[];
  int n = fileglob(cfiles, "cam/*");
  printf("The first file in EAGLE's cam folder is %s.\n",
         cfiles[0]);
}
```

As its name implies, `fileread` reads data from a file. It's particularly interesting because its operation depends on its first argument, which can take one of three datatypes:

- `char[]`—Places the file's bytes into the array and returns the number of bytes read
- `string[]`—Places the file's lines of text into the array and returns the number of lines read

- `string`—Places the file's complete text into the string and returns the length of the string

As with `fileglob`, the initial directory is EAGLE's top-level folder. The following code reads the lines of the $EAGLE/scr/eagle.scr and prints the number of lines read.

```
output("output.txt") {
  string lines[];
  int numlines = fileread(lines, "scr/eagle.scr");
  printf("There are %d lines in EAGLE's default script.\n",
          numlines);
}
```

Unit Conversion Functions

EAGLE uses its own set of units to locate coordinates in an editor's grid. To access these coordinates in a ULP, the program must be able to convert these units to and from traditional units. Table 11.7 lists the eight functions that make this possible.

Table 11.7

Unit Conversion Functions

Function Name	Description
`real u2inch(int n)`	Converts grid units to inches
`real u2mic(int n)`	Converts grid units to microns
`real u2mil(int n)`	Converts grid units to mils
`real u2mm(int n)`	Converts grid units to millimeters
`int inch2u(real n)`	Converts inches to grid units
`int mic2u(real n)`	Converts microns to grid units
`int mil2u(real n)`	Converts mils to grid units
`int mm2u(real n)`	Converts millimeters to grid units

As shown in the table, EAGLE's internal units are given as integers and traditional measurements are given as reals. The following code converts the coordinate (0.5, –0.5) from millimeters to internal units:

```
int internal_x = mm2u(0.5);
int internal_y = mm2u(-0.5);
```

Chapter 12 discusses the UL data structures used in schematic and board designs. Many of these have fields named `x` and `y` that represent coordinates. It's important to remember that functions are available to convert these coordinates to useful units.

11.4 Control Structures

Every high-level language has statements that control how other statements are executed. These control structures fall into one of two categories: those that specify whether a set of statements should be executed and those that specify how many times a set of statements should be repeated. In the UL, the `if..else` and `switch..case` statements fall into the first category. The `while` and `for` statements fall into the second category.

11.4.1 The if..else Statement

The simplest of the control structures is the `if..else` statement, which evaluates a logical expression to decide whether to execute one set of statements or another. This statement is exactly like the `if..else` statement in regular C, which means the following rules apply:

- The logical expression must be surrounded by parentheses.
- The logical expression may use any of C's operators (`==`, `!=`, `>`, `>=`, `<`, `<=`, `&&`, `||`) to compare the two values.
- If statements should be executed together, they must be surrounded by curly braces.

The following code demonstrates how `if..else` works. If the ULP is invoked with multiple arguments, it will execute the two statements in curly braces. Otherwise it executes a single statement.

```
output("args.txt") {
  if (argc > 1) {
    printf("%d additional arguments.\n", argc);
    printf("The first argument is %s.\n", argv[0]);
  }
  else
    printf("No additional arguments.\n");
}
```

11.4.2 The switch..case Statement

The `switch..case` statement makes it easy to select from multiple options. Each option is represented by a `case` statement, and each option consists of four parts:

- The `case` keyword
- An integer or character followed by a colon
- One or more statements to be executed if the `switch` value matches the `case` value
- The `break` keyword

If the `switch` value doesn't match any of the `case` options, the statement will execute the `default` option if it's available. This is shown in the following code:

```
switch(argc) {
  case 1: printf("Case 1\n"); break;
  case 2: printf("Case 2\n"); break;
  case 3: printf("Case 3\n"); break;
  default: printf("More than three arguments.");
}
```

Note that the `case` options can't use strings. Each case value must be an integer or a character.

11.4.3 The while and do..while Loops

The `while` statement is similar to the `if` statement, but if the logical condition evaluates to a nonzero value (such as true), the selected statements execute until the condition evaluates to zero (such as false). This is shown in the following code:

```
output("x.txt") {
  int x = 0;
  while(x < 5) {
    printf("x = %d.\n", x);
    x++;
  }
}
```

The loop executes until the `x < 5` condition becomes false. That is, it continues executing the `printf` statement until x equals 5. If x remains less than 5, the loop continues forever.

If x starts at a value greater than or equal to 5, the example loop never executes. But in some cases, the loop should always execute once. That is, the loop should check the condition *after* the statements are executed instead of *before*. For this reason, UL provides the `do..while` loop. The following code shows how it works.

```
int x = 5;
do {
  x++;
} while(x < 5)
```

11.4.4 The for Loop

The `for` loop is commonly used to repeat statements according to an index variable. Its structure is given as follows:

```
for(init_statement; test_statement; inc_statement) {
  statement_block
}
```

In the first line, each of the three statements is optional. Their purposes are given as follows:

- `init_statement`—Executes first. Usually initializes a variable called the index.
- `test_statement`—Executes after `init_statement`. Usually tests the index's value.
- `inc_statement`—If `test_statement` evaluated to true, this executes after the statements in `statement_block`. Usually modifies the index's value.

The loop repeats until the condition in `test_statement` evaluates to false. One interesting point about UL is that `test_statement` can be set to an array. In this case, the loop executes for each element in the array.

As an example, the following code creates an array of integers and uses a `for` loop to print the value of each.

```
output("num.txt") {
  real num_array[] = {1.1, 2.2, 3.3, 4.4};
  for (int i=0; num_array[i]; i++)
    printf("The value at index %d is %f.\n", i, num_array[i]);
}
```

In this example, the statement executed by the for loop isn't surrounded by curly braces. Curly braces are required only when two or more statements are to be executed together. This holds true for each of the control structures discussed in this section.

There are two last statements that deserve attention: `break` and `continue`. The `break` statement ends the execution of the current loop iteration. Execution resumes after the loop statement. `continue` also ends the execution of the current loop iteration, but it doesn't end execution for the entire loop. Instead, execution resumes with a new loop iteration (if possible).

11.5 The exit Statement

As in regular C, the `exit` statement tells EAGLE to halt execution of the program. This isn't necessary, and if this statement isn't present, EAGLE will stop executing the ULP when it reaches the end of the file.

But besides telling EAGLE to halt execution, `exit` can provide one of two pieces of information. These are given by the two usages of `exit`:

- `exit(int code)`—Returns an error code to identify execution status
- `exit(string cmd)`—Returns a command string for execution

The second usage is critically important. Up to now, all the UL code presented in this chapter has focused on printing text to a file. This is a common use of ULPs, but with the second usage of `exit`, ULPs can execute the editor commands discussed in Chapter 10, "Editor Commands."

As a simple example, the code in Listing 11.2 examines the name of the active board design and saves the design to a file with a slightly modified name. To save the design, the ULP executes the `write` command discussed in Chapter 10.

Listing 11.2: change_name.ulp

```
board(b) {
  string file_name = filename(b.name);
  int name_length = strlen(file_name);
  string new_name = strsub(file_name, 0,
                           name_length - 4) + "_new.brd";
  string cmd = "write @" + new_name;
  exit(cmd);
}
```

This code uses capabilities that won't be explained until the next chapter, but the important point is the `exit` function, which accepts a string called `cmd`. This string is initialized in the following way:

```
string cmd = "write @" + new_name;
```

This line of code uses UL's string handling to create a command that can be executed by EAGLE. Specifically, it appends the `write` command to the new filename. As explained in Chapter 10, `write` saves the design to the named file. The `exit` function directs the command string (`cmd`) to EAGLE, which executes the `write` statement as if it had been part of a regular script.

11.6 Conclusion

The User Language was designed to resemble C, and if you're already familiar with C programming, this chapter should present little difficulty. The data types, control structures, and functions generally have the same names and behavior as their counterparts in C.

One important difference between C and the UL is that ULPs are interpreted, not compiled. That is, a ULP executes line-by-line, so there's no need for a `main` function. In addition, UL has special functions like `output` that define blocks of code. These are called executable statements. In the case of `output`, every `printf` statement in the block directs its data to `output`'s file argument.

UL's `exit` function is particularly important to know. Like the `exit` function in C, it can accept an integer code that returns the program's status. Unlike C's `exit` function, it can also accept a string containing editor commands to be executed when the ULP terminates. This combination of editor commands and UL makes it possible to code powerful programs.

The next chapter continues the discussion of UL by explaining how its special data types make it possible to examine schematic designs and board designs. As you proceed, remember that the capabilities discussed in this and the previous chapter are available in UL programs.

Chapter 12

Examining Designs with the User Language

Chapter 11, "Introduction to the User Language (UL)," introduced the UL and showed how to code and execute UL programs (ULPs). Now we're going to put UL to practical use. Specifically, this chapter discusses how to use UL to examine schematic designs and board designs. I'll explain how to use UL to generate a Bill of Materials (BOM) or modify components in a design. Chapter 13, "Creating Dialogs and Menu Items," will go even further and show how ULPs can access elements of the EAGLE user interface.

Chapter 11 showed a number of ways that UL resembles the C programming language. But it left out one important point: data structures. A data structure may be thought of as a data type made up of other data types. UL's structures are similar to those used in C except UL doesn't allow for user-defined structures. Instead, it provides a vast array of builtin structures representing designs and elements of designs.

To examine a design in a UL, it's important to understand how these structures work and how to access their fields. This chapter presents a large number of these structures, with emphasis on the two most important, `UL_SCHEMATIC` and `UL_BOARD`. As will be shown, these top-level structures define programming blocks in a manner similar to the `output` block presented in Chapter 11.

This chapter presents most of the structures associated with schematic and board designs, but not all of them. For a more complete listing of structures and fields, I recommend the *User Language Manual*, which can be freely downloaded from CadSoft.

12.1 UL-Specific Data Types

In addition to the C-like data types discussed in Chapter 11, UL supports a set of UL-specific datatypes whose name start with `UL_`. For example, a schematic design is represented by a `UL_SCHEMATIC` and a board design is represented by a `UL_BOARD`. Throughout this book, I'll refer to these UL-specific data types as *UL types*.

I'll refer to an instance of a UL type as a *UL structure*. That is, if `b` is the name of an instance of `UL_BOARD`, I'll refer to `b` as a structure of the `UL_BOARD` type.

12.1.1 Members

A UL type is similar to a `struct` definition in C. That is, each type may be composed of multiple elements of different types. For example, a `UL_PAD` contains a series of `int`s, `real`s, and `string`s, and each element can be accessed with a specific name.

These subelements or fields of a UL type are called *members*. Each member has a name and the member can be accessed using dot notation, in which a dot separates the structure name from the member name. For example, if `x` is a UL structure and `y` is one of its members, `y` can be accessed as `x.y`.

The members of a UL structure can be divided into two categories: data members and loop members. If there can be only one member of a given type, the element is called a data member. For example, a schematic can have only one name, so `name` is a data member of `UL_SCHEMATIC`. If the schematic is called `s`, its name is accessed as `s.name`.

12.1.2 Loop Members

If a structure can contain multiple members of a given type, the type is called a loop member. For example, a schematic may contain multiple sheets, so `sheets` is a loop member of the `UL_SCHEMATIC` datatype. Names of loop members are always given in plural form.

Loop members are accessed using block statements similar to this:

```
loop_member(id) {
  ...
}
```

The statements in this block will be executed for each `loop_member` in the structure. With each iteration, the statements inside the block can refer to the current `loop_member` using the `id` variable. An example will make this clear.

```
schematic(s) {
  printf("Schematic name: %s\n", s.name);
  s.sheets(sh) {
    printf("Sheet number: %d\n", sh.number);
  }
}
```

The outer block assigns the schematic an identifier of `s` and the inner block assigns each of the schematic's sheets an identifier of `sh`. The statements in the inner block execute once for each sheet, and with each iteration, the current sheet is assigned an identifier of `sh`. If the schematic has five sheets, the second `printf` statement will execute five times and print the number of each sheet.

12.1.3 Top-Level Structures and Their Executable Statements

In general, a ULP can access three top-level structures: `UL_SCHEMATICs`, `UL_BOARDs`, and `UL_LIBRARY` structures. The `UL_SCHEMATIC` type is discussed in Section 12.2 and the `UL_BOARD` type is discussed in Section 12.3.

Right now, I want to introduce the three executable statements that relate to these structures. As discussed in Chapter 11, an executable statement is an EAGLE-defined function block that is executed from top to bottom in a ULP.

So far, the only executable statement we've seen is `output()`. The following executable statements relate to the three top-level structures:

- `schematic(id) {}`—Examines the content of the current schematic
- `board(id) {}`—Examines the content of the current board design
- `library(id) {}`—Examines the content of the current library

These executable statements make it possible to access information about EAGLE's designs and libraries. Only one schematic design, board design, and library can be open for editing in an EAGLE instance, so a ULP can't iterate through multiple designs or libraries.

> **NOTE** These functions work properly only if the corresponding editor is open. That is, `schematic()` executes only if the schematic editor is open, `board()` executes only if the board editor is open, and `library()` executes only if the library editor is open. If the corresponding editor isn't open, the ULP will produce an error condition and stop executing.

12.2 Schematic Designs (UL_SCHEMATIC)

A `UL_SCHEMATIC` structure represents a schematic design. To examine a schematic in a ULP, it's important to understand the members of the structure. Table 12.1 lists the data members of the `UL_SCHEMATIC`.

Table 12.1

Data Members of UL_SCHEMATIC

Data Member Name	Type	Description
alwaysvectorfont	int	Specifies whether the fonts are always vectorized or configured in the user interface
description	string	The textual description of the schematic
grid	UL_GRID	The grid associated with the schematic
headline	string	Short description of the schematic
name	string	The schematic's name
verticaltext	int	Reading direction for vertical text
xreflabel	string	The format string used for cross-references

Table 12.2 lists the loop members of the UL_SCHEMATIC structure. Remember that a loop member is essentially similar to an array. That is, the nets() function makes it possible to iterate through the UL_NET structures contained in the UL_SCHEMATIC.

Table 12.2

Loop Members of UL_SCHEMATIC

Loop Member Name	Type	Description
attributes()	UL_ATTRIBUTE	Array of name-value pairs (global variables)
classes()	UL_CLASS	Net classes associated with the schematic
layers()	UL_LAYER	Layers used in the schematic
libraries()	UL_LIBRARY	Libraries used in the schematic
nets()	UL_NET	Electrical connections in the schematic
parts()	UL_PART	Components of the schematic
sheets()	UL_SHEET	Sheets of the schematic
variantdefs()	UL_VARIANTDEF	Assembly variants based on the schematic

The UL_LAYER structure is discussed in the next section. This section looks at all the other members of UL_SCHEMATIC. I'll start with the data members.

12.2.1 Data Members of UL_SCHEMATIC

Most of the data members in Table 12.1 are easy to understand. But three of them deserve extra attention: alwaysvectorfont, verticaltext, and grid.

Schematic Text Members

The text in a schematic can be vectorized (EAGLE-specific), proportional (usually Helvetica), or monospace (usually Courier). Schematic designs use vectorized fonts by default, but you can change this in the editor by going to Options > User Interface...

in the main menu. You can also decide whether this decision should be persistent throughout the design.

The `alwaysvectorfont` identifies whether the font choice has been set in the user interface (`ALWAYS_VECTOR_FONT_GUI`) and whether the font choice should be persistent (`ALWAYS_VECTOR_FONT_PERSISTENT`). The following code shows how this can be accessed.

```
output("font.txt") {
  schematic(s) {
    if(s.alwaysvectorfont & ALWAYS_VECTOR_FONT_GUI)
      printf("The font choice is set in the user interface.");
    else
      printf("The schematic always uses vectorized text.");
  }
}
```

By default, vertical text in a schematic is printed from bottom to up. In the schematic editor, this can be set by going to Options > User Interface... in the main menu. To check the setting in a ULP, `verticaltext` can be compared to `VERTICAL_TEXT_UP` or `VERTICAL_TEXT_DOWN`. These constants work in the same way as the constants used with `alwaysvectorfont`.

The Grid (UL_GRID)

The grid settings of a schematic design can be accessed through the `UL_GRID` structure associated with the `UL_SCHEMATIC`. This structure has members of its own, and Table 12.3 lists each of them with their data type.

Table 12.3

Members of UL_GRID

Member Name	Type	Description
`distance`	`real`	The spacing between grid points
`dots`	`int`	Specifies if the editor displays the grid using lines (0) or dots (1)
`multiple`	`int`	The spacing of the displayed grid as a multiple of the distance
`on`	`int`	Specifies whether the grid is displayed (0) or not (1)
`unit`	`int`	The units used to display text
`unitdist`	`int`	The units used by the grid distance

The most important of these members is `distance`, which identifies the editor's grid spacing. But this value is useless without knowing the grid's units, which are given by `unitdist`. This may be take one of four values:

- `GRID_UNIT_MIL`—The grid uses mils for measurement.
- `GRID_UNIT_INCH`—The grid uses inches for measurement.
- `GRID_UNIT_MM`—The grid uses millimeters for measurement.
- `GRID_UNIT_MIC`—The grid uses microns for measurement.

The following code prints out the grid spacing and the units used for the spacing.

```
output("grid.txt") {
  schematic(s) {
    printf("The grid spacing is %f ", s.grid.distance);
    switch(s.grid.unitdist) {
      case GRID_UNIT_MIL: printf("mils.\n"); break;
      case GRID_UNIT_INCH: printf("inches.\n"); break;
      case GRID_UNIT_MM: printf("millimeters.\n"); break;
      case GRID_UNIT_MIC: printf("microns.\n"); break;
    }
  }
}
```

The `unitdist` and `unit` members have similar names but different functions. `unitdist` identifies the units used by the primary grid spacing. In contrast, `unit` identifies the units used for the displayed text, which is usually set with the alternate grid. Oddly enough, `UL_GRID` has no member that provides the spacing of the editor's alternate grid.

12.2.2 Creating a Bill of Materials: Parts and Attributes

A common use of ULPs involves searching through a schematic's components and using their attributes to create a BOM. To see how this can be done, it's important to understand the `UL_PART` and `UL_ATTRIBUTE` structures.

Schematic Components (UL_PART)

The `UL_SCHEMATIC` structure provides the `parts()` member for iterating through the components of a schematic. Each component is represented by a `UL_PART` structure. Table 12.4 lists the different members of this structure.

Table 12.4

Members of UL_PART

Member Name	Type	Description
`attribute[]`	`string`	Returns the value of a specific attribute
`device`	`UL_DEVICE`	The part's underlying device
`deviceset`	`UL_DEVICESET`	The deviceset containing the part's device
`name`	`string`	The part's name
`populate`	`int`	Identifies if the part is part of the current variant
`value`	`string`	The value data defined in the schematic
`attributes()`	`UL_ATTRIBUTE`	The collection of attributes associated with the part
`instances()`	`UL_INSTANCE`	Information about the specific instance of the part
`variants()`	`UL_VARIANT`	The assembly variants containing the part

An attribute is simply a combination of a name and value, both given as strings. The `attribute[]` member accepts the name string and returns the value. The following code shows how this works. It checks if any part in the schematic has an attribute whose name is `VENDORNAME`. If so, it prints the part's name and the attribute's value.

```
output("attr.txt") {
  schematic(s) {
    s.parts(p) {
      if(p.attribute["VENDORNAME"])
        printf("Part %s comes from vendor %s.\n",
          p.name, p.attribute["VENDORNAME"]);
    }
  }
}
```

The `attributes()` member provides the entire list of attributes associated with the part. Each attribute is represented by a `UL_ATTRIBUTE` structure. This will be discussed next.

Design Attributes (UL_ATTRIBUTE)

Chapter 4, "Designing the Femtoduino Schematic," explained how to set attributes in the schematic editor. Attributes can be associated with a part or the overall design (see the `attributes()` member in Table 12.2). Each attribute is represented by a `UL_ATTRIBUTE` structure and Table 12.5 lists its members.

Table 12.5

Members of UL_ATTRIBUTE

Member Name	Type	Description
constant	int	Identifies whether the attribute can be overwritten (0) or not (1)
defaultvalue	string	The default value of the attribute
display	int	Identifies what aspect of the attribute should be displayed
name	string	The attribute's name
text	UL_TEXT	Properties of the displayed attribute text
value	string	The attribute's value

Component attributes can be set with the `attribute` command presented in Chapter 10, "Editor Commands." This accepts an optional default value for the attribute, which is used if the attribute is otherwise undefined. This is provided by the `defaultvalue` member and the regular value is provided by `value`.

Many different types of structures have attributes. The `display` and `text` members are only available if the containing structure is a `UL_INSTANCE` or `UL_ELEMENT`. The first identifies whether the attribute's name and/or value is displayed or if nothing is displayed. The second provides information about how the displayed text is depicted.

As an example, the following code looks for a part named `U5` and iterates through its attributes. For each attribute whose `constant` member is set to 0, the code prints its name and value to the attr.txt file.

```
output("attr.txt") {
  schematic(s) {
    s.parts(p) {
      if(p.name == "U5") {
        p.attributes(a) {
          if(!a.constant)
            printf("Attribute %s has value %s.\n",
              a.name, a.value);
        }
      }
    }
  }
}
```

Now that you understand how parts and attributes work, you're ready to code a ULP capable of generating a BOM.

Generating a BOM

If you go to the CadSoft web site and click Downloads, you'll find an option for User Language programs. This takes you to CadSoft's free ULPs, and one of the most popular has a name that begins with bom-ex. This ULP generates a BOM by examining each component in a design and searching through its attributes.

To use bom-ex, the attributes of the schematic's components need to be assigned specific names. For example, the attribute for a component's part number must be named `PARTNO`, and the attribute representing its tolerance must be named `TOL`.

The bom-ex program is impressively long and I'm not going to present its code here. Instead, Listing 12.1 presents the code for bom_short.ulp, which can be found in the ulp folder of this book's example archive. This creates a simple BOM that examines each component in the schematic and lists the name, value, manufacturer, and cost of its components. This data is formatted using comma-separated values, so it should be easy to open in a spreadsheet application.

The four fields of bom_short.ulp are acquired in the following way:

- **Name**—Obtained from the part's `name` member
- **Value**—Obtained from the part's `value` member
- **Manufacturer**—Obtained from the part's attribute with the name `MFR`
- **Cost**—Obtained from the part's attribute with the name `COST`

Listing 12.1: bom_small.ulp

```
// Check to see if the schematic editor is open
if (!schematic) {
  dlgMessageBox("This can be executed only if
                the schematic editor is open.");
  exit(1);
}
else {
  schematic(s) {
    output(filesetext(s.name, ".csv")) {

      // Print the header for the CSV file
      printf("Name,Value,Manufacturer,Cost\n");

      // Print the name, value, manufacturer, and cost
      s.parts(p) {
        printf("\"%s\",\"%s\",", p.name, p.value);

        // Print the name of the manufacturer
        if(p.attribute["MFR"])
          printf("\"%s\",", p.attribute["MFR"]);
        else
          printf("\"\",");

        // Print the cost
        if(p.attribute["COST"])
          printf("\"%s\"", p.attribute["COST"]);
        else
          printf("\"\"");

        printf("\n");
      }
    }
  }
}
```

This should be straightforward, but there are three points I want to make:

- Comments in a ULP are exactly like those in C/C++ programming. That is, a comment may be delimited by `/*` and `*/`. An entire line can be commented if preceded by `//`.
- The program starts with an `if` statement that checks if the schematic editor is open. If the editor isn't open, the program displays a helpful dialog. If the check isn't made, an error will result and EAGLE will display a less-helpful dialog.
- The `output` executable statement is inside the `schematic` statement. This is important because the name of the output file depends on the schematic's name.

Component Instances (UL_INSTANCE)

An important loop member of UL_PART is instances(), which provides the UL_INSTANCEs associated with the part. A UL_INSTANCE represents a specific instantiation of a part and Table 12.6 lists its members.

Table 12.6

Members of UL_INSTANCE

Member Name	Type	Description
angle	real	The instance's angle - 0, 90, 80, or 270
column	string	The column index containing the instance
gate	UL_GATE	The gate corresponding to the instance
mirror	int	Indicates whether the part is on the front (0) or rear (1) side of the board
part	UL_PART	The part corresponding to the instance
row	string	The row index containing the instance
sheet	int	The number of the sheet containing the part
smashed	int	Whether the instance's name/value can be moved separately from the instance
value	string	The value of the instance
x, y	int	Coordinates of the instance's origin
attributes()	UL_ATTRIBUTE	The attributes associated with the instance
texts()	UL_TEXT	The text associated with the instance
xrefs()	UL_GATE	The cross-reference gates of this instance

Most of these members relate to the instance's position and orientation within the schematic. More information about the instance can be obtained by accessing the structure's UL_GATE member and the UL_SYMBOL of the UL_GATE.

Component Gates (UL_GATE) and Symbols (UL_SYMBOL)

Chapter 4 explained that a component in the schematic editor may be split into multiple gates that can be moved separately. For example, if a component contains an array of eight resistors, each resistor in the schematic is represented by a gate that can be positioned independently of the others.

Table 12.7 lists the members of the UL_GATE structure. Note that if the component has only one gate, the position of the UL_GATE will correspond to the position of the UL_INSTANCE.

Table 12.7

Members of UL_GATE

Member Name	Type	Description
addlevel	int	Behavior of the gate when the instance is added to the design
name	string	The name of the gate
swaplevel	int	The value identifying other gates this gate can be swapped with (0 if the gate can't be swapped)
symbol	UL_SYMBOL	The symbol corresponding to the gate
x, y	int	Coordinates of the gate's origin

When an instance containing multiple gates is added to a schematic, the gates behave differently depending on addlevel. This can take one of five values:

- `GATE_ADDLEVEL_MUST`—The gate must be added whenever any gate is added. Delete removes the entire device.
- `GATE_ADDLEVEL_CAN`—If other gates have an addlevel set to Next, these gates can only be accessed with Invoke.
- `GATE_ADDLEVEL_NEXT`—If a component has multiple gates, successive gates are made ready as each is placed in the schematic.
- `GATE_ADDLEVEL_REQUEST`—Can only be accessed through Invoke.
- `GATE_ADDLEVEL_ALWAYS`—The gate must be added whenever any gate is added. Delete removes only the given gate, which can be added with Invoke.

In the schematic editor, each gate is graphically represented by a symbol. In UL, a symbol is represented by a `UL_SYMBOL`. Table 12.8 lists the members of this structure.

Table 12.8

Members of UL_SYMBOL

Member Name	Type	Description
area	UL_AREA	Opposite corners of the symbol
description	string	Full description of the symbol
headline	string	Brief description of the symbol
library	string	Name of the library containing the symbol
name	string	Name of the symbol
circles()	UL_CIRCLE	The circle shapes contained in the symbol
dimensions()	UL_DIMENSION	The dimensions of the symbol
frames()	UL_FRAME	The frames contained in the symbol
rectangles()	UL_RECTANGLE	The rectangle shapes contained in the symbol
pins()	UL_PIN	The pins of the symbol

polygons()	UL_POLYGON	The polygons contained in the symbol
texts()	UL_TEXT	The text objects contained in the symbol
wires()	UL_WIRE	The lines contained in the symbol

With this information, we can construct a ULP that finds the symbols furthest to the left and right. This can be useful when printing a schematic or drawing a frame around it. The code in Listing 12.2 shows how this works.

Listing 12.2: bounds.ulp

```
schematic(s) {

  // Initialize bounds
  int left = 200, right = -200;

  string left_instance, right_instance;
  s.parts(p) {
    p.instances(i) {

      // Find leftmost instance
      if(i.gate.symbol.area.x1 < left) {
        left = i.gate.symbol.area.x1;
        left_instance = i.name;
      }

      // Find rightmost instance
      if(i.gate.symbol.area.x2 > right) {
        right = i.gate.symbol.area.x2;
        right_instance = i.name;
      }
    }
  }

  output("bounds.txt") {
    printf("leftmost = %s, rightmost = %s\n",
      left_instance, right_instance);
  }
}
```

12.2.3 Nets and Net Classes

In addition to searching through parts, the `UL_SCHEMATIC` structure makes it easy to examine the parts' connections (nets) and their net classes. In a ULP, nets are represented by `UL_NET` structures and net classes are represented by `UL_CLASS` structures. This section explains both and demonstrates how they can be used in code.

Schematic Connections (UL_NET)

A net represents a single electrical connection in the schematic editor. Each net is represented by a `UL_NET` and Table 12.9 lists the structure's members.

Table 12.9

Members of UL_NET

Member Name	Type	Description
`class`	`UL_CLASS`	The net class
`column`	`string`	The left and right columns spanned by the net
`name`	`string`	The name of the net
`row`	`string`	The top and bottom rows spanned by the net
`pinrefs()`	`UL_PINREF`	The pins to which the net is connected
`segments()`	`UL_SEGMENT`	The segments that make up the net

The `row` and `column` members don't return integers. Instead, each returns a string containing two values. For `row`, the string contains the indices for the top and bottom rows spanned by the net. For `column`, the string contains the indices for the left and right columns spanned by the net. The rows and columns are determined by the schematic's frame, which is discussed in Chapter 4.

Suppose you want to know which components in the schematic are connected by a given net. In this case, you need to access the `pinrefs()` member of the `UL_NET`. This returns an array of `UL_PINREF` structures.

Pins Connected to Nets (UL_PINREF and UL_PIN)

In a schematic, a component's connection points are called pins. In a ULP, parts access pins through the `UL_PINREF` structure. Table 12.10 lists its three members.

Table 12.10

Members of UL_PINREF

Member Name	Type	Description
`instance`	`UL_INSTANCE`	The component instance containing the pin
`part`	`UL_PART`	The component containing the pin
`pin`	`UL_PIN`	The pin corresponding to the reference

If you want to know only which component contains a pin, the `UL_PINREF` is all you need. But if you want more details such as the pin's position or name, you need to access the corresponding `UL_PIN`. Table 12.11 lists each member of `UL_PIN`.

Table 12.11

Members of UL_PIN

Member Name	Type	Description
`direction`	`int`	The pin's direction
`length`	`int`	The length of the pin (long, medium, short, or point)

`visible`	`int`	Specifies the pin's visibility
`function`	`int`	Identifies a special aspect of the pin (clock, inverted)
`angle`	`real`	The pin's angle - 0, 90, 80, or 270
`name`	`string`	The pin's name
`net`	`string`	The name of the net to which the pin is connected
`route`	`int`	Identifies whether contacts must be routed to one another
`swaplevel`	`int`	The value identifying other pins this pin can be swapped with (0 if the pin can't be swapped)
`x, y`	`int`	The location of the pin's connection point
`circles()`	`UL_CIRCLE`	The circle or circles making up the pin's shape
`texts()`	`UL_TEXT`	The displayed text associated with the pin
`wires()`	`UL_WIRE`	The line or lines making up the pin's shape

The `direction`, `length`, `visible`, and `function` members are set with special enumerated types. To understand this, it helps to look at Tables 10.4–10.7 in Chapter 10.

- `direction`—As shown in Table 10.4, a pin's direction can be set to `io`, `in`, `out`, `oc`, and others. Similarly, `direction` can be set to `PIN_DIRECTION_IO`, `PIN_DIRECTION_IO`, `PIN_DIRECTION_OUT`, `PIN_DIRECTION_OC`, and others.
- `length`—As given in Table 10.5, a pin's length can be set to `point`, `short`, `middle`, or `long`. Similarly, `length` can be set to `PIN_LENGTH_POINT`, `PIN_LENGTH_SHORT`, `PIN_LENGTH_MIDDLE`, or `PIN_LENGTH_LONG`.
- `visible`—Table 10.6 lists the values a pin's visibility can take: `both`, `pin`, `pad`, or `off`. Similarly, `length` can be set to `PIN_VISIBILITY_FLAG_PIN`, `PIN_VISIBILITY_FLAG_PAD`, or `PIN_VISIBILITY_FLAG_OFF`.
- `function`—As listed in Table 10.7, a pin's function can be identified by `dot`, `clk`, or `dotclk`. The function member can be set to `PIN_FUNCTION_FLAG_NONE`, `PIN_FUNCTION_FLAG_DOT`, or `PIN_FUNCTION_FLAG_CLK`.

By combining the `UL_PART`, `UL_INSTANCE`, and `UL_PIN` structures, it's possible to code ULPs that automatically connect pins in the schematic. To demonstrate how this can be done, this section constructs a ULP that performs the following tasks:

- Iterate through each instance in the schematic and its pins.
- If the instance is IC1 and the pin is GND, set (x1, y1) equal to the pin's location.
- If the instance is IC2 and the pin is GND, set (x2, y2) equal to the pin's location.
- Using the `net` command, draw a net between (x1, y1) and (x2, y2) and assign it the name `netgnd`.

Listing 12.3 shows how these steps can be implemented in code. It uses UL to generate a `net` command and executes it by making the command an argument of the `exit` function.

Listing 12.3: net_connect.ulp

```
// Convert internal units to grid units
real convert_to_grid(int dim, int units) {
  real ans;
  switch(units) {
    case GRID_UNIT_MIC:
      ans = u2mic(dim); break;
    case GRID_UNIT_MM:
      ans = u2mm(dim); break;
    case GRID_UNIT_MIL:
      ans = u2mil(dim); break;
    case GRID_UNIT_INCH:
      ans = u2inch(dim); break;
  }
  return ans;
}

schematic(s) {

  // Declare variables
  real x1, y1, x2, y2;
  string loc1 = "", loc2 = "", cmd = "";

  s.parts(p) {
    p.instances(i) {
      i.gate.symbol.pins(pin) {

        // Search for the first pin connection
        if((i.name == "IC1") && (pin.name == "GND")) {
          x1 = convert_to_grid(pin.x, s.grid.unitdist);
          y1 = convert_to_grid(pin.y, s.grid.unitdist);
          sprintf(loc1, "(%f %f)", x1, y1);
        }

        // Search for the second pin connection
        else if((i.name == "IC2") && (pin.name == "GND")) {
          x2 = convert_to_grid(pin.x, s.grid.unitdist);
          y2 = convert_to_grid(pin.y, s.grid.unitdist);
          sprintf(loc2, "(%f %f)", x2, y2);
        }
      }
    }
  }

  // Create the net command
  if((loc1 != "") && (loc2 != "")) {
    cmd = "net netgnd " + loc1 + " " + loc2;
  }
  exit(cmd);
}
```

The ULP's main processing is performed by the two `if` statements. Both check for a specific part name and pin name. If the check is true, they find the coordinates of the pin's connection point and call `convert_to_grid` to convert the coordinates to the grid's units. After the conversions are made, the converted coordinates are combined into a string that will be used as a parameter of the generated `net` command.

The `convert_to_grid` function is important to understand. It accepts two parameters: a dimension given in EAGLE's internal units and a value identifying the units of the schematic's grid. The second parameter specifies which conversion function is needed to convert the first parameter.

Net Classes (UL_CLASS)

Each net begins as part of the default net class. But new classes can be created by going to Edit > Net classes... in the schematic editor's main menu. Each class has four properties that can be configured: the name, trace width, drill diameter, and clearance. These properties are reflected by the members of the `UL_CLASS` structure. Table 12.12 lists each of these members along with their data types and a description.

Table 12.12

Members of UL_CLASS

Member Name	Type	Description
`clearance`	`int`	Minimum distance between objects in the signal layers (traces, pads, vias, and so on)
`drill`	`int`	Drill diameter for the net class
`name`	`string`	The name of the net class
`number`	`int`	The ID number of the net class
`width`	`int`	The width of the traces corresponding to nets in the net class

These members are all easy to understand. The following code iterates through the nets of a schematic and prints the name and trace width of their net classes.

```
output("classes.txt") {
  schematic(s) {
    s.nets(n) {
      printf("Net %s belongs to class %s,
              whose trace width is %u.\n",
              n.name, n.class.name, n.class.width);
    }
  }
}
```

12.2.4 Sheets and Frames

In Chapter 4, the Femtoduino schematic is divided into separate subcircuits for simplicity. When designs get especially large, the schematic needs to be split into separate sheets. In a ULP, each sheet of a schematic is represented by a `UL_SHEET` structure. Table 12.13 lists the structure's members.

Table 12.13

Members of UL_SHEET

Member Name	Type	Description
area	UL_AREA	Coordinates of the sheet's lower-left corner and upper-left corner
description	string	Text description of the sheet
headline	string	Brief description of the sheet
number	int	The sheet's identifying number
busses()	UL_BUS	The buses contained in the sheet
circles()	UL_CIRCLE	The circle shapes in the sheet
dimensions()	UL_DIMENSION	The dimensions of the sheet
frames()	UL_FRAME	The frames used to outline the sheet
instances()	UL_INSTANCE	The part instances contained in the sheet
nets()	UL_NET	The nets contained in the sheet
polygons()	UL_POLYGON	The polygons contained in the sheet
rectangles()	UL_RECTANGLE	The rectangular shapes in the sheet
texts()	UL_TEXT	The text objects of the sheet
wires()	UL_WIRE	The lines contained in the sheet

Many of these members relate to the sheet's geometry: its area, dimensions, and shapes. One interesting point is that `UL_SHEET` has a member representing the frame and `UL_SCHEMATIC` does not.

In my ULPs, I've found it helpful to know how many rows and columns are in the schematic's frame. The `UL_FRAME` structure provides this information and Table 12.14 lists its members.

Table 12.14

Members of UL_FRAME

Member Name	Type	Description
columns	int	Number of columns in the frame
rows	int	Number of rows in the frame
border	int	The type of border used by the frame
layer	int	The number of the layer containing the frame
x1, y1	int	Coordinates of the frame's lower-left corner
x2, y2	int	Coordinates of the frame's lower-left corner
text()	UL_TEXT	The text presented in the frame
wires()	UL_WIRE	The wires contained in the frame

The following code demonstrates how the UL_SHEET and UL_FRAME can be accessed in a ULP. For each sheet of the schematic, it prints the number of rows and columns in the frame.

```
output("sheets.txt") {
  schematic(s) {
    s.sheets(sh) {
      sh.frames(f) {
        printf("The frame of Sheet %d
                has %d rows and %d columns.\n",
                sh.number, f.rows, f.columns);
      }
    }
  }
}
```

12.2.5 Variants

Chapter 4 explained that an assembly variant is essentially a clone of a schematic that can be modified in three ways:

- Components from the original schematic can be excluded from the variant.
- Component values can be changed, such as resistances and capacitances.
- Component technologies can be changed, such as changing from surface-mount to ball grid arrays.

In the schematic editor, variants can be configured by going to Edit > Assembly Variants... on the main menu. In a ULP, there are two structures related to variants: UL_VARIANTDEF and UL_VARIANT. The first structure can be accessed from the UL_SCHEMATIC and it has a single member: name.

The second structure, UL_VARIANT, can't be accessed from the UL_SCHEMATIC. I thought this was odd at first, but the schematic itself is an implementation of a variant. So it makes sense that the schematic structure can't access other variants.

Instead, ULPs access UL_VARIANTs through UL_PARTs. That is, each UL_PART can access each variant in which it's included. Table 12.15 lists the members of UL_VARIANT.

Table 12.15

Members of UL_VARIANT

Member Name	Type	Description
populate	int	Whether the part is included in the variant (1) or not (0)
value	string	The part's value in the variant
technology	string	The part's technology in the variant
variantdef	UL_VARIANTDEF	The variant definition corresponding to the variant

The following code shows how ULPs can access variants. It iterates through each part of the schematic. For each part, it prints the variants' names and whether the part is populated in the variant.

```
output("variant.txt") {
  schematic(s) {
    s.parts(p) {
      p.variants(v) {
        if(v.populate)
          printf("Variant %s contains Part %s.\n",
                 v.variantdef.name, p.name);
        else
          printf("Variant %s doesn't contain Part %s.\n",
                 v.variantdef.name, p.name);
      }
    }
  }
}
```

When I execute this code, `v.populate` is only true if the component has been modified in the variant. If the part is included in the variant but it hasn't been modified, nothing is printed.

12.3 Board Designs (UL_BOARD)

Just as a `UL_SCHEMATIC` structure represents a schematic design, a `UL_BOARD` represents a board design. The board design's information can be accessed by calling the `board()` executable statement. Inside this statement, the members of `UL_BOARD` can be accessed using dot notation.

Table 12.16 lists the members of `UL_BOARD`. Note that the names of data members are not followed by parentheses and the names of the loop members are.

Table 12.16

Members of UL_BOARD

Member Name	Type	Description
`alwaysvectorfont`	`int`	Identifies how the text is configured with regard to vectorized font
`area`	`UL_AREA`	Coordinates of the design's lower-left and upper-right corners
`description`	`string`	Description of the board design
`grid`	`UL_GRID`	Properties of the board editor's grid
`headline`	`string`	Brief description of the board design
`name`	`string`	The name of the board design
`verticaltext`	`int`	Identifies whether vertical text is printed from bottom to top or top to bottom

attributes()	UL_ATTRIBUTE	The design's global attributes
circles()	UL_CIRCLE	Circular shapes included in the design
classes()	UL_CLASS	Net classes used in the board design
dimensions()	UL_DIMENSION	Dimensions of the board
elements()	UL_ELEMENT	Components in the board design
frames()	UL_FRAME	Frames associated with the board
holes()	UL_HOLE	Drill holes
layers()	UL_LAYER	Layers used in the board design
libraries()	UL_LIBRARY	Component libraries used in the board design
polygons()	UL_POLYGON	Polygons (copper pours) in the design
rectangles()	UL_RECTANGLE	Rectangular shapes in the design
signals()	UL_SIGNAL	Named nets (signals) in the board design
texts()	UL_TEXT	Text depicted in the board design
variantdefs()	UL_VARIANTDEF	Variant definitions in the board design
wires()	UL_WIRE	Lines drawn in the board design

Many of these have the same types as those in `UL_SCHEMATIC` and serve the same purposes. One major difference is that components in a `UL_SCHEMATIC` are represented by `UL_PART`s and components in a `UL_BOARD` are represented by `UL_ELEMENT`s.

12.3.1 Board Design Components (UL_ELEMENT and UL_PACKAGE)

As discussed in Chapter 5, "Layout and Design Rules," the components in a board design are referred to as packages. But in the UL, a component of a `UL_BOARD` is accessed as a `UL_ELEMENT` structure. Table 12.17 lists its members.

Table 12.17

Members of UL_ELEMENT

Member Name	Type	Description
angle	real	Counterclockwise angle (0.0–359.9)
attribute[]	string	Value of a specific attribute
column	string	The element's column index in the frame
locked	int	Indicates whether the user can move/copy the element (0) or not (1)
mirror	int	Indicates whether the part is on the front (0) or rear (1) side of the board
name	string	The element's name
package	UL_PACKAGE	The package associated with the element
populate	int	Indicates whether the element is populated
row	string	The element's row index in the frame

smashed	int	Whether text has been separated from the element for the purpose of translation/rotation
spin	int	Indicates whether the element's text can be printed in any direction
value	string	The element's value (resistance, capacitance)
x, y	int	The coordinates of the element's origin
attributes()	UL_ATTRIBUTE	The array containing the element's attributes
texts()	UL_TEXT	The element's smashed text
variants()	UL_VARIANT	The element's assembly variants

It's important to understand the difference between a UL_ELEMENT and its associated UL_PACKAGE. A UL_PACKAGE is defined in a library along with its pads, holes, and shape. A UL_ELEMENT represents one instance of the UL_PACKAGE in the design. Table 12.18 presents the name and type of each number of UL_PACKAGE.

An example will make this clear. Suppose three resistors are added to a design, each having the package SMD1206. The UL_ELEMENT names might be R1, R2, and R3, but in each case, the UL_PACKAGE name is SMD1206.

Table 12.18

Members of UL_PACKAGE

Member Name	Type	Description
area	UL_AREA	The lower-left and upper-right corners containing the package
description	string	The package's full description
headline	string	A brief description of the package
library	string	The library containing the package
name	string	The package's name
circles()	UL_CIRCLE	The circle shapes that make up the package
contacts()	UL_CONTACT	The package's pads
dimensions()	UL_DIMENSION	The package's dimensions
frames()	UL_FRAME	The frames contained in the package
holes()	UL_HOLE	The holes contained in the package
polygons()	UL_POLYGON	The polygons contained in the package
rectangles()	UL_RECTANGLE	The rectangle shapes contained in the package
texts()	UL_TEXT	The text objects contained in the package
wires()	UL_WIRE	The lines contained in the package

Most of these members have the same names and purposes as those found in UL_SYMBOL. Two of them deserve added explanation.

- contacts()—Contains the array of the package's pads
- holes()—Contains the drill holes and vias that make up the package

A simple example will show how UL_ELEMENT and UL_PACKAGE are used. The following code computes and prints the number of pads in the design.

```
output("numpads.txt") {
  int num_pads = 0;
  board(b) {
    b.elements(e) {
      e.package.contacts(c) {
        num_pads += 1;
      }
    }
    printf("Number of pads in the design: %d\n", num_pads);
  }
}
```

12.3.2 Layers and Polygons

Chapter 5 explained how to arrange packages in the Femtoduino's board design. Two important tasks involved changing the layers of the bottom components and creating polygons to serve as ground planes on the top and bottom layers. In UL, the layers of a board are represented by UL_LAYER structures and polygons are represented by UL_POLYGON structures.

Layers (UL_LAYER)

When positioning objects in a board design, it's important to keep track of which layers are used and what features can be found in each layer. In a ULP, this information can be obtained by examining the members of each UL_LAYER in the UL_BOARD. Table 12.19 lists each of these members.

Table 12.19

Members of UL_LAYER

Member Name	Type	Description
color	int	Color associated with the layer
fill	int	Identifies whether the layer is displayed as filled or in outline
name	string	Name of the layer
number	int	Number of the layer
used	int	Whether the layer contains objects (1) or not (0)
visible	int	Whether the layer is visible (1) or not (0)

These members are straightforward to understand. Personally, I like to use this structure to verify that the board's features are only on the layers I intend to use. The following code prints the name and number of each layer containing objects.

```
output("layers.txt") {
  board(b) {
    b.layers(l) {
      if(l.used)
        printf("Layer #%d (%s) contains objects.\n",
                l.number, l.name);
    }
  }
}
```

Polygons (UL_POLYGON)

In Chapter 5, the ground planes in the Femtoduino design were created by drawing polygons in the board design. A polygon identifies a region of copper, and like a trace or airwire, each polygon carries a signal.

In the UL, a `UL_POLYGON` represents a conductive region in a `UL_BOARD`. Table 12.20 lists the members of the `UL_POLYGON` structure.

Table 12.20

Members of UL_POLYGON

Member Name	Type	Description
`isolate`	`int`	Minimum distance between the polygon and other objects
`layer`	`int`	The number of the layer containing the polygon
`orphans`	`int`	Indicates whether conductive islands should be created (1) or not (0)
`pour`	`int`	Indicates the pattern of the copper pour
`rank`	`int`	If polygons overlap, identifies which should be given priority
`spacing`	`int`	Distance between lines if the copper pour pattern is hatched
`thermals`	`int`	Indicates whether pads should connect to the polygon through thermal pads (1) or not (0)
`width`	`int`	Width of the polygon's fill lines
`wires()`	`UL_WIRE`	The lines that make up the uncomputed polygon
`contours()`	`UL_WIRE`	The lines that make up the computed polygon
`fillings()`	`UL_WIRE`	The fill lines that make up the polygon

The last three members provide arrays of `UL_WIRE` structures and can be used to identify the lines that make up a `UL_POLYGON`. As explained in Chapter 4, if a polygon is used as a copper pour and other conductive elements lie in the same layer, EAGLE will compute a new shape for the polygon to prevent intersection.

The `wires()` member provides the original lines set by the user while the `contours()` and `fillings()` members relate to the computed polygon. `contours()` returns the oriented lines that make up the computed polygon while `fillings()` provides the fill wires that make up the polygon.

The `pour` member identifies the pattern of copper inside the copper pour. This can take one of two values:

- `POLYGON_POUR_SOLID`—No pattern, the polygon is filled with copper.
- `POLYGON_POUR_HATCH`—Copper is patterned in a cross-hatched manner.

12.3.3 Signals and Contacts

Of the circuit design tasks that need to be automated, routing is the most important. Certain versions of EAGLE provide an autorouter, but if your version doesn't have this (or you simply don't like it), the UL makes it possible to code your own.

The editor command that makes routing possible is `route`, discussed in Chapter 10. To use this properly requires examining the signals and pads in the board design. These features can be accessed using five types of UL structures:

1. `UL_SIGNAL`—Represents an electrical signal in the board editor
2. `UL_CONTACTREF`—A reference to a pad
3. `UL_CONTACT`—A through-hole pad or SMD pad
4. `UL_SMD`—A pad connected to a surface-mount device (SMD) lead
5. `UL_PAD`—A through-hole pad

Signals (UL_SIGNAL)

In a UL program, the signals of a board design can be accessed by iterating through the `UL_SIGNAL` structures provided by the `signals()` member of `UL_BOARD`. Table 12.21 lists the members of the `UL_SIGNAL` type.

Table 12.21

Members of UL_SIGNAL

Member Name	Type	Description
`airwireshidden`	`int`	Identifies whether airwires are hidden (1) or not (0)
`name`	`string`	The signal name
`contactrefs()`	`UL_CONTACTREF`	References to the contacts (through-hole pads and SMD pads) connected to the signal
`polygons()`	`UL_POLYGON`	The polygons carrying the signal
`vias()`	`UL_VIA`	The vias carrying the signal
`wires()`	`UL_WIRE`	Airwires and traces carrying the signal

You may notice that UL doesn't provide structures for airwires or routed traces in a board design. Instead, both types of connections are provided by the `wires()` member. Up to this point, `UL_WIRE`s have been used only for graphical lines, but in Table 12.21, `wires()` provides all the electrical connections (airwires and routed traces) for a given signal. If a `UL_WIRE` has a width of 0, it represents an airwire. If the width is greater than 0, it represents a routed trace.

The code in Listing 12.4 shows how this works. It searches for the signal named VCC and uses the `wires()` method to print out its airwires and routed traces.

Listing 12.4: signal_search.ulp

```
output("wires.txt") {
  board(b) {
    b.signals(s) {

      // Searches for the signal named VCC
      if(s.name == "VCC") {
        s.wires(w) {

          // Checks for airwire or routed trace
          if(w.width == 0)
            printf("Airwire ");
          else
            printf("Routed trace ");
          printf("from (%f, %f) to (%f, %f).\n",
            convert_to_grid(w.x1, b.grid.unitdist),
            convert_to_grid(w.y1, b.grid.unitdist),
            convert_to_grid(w.x2, b.grid.unitdist),
            convert_to_grid(w.y2, b.grid.unitdist));
        }
      }
    }
  }
}
```

The `airwireshidden` member doesn't refer to the visibility of the airwire layer (19). It identifies whether the user selected the signal's airwires to be hidden.

Contact References (UL_CONTACTREF)

The UL refers to pads (through-hole and SMD) as contacts. Each contact is represented by a `UL_CONTACT` structure, but this can't be accessed directly from a `UL_SIGNAL`. Instead, the `contactrefs()` member in Table 12.22 provides *references* to contacts. These references are given as `UL_CONTACTREF` structures and Table 12.22 lists its members.

Table 12.22

Members of UL_CONTACTREF

Member Name	Type	Description
contact	UL_CONTACT	The UL_CONTACT representing a pad
element	UL_ELEMENT	The associated component
route	int	Identifies whether the contact must be routed to any contact or all contacts
routetag	string	Describes the group of contacts connected to the pin

The route member may be set to one of two values:

- CONTACT_ROUTE_ALL—The contact must be explicitly routed to all contacts.
- CONTACT_ROUTE_ANY—The contact may be routed to any contact.

Contacts (UL_CONTACT)

A contact represents a generic pad independent of technology. In UL a contact is represented by a UL_CONTACT. The structure's members provide top-level information about the contact and Table 12.23 lists each of them.

Table 12.23

Members of UL_CONTACT

Member Name	Type	Description
name	string	The name of the contact
pad	UL_PAD	The through-hole pad corresponding to the contact
signal	string	The name of the signal connected to the contact
smd	UL_SMD	The SMD pad corresponding to the contact
x, y	int	The coordinates of the contact's origin

The pad and smd members are important to understand. if the contact represents a SMD pad, smd will provide a UL_SMD structure and pad will provide nothing. Similarly, if the contact represents a through-hole pad, pad will provide a UL_PAD structure and smd will provide nothing. The following discussion explains the UL_SMD and UL_PAD structures in detail.

SMDs (UL_SMD)

In general, a SMD pad is a conductive region with a rectangular or rounded-rectangular shape. In the UL, SMD pads are represented by UL_SMDs. Table 12.24 lists the structure's members.

Table 12.24

Members of UL_SMD

Member Name	Type	Description
angle	real	Angular orientation of the SMD pad
dx[layer], dy[layer]	int	Dimensions of the pad for a given layer
flags	int	Specifies whether stop/cream masks are present, thermals
name	string	Name of the SMD pad
roundness	int	Degree of roundness of the pad
signal	string	Name of the signal carried by the pad
x, y	int	Coordinates of the pad's origin

It's important to see the difference between the x, y, dx, and dy members. x and y identify the coordinates of the pad's origin. dx and dy provide the pad's dimensions on different layers. By default, when EAGLE creates an SMD pad, it creates regions (masks)in the tStop (29) and tCream (31) layers. The tStop region is generally slightly larger than the pad, so dx[29] and dy[29] are usually larger than dx[1] and dy[1]. The tCream region is frequently the same size as the pad, so dx[31] = dx[1] and dy[31] = dy[1].

flags identifies whether the pad has an associated stop mask and cream mask, and if the possibility exists that it could be a thermal pad. Its possible values are given as follows:

- SMD_FLAG_STOP—Stop mask created for the pad.
- SMD_FLAG_CREAM—Cream mask created for the pad.
- SMD_FLAG_THERMALS—The pad may be a thermal pad if located on or near a polygon.

Pads (UL_PAD)

In UL, each through-hole pad in a design is represented by a UL_PAD structure. Table 12.25 lists each of the members of this structure.

Table 12.25

Members of UL_PAD

Member Name	Type	Description
angle	real	Angular orientation of the through-hole pad
diameter[layer]	int	Pad diameter at the given layer
drill	int	Number of the drill
drillsymbol	int	Number of the drill symbol for this diameter
elongation	int	Amount of elongation
flags	int	Specifies whether stop/cream masks are present, thermals
name	string	Name of the through-hole pad
shape[layer]	int	Shape of the pad for a given layer
signal	string	Name of the signal carried by the pad
x, y	int	Coordinates of the pad's origin

The `shape` member accepts a layer number and returns a value representing the shape. This may be set to `PAD_SHAPE_SQUARE`, `PAD_SHAPE_ROUND`, `PAD_SHAPE_LONG`, `PAD_SHAPE_OCTAGON`, or `PAD_SHAPE_OFFSET`.

If the `PAD_SHAPE_OFFSET` or `PAD_SHAPE_OFFSET` shapes are selected, the `elongation` member identifies the extent to which the pad's shape has been stretched. More precisely, it identifies the ratio of the pad's long side to its short side as an integer percentage.

The flags member is similar to the flags member of the `UL_SMD`, but because through-hole pads don't require solder paste, there are no cream masks created. Also, many boards assign a special shape to identify the first pad of a device with multiple through-hole leads. The flags member can take the following values:

- `PAD_FLAG_STOP`—Stop mask created for the pad.
- `PAD_FLAG_FIRST`—The pad should be assigned a special shape.
- `PAD_FLAG_THERMALS`—The pad may be a thermal pad if located on or near a polygon.

12.3.4 Vias and Holes

The last aspects of board designs we'll look at are the vias and holes. These are represented by `UL_VIA` and `UL_HOLE` structures, whose members resemble those of the `UL_PAD` structure presented earlier. This discussion presents both structures and shows how they can be used to generate a custom drill file.

Vias (UL_VIA)

A via is a conductive hole connecting two or more layers. Accessing the `UL_VIAs` in a board design requires two steps:

1. For the `UL_BOARD`, invoke `signals()` to access the design's `UL_SIGNAL` structures.
2. For each `UL_SIGNAL`, invoke `vias()` to access the `UL_VIAs` carrying the signal.

Table 12.26 lists the members of the `UL_VIA` structure. The primary difference between these members and those of 12.25 is the `start` and `end` members.

Table 12.26

Members of UL_VIA

Member Name	Type	Description
`diameter[layer]`	`string`	The via's diameter at the given layer
`drill`	`int`	Number of the drill
`drillsymbol`	`int`	Number of the drill symbol for this diameter
`end`	`int`	The number of the via's ending layer
`flags`	`int`	Configuration options for the via
`shape[layer]`	`int`	Shape of the pad for the given layer
`start`	`int`	The number of the via's starting layer
`x, y`	`int`	The coordinates of the via's origin

Vias don't require solder paste and they can't be implemented with thermal pads. Therefore, the only configuration option that can be set with flags involves the stop mask, which can be set with the value `VIA_FLAG_STOP`.

As with through-hole pads, a via's shape can be configured by the designer. The shape member may take one of three values: `VIA_SHAPE_ROUND`, `VIA_SHAPE_SQUARE`, and `VIA_SHAPE_OCTAGON`.

Holes (UL_HOLE)

Regular drill holes don't have pads or conductive material. Therefore, the `UL_HOLE` structure is simpler than the `UL_PAD` or `UL_VIA` structures. Table 12.27 lists its members.

Table 12.27

Members of UL_HOLE

Member Name	Type	Description
diameter[layer]	string	The hole's diameter at the given layer
drill	int	Number of the drill
drillsymbol	int	Number of the drill symbol for this diameter
x, y	int	The coordinates of the hole's origin

These members have the same names and roles as those described earlier. All drill holes are round, so there is no `shape` member. Drill holes don't have stop masks, cream masks, or thermal pads, so there is no `flags` member.

Generating a Drill Rack

Chapter 7, "Generating and Submitting Output Files," discussed drill files, which EAGLE generates to provide information about the holes that need to be drilled into the board. In particular, EAGLE has a file called drillcfg.ulp that creates a file called a drill rack.

This ULP looks at all the holes that need to be drilled and lists each required drill diameter along with a unique identifier starting with `T`. For example, the Femtoduino circuit introduced in Chapter 4 requires two drills: 0.016" and 0.040". Its drill rack file is given as follows:

```
T01 0.016in
T02 0.040in
```

The code in Listing 12.5 accomplishes a similar result. It searches through the board design and fills an array with drill values. This search is conducted in three steps:

- Iterate through the board's `UL_HOLE` structures and store their drill values.
- For each signal, iterate through the `UL_VIA` structures and store their drill values.
- For each package, iterate through the `UL_CONTACTS`. If the `UL_CONTACT` has an associated `UL_PAD`, store its drill values.

Listing 12.5: drill_rack.ulp

```
// The array of drill values
int drills[];

// Check whether the drill is already present
int new_drill(int check_drill) {

  // Search for drill value in array
  for(int i=0; drills[i]; i++) {
    if(drills[i] == check_drill) {
      return 0;
    }
  }
  return 1;
}

board(b) {
  output(filesetext(b.name, ".drl")) {

    // Array index
    int i = 0;

    // Find the UL_HOLE drills
    b.holes(h) {
      if(new_drill(h.drill))
        drills[i++] = h.drill;
    }

    // Find the UL_VIA drills
    b.signals(s) s.vias(v) {
      if(new_drill(v.drill))
        drills[i++] = v.drill;
    }

    // Find the UL_PAD drills
    b.elements(e) e.package.contacts(c) {
      if(c.pad && new_drill(c.pad.drill)) {
        drills[i++] = c.pad.drill;
      }
    }

    for(int j=0; drills[j]; j++)
      printf("T%02d %f\n", j+1,
        convert_to_grid(drills[j], b.grid.unitdist));
  }
}
```

There are four points I'd like to mention about this program:

- UL doesn't provide a boolean data type, so the `if`/`while` statements consider 0 to be false and a nonzero value to be true. This is why `new_drill` returns 0 if the drill value is already in the array and 1 otherwise.

- The resulting drill rack file has the name of the board design and the *.drl suffix. It is stored in the current project directory.
- This program assumes that the grid's units are the desired units for the drill rack. This may not be a safe assumption.
- This program doesn't provide as much customization or precision checking as drillcfg.ulp. Therefore, if you need to generate a professional drill rack file, I recommend that you use drillcfg.ulp.

12.4 Conclusion

To code a nontrivial UL program, you need to be familiar with a wide range of data structures. At the very least, you should understand the two top-level structures: `UL_SCHEMATIC` and `UL_BOARD`.

The first part of this chapter discussed the `UL_SCHEMATIC` structure, which represents the currently active schematic design. This provides access to the components of the design, which are represented by `UL_PART` structures. The components' symbols are represented by `UL_SYMBOL` structures, so if you're interested in a component's geometry or pins, this is the structure to know.

The second part of this chapter presented the `UL_BOARD` and its many members. In a board design, components are represented in UL by `UL_ELEMENT` structures. To obtain information about an element's geometry or pads, you need to access its associated `UL_PACKAGE`.

UL doesn't have structures that represent airwires or routed traces. Instead, each `UL_SIGNAL` can access an array of `UL_WIRE` structures. These `UL_WIRE`s represent all connections carrying the given signal. If an `UL_WIRE` has a width of 0, it corresponds to an airwire in the design. Otherwise, it represents a routed trace.

Chapter 13

Creating Dialogs and Menu Items

UL programs can receive user input through command-line arguments, but it's easier for the user to enter data using dialogs. A good example of this is the drillcfg.ulp program discussed in Chapter 7, "Generating and Submitting Output Files." To create a drill rack file, this program displays dialogs asking for unit selection, drill rack verification, and file storage. This information would be difficult to obtain using command-line arguments only.

Thankfully, the User Language provides many ways to create dialogs for EAGLE. Most of this chapter is devoted to explaining how these dialogs can be configured, and I'll focus on three aspects:

1. **Predefined dialogs**—Four dialog types that can be quickly inserted into a program
2. **Widgets**—Different graphical elements that can be added to a dialog
3. **Layout**—Methods of arranging widgets in the dialog space

The last section of this chapter presents the MENU command, which makes it possible to add custom actions to the main menu of an EAGLE window. This enables the user to execute scripts and ULPs without having to enter text in the command box.

13.1 Predefined Dialogs

Before we look at custom dialogs, I'd like to describe UL's predefined dialogs, which are much easier to implement in code. The UL provides four different predefined dialogs.

1. **Message box**—Presents a notification to the user
2. **Directory dialog**—Allows the user to select a directory
3. **File open dialog**—Allows the user to open an existing file
4. **File save dialog**—Allows the user to select the name and location of a file to which data should be saved

In each case, the dialog is created by a UL function whose arguments configure the dialog's appearance. This section discusses these functions and shows how they're used in ULPs.

13.1.1 Message Box

A message box is a simple dialog that presents information to the user and closes when a button is pressed. The `dlgMessageBox` function creates a new message box and opens it in the current window. Its signature is given as follows:

```
int dlgMessageBox(string message[, button_strings])
```

The function accepts a message string and a list of strings to be printed in buttons along the dialog's bottom. Its return value identifies which of the buttons were pressed. For example, `dlgMessageBox` returns 0 if the first button is clicked, 1 if the second button is clicked, and so on.

As an example, consider the following function call:

```
int button = dlgMessageBox(";Hi there, <i>user</i>!",
                           "&Option1", "O&ption2");
```

This produces the dialog box shown in Figure 13.1.

Figure 13.1: An Example Message Box

As shown, the dialog's title is determined by the name of the ULP file. The buttons are positioned to occupy the bottom of the dialog. There are four additional points that deserve attention:

- The message is preceded by an icon that identifies the dialog as providing information as opposed to a warning or error.
- The message string uses HTML formatting to print `user!` in italics.
- If the user presses Enter, the first button will be clicked and the dialog will close. If the user presses the Escape key, the second button will be clicked.
- Each button has an associated accelerator key that allows the user to select it with the appropriate keystroke.

The rest of this section elaborates on these points and shows how they can be configured in code. Much of this discussion applies to all UL dialogs, not just message box dialogs.

Dialog Icons

Without special characters, message boxes are displayed without icons. But UL provides three different icons that indicate the nature of the message. In each case, the icon can be selected by preceding the message string with a punctuation mark.

- **Information**—If the dialog provides information to the user, the information icon can be included by preceding the message string with the semicolon (;). This is the icon displayed in Figure 13.1.
- **Warning**—If the dialog is intended to caution the user, the warning icon can be included by preceding the message string with an exclamation point (!).
- **Error**—If the dialog is intended to show an error condition, the error icon can be included by preceding the message string with a colon (:).

The dialog's appearance can be further customized by including HTML formatting tags. This topic is discussed next.

HTML Formatting

Text in dialogs, descriptions, and `#usage` directives can be formatted with a subset of the tags used by the Hypertext Markup Language (HTML). Table 13.1 lists many of the formatting tags recognized by the UL.

Table 13.1

Formatting Tags Available for UL Dialogs (Abridged)

HTML Tag	Description
<i>..</i>	Italicized text
<b>..</b>	Boldface text
<u>..</u>	Underlined text
<code>..</code>	Text printed as code (monospace, Courier)
<center>..</center>	Centered text
<p>..</p>	Left-aligned paragraph
<h1>..</h1> <h2>..</h2> <h3>..</h3>	Heading styles (the lower the number, the larger the font)
<ol> <li>..</li> <li>..</li> </ol>	Ordered list of items
<ul> <li>..</li> <li>..</li> </ul>	Bulleted list of items
 	Line break
<hr>	Horizontal line
	Nonbreaking space
<	The less-than sign (<)
>	The greater-than sign (>)

For the full list of tags, download the User Language Manual from CadSoft's web site. For a thorough discussion of HTML, I recommend the free online HTML tutorial at http://www.w3schools.com/html/default.asp.

The following code demonstrates how many of these tags are used. The message consists of a bulleted list of three items, each printed in a different heading.

```
string s1 = "<li><h1>An <u>ERROR</u> has occurred</h1></li>";
string s2 = "<li><h2>And you'd better do something</h2></li>";
string s3 = "<li>Or something <b>bad</b> will happen.</li>";
int answer = dlgMessageBox(":<ul>" + s1 + s2 + s3 + "</ul>",
                           "I Understand");
```

Figure 13.2 shows what the resulting dialog looks like. It displays an error icon on the left because the message string starts with a colon.

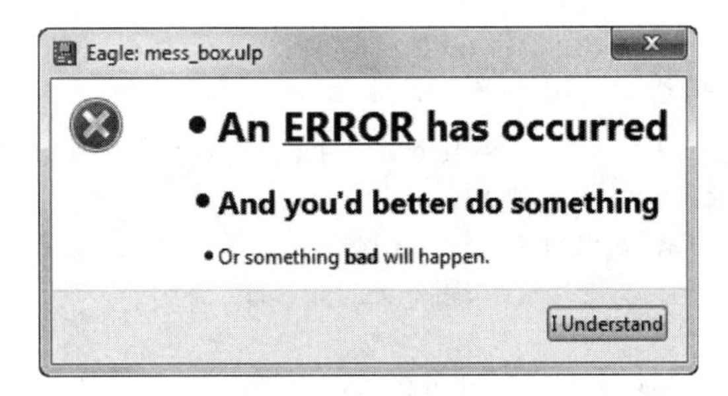

Figure 13.2: HTML Formatting

The last two items in Table 13.1 deserve additional explanation. Because UL text can be formatted with HTML, EAGLE interprets any < and > characters as parts of HTML tags. Therefore, if a message contains these characters without formatting, they need to be specially marked. That is, `<` should be used instead of < and `>` should be used instead of >.

It's interesting to compare the buttons in Figure 13.1 to the button in Figure 13.2. In Figure 13.1, the buttons are expanded to occupy the width of the dialog. The button in Figure 13.2 has more text, but its size is reduced to match that of the text.

Default and Cancel Buttons

In Figure 13.1, the left button is outlined and shaded. This means the left button is the dialog's default button, and if the user presses Enter, this button will be clicked.

Without configuration, the first button listed in `dlgMessageBox` is made the default button. This behavior can be modified by preceding the intended default button with +. For example, the following code looks like the code shown earlier, but makes the right button the default button.

```
int button = dlgMessageBox(";Hi there, <i>user</i>!",
                           "&Option1", "+O&ption2");
```

In addition to the default button, dialogs have cancel buttons that are clicked when the user presses the Escape key. Without configuration, the last button is always made the cancel button. This can be configured by preceding the intended cancel button with -. This is shown in the following code, which makes the second button the cancel button.

```
int button = dlgMessageBox(msg, "Button", "-Cancel", "+Default");
```

Accelerator Keys and Escape Characters

In Figure 13.1, the text of both buttons contains an underlined character. That is, the O in the left button is underlined and the p in the right button is underlined. When the user presses either the `o` or `p` keys, the corresponding button is pressed. These special keys are called accelerator keys, and they can be configured by preceding the desired character with an ampersand (&).

The following code creates a message box with three buttons named ABC. With accelerator keys, the first button is clicked when the `a` key is pressed, the second is clicked when the `b` key is pressed, and the third is clicked when the `c` key is pressed.

```
int button = dlgMessageBox(msg, "&ABC", "A&BC", "AB&C");
```

HTML formatting isn't available for button text, so the & character can't be used in place of an ampersand. But an ampersand can be displayed in a button by preceding the character with two backslashes. These escape characters tell EAGLE that the following character should be interpreted and displayed like a regular character.

For example, the following code creates a dialog whose buttons use escape characters to display ampersands.

```
int button = dlgMessageBox("Favorite pair of letters?",
                           "A \\& B", "P \\& Q", "X \\& Y");
```

13.1.2 Directory Dialogs

Unlike message boxes, which may serve a wide range of purposes, a directory dialog has only one role: to allow the user to select a directory. This is easy to understand, and the required function, `dlgDirectory()`, is easy to use. Its signature is given as follows:

```
string dlgDirectory(string title[, string start])
```

The only required argument of `dlgDirectory` is a title for the dialog. This will be printed along the dialog's title bar.

By default, the dialog opens the top-level EAGLE directory for searching. If a different path is provided after the title, that directory will be used instead. If a folder under EAGLE's top-level directory is provided, that will be the starting point.

The following code demonstrates how this works. It launches a directory dialog that starts the search in the examples folder under the $EAGLE/projects directory.

```
string folder = dlgDirectory("Select a project folder",
                             "projects/examples");
```

Figure 13.3 shows what the resulting dialog looks like on my Windows 7 system.

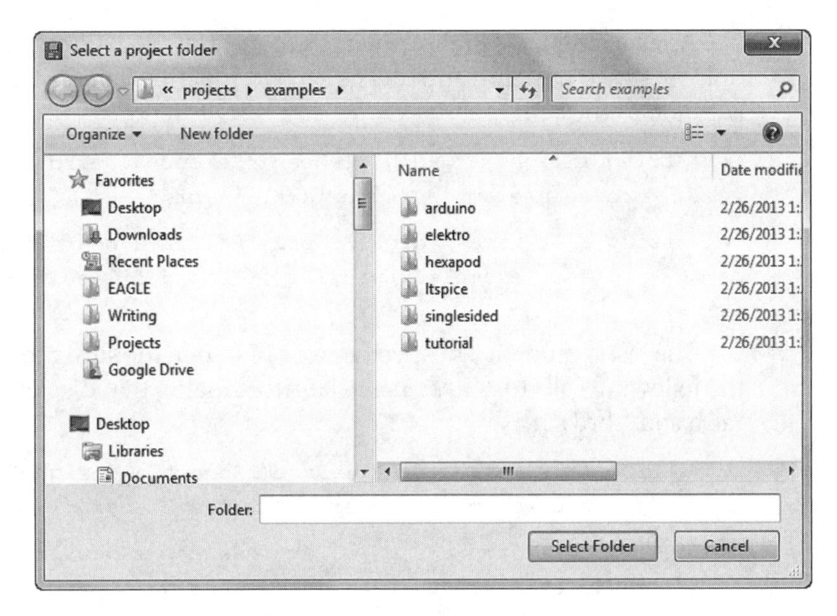

Figure 13.3: An Example Directory Dialog

13.1.3 File Open and File Save Dialogs

The last two types of predefined dialogs involve file selection. A dialog of the first type allows the user to select a file for editing and is launched in a ULP by the `dlgFileOpen` function. A dialog of the second type allows the user to select a file for storing content. This is launched by `dlgFileSave`.

These functions are called in nearly the same way. This is reflected by the similarity of their signatures, given here:

- ```
  string dlgFileOpen(string title[, string start[,
                     string filter]])
  ```
- ```
  string dlgFileSave(string title[, string start[,
                     string filter]])
  ```

The first two arguments serve the same roles as those of `dlgDirectory`. The first string serves as the dialog's title, and the optional second string sets the initial folder opened by the dialog.

The third argument provides the suffix or suffixes that identifies which files are potential matches. For example, if a file open dialog should open only *.brd files, the third argument should be configured so that the dialog displays only files with the *.brd suffix.

This suffix filter can be given using one of two methods. The first identifies only the suffix preceded by the * wildcard. The second method provides a description along with the suffix.

As an example, the following calls to `dlgFileOpen` filter the file dialog to show only files with the *.brd suffix. The second function call provides a description in addition to the suffix.

- `string file_name = dlgFileOpen("Select a board file", "projects/examples/tutorial", "*.brd");`
- `string file_name = dlgFileOpen("Select a board file", "projects/examples/tutorial", "Board files (*.brd)");`

The last argument can specify more than one file suffix to be filtered. For example, the following calls to `dlgFileSave` launch a dialog that displays only files with the *.sch and *.brd suffixes.

- `string file_name = dlgFileSave("Select a design file", "projects/examples/tutorial", "*.sch *.brd");`
- `string file_name = dlgFileSave("Select a design file", "projects/examples/tutorial", "Design files (*.sch *.brd)");`

`dlgFileOpen` and `dlgFileSave` both return strings that identify the path of the file selected by the user. If the user presses Cancel, the return string will be empty.

13.2 Custom Dialogs and Widgets

If the four types of predefined dialogs won't be sufficient, you need to create a custom dialog. Thankfully, the UL provides a wide range of graphical elements that can be added to a dialog window. These are called *widgets* and a large portion of this section is devoted to explaining the different widgets available.

But before you can add widgets to a dialog, you need to know how to create custom dialogs. This is made possible by the `dlgDialog` function, so this is the best place to start the discussion.

13.2.1 Creating a New Dialog

The `dlgDialog` function configures and launches a new dialog. Unlike the other functions discussed in this chapter, this defines a block of statements similar to the `board(){}` or `schematic(){}` functions from Chapter 11, "Introduction to the User Language (UL)." Its usage is given as follows:

```
int result = dlgDialog("Dialog Title") {
  ..
}
```

The return value of `dlgDialog` identifies the result of the dialog's operation. This can be defined in code or configured when the user presses a button. If the user closes the dialog without pressing a button, the return value is –1.

The statements inside the block configure the dialog's appearance and behavior. Both aspects depend on the dialog's widgets. The rest of this section presents the widgets available for UL dialogs by dividing them into five categories:

- Labels and text views
- Buttons
- Edit widgets
- Checkboxes and combo boxes
- List widgets

13.2.2 Labels and Text View Widgets

In my opinion, the easiest dialog widgets are those that present text to the user and receive the characters entered by the user. Widgets of the first type are called *labels* and widgets of the second type are called *text views*. This discussion presents both kinds and demonstrates how they can be created and configured.

Labels

A label is a widget that presents text in the dialog. This is reflected by the signature of `dlgLabel`, whose only required argument is a string.

```
void dlgLabel(string text[, int update])
```

The `text` argument defines the label's message. This can use the same HTML formatting tags listed in Table 13.1, so characters can be displayed in italics, boldface, or monospace font. The same rules also apply for special characters.

As an example, the following code creates a dialog that presents a simple welcome message.

```
dlgDialog("Welcome Message") {
  dlgLabel("<i>Good day,</i> <b>user!</b>");
};
```

Figure 13.4 shows what this dialog looks like. As shown, the dialog's width is determined by the label's width, not the dialog's title.

Figure 13.4: A Dialog with a Label

By default, the label's text is static, and can't be changed by the user or the program. But if the optional second argument of `dlgLabel` is set to a positive integer and the label's text is set to a string variable (not a string literal like `"Hello world"`), the label's message can be modified by the program.

In addition to displaying HTML-formatted text, labels can also provide hyperlinks to uniform resource identifiers (URIs). For example, the following code presents a hyperlink pointing to CadSoft's web site in the USA.

```
dlgDialog("Label with a Hyperlink") {
 dlgLabel("<a href=\"http://www.cadsoftusa.com\">CadSoft</a>");
};
```

When this hyperlink is clicked, a browser will open to CadSoft's web site. As shown, the double-quotes in the `<a>..</a>` tag must be escaped with a backslash.

If the URI points to a directory, a file explorer application will open to the appropriate folder. On a Windows system, the following label's hyperlink opens the C:\test directory.

```
dlgLabel("<a href=\"file:///C:/test\">Test folder</a>");
```

Text Views

A label widget can't display text on multiple lines. That is, if a line-break (`\n`) is inserted in the text, it will be ignored.

To make up for this, UL provides the text view, which presents a message to the user that may occupy multiple lines. It's created by the `dlgTextView` function, which has two signatures:

```
dlgTextView(string text)
dlgTextView(string text, string &link) {statement_block}
```

The first signature creates the text view as a multiline label. It can use the same HTML formatting as a regular label and it creates line breaks for `\n` characters. For example, the following dialog presents text occupying three lines.

```
dlgDialog("Text View with Three Lines") {
  dlgTextView("Line 1\nLine 2\nLine 3");
};
```

In addition to regular hyperlinks, text view can display special links that execute code in a ULP. This is where the second signature of `dlgTextView` comes in. When a special link is clicked, the program executes the statement or statements following the `dlgTextView` function call. These statements can access the text specified by the link's `href` attribute.

Configuring a special hyperlink for `dlgTextView` requires three steps:

1. In the text view's message, use the `<a>..</a>` tag to create the hyperlink.
2. Set the second argument of `dlgTextView` equal to a string variable.
3. Follow `dlgTextView` with one or more statements that should be executed when the link is clicked. These statements can access the second argument of `dlgTextView`, which will be set equal to the value of `href`.

The code in Listing 13.1 demonstrates how this works. The text view contains a bulleted list of four items. When a link is clicked, the program writes a statement to a file that identifies which link was clicked.

There are two points I'd like to mention about this code. First, the `target` variable must be declared as a string, but its value is set when the hyperlink is clicked. Second, it's usually a good idea to call `exit()` after the hyperlink is clicked. This is because, from what I've seen, the widget becomes blank as soon as the link is clicked.

Listing 13.1: text_view.ulp

```
output("example.txt") {
  string list = "<ul>"
  "<li><a href=\"first link\">First item</a></li>"
  "<li><a href=\"second link\">Second item</a></li>"
  "<li><a href=\"third link\">Third item</a></li>"
  "<li><a href=\"fourth link\">Fourth item</a></li>"
  "</ul>";
  dlgDialog("Text View Example") {
    string target;
    dlgTextView(list, target) {
      printf("The %s was clicked.", target);
      exit(0);
    }
  };
};
```

13.2.3 Buttons

The UL provides two types of buttons: push buttons and radio buttons. Push buttons tell the application to do something and radio buttons allow the user to select one option from multiple choices. This discussion shows how to create dialogs with both types of buttons.

Push Buttons

The push button, sometimes called a switch, presents text to the user and performs an operation when clicked. Both aspects are defined by the `dlgPushButton` function, whose signature is given as follows:

```
dlgPushButton(string text) {statement_block}
```

The `text` argument specifies the string to be displayed on the button. HTML formatting isn't available, but an accelerator key can be associated with the button by preceding a character of its text with `&`. If an accelerator key is available, the user can effectively click the button by pressing Alt and the given character.

As an example, the following code creates a push button that displays `Hello`. The `H` is preceded by `&`, so the user can effectively click the button by pressing Alt-H.

```
dlgPushButton("&Hello") {}
```

In addition to setting accelerator keys, a button can be made the default button by preceding its name with `+`. This will be clicked when the user presses Enter. A button can also be preceded with `-`, which means the button will be clicked when Esc is pressed.

The statements in the block following `dlgPushButton` are executed when the button is pressed. When coding these blocks, it's important to be familiar with four functions:

- `dlgAccept()`—Closes the dialog; `dlgDialog` returns a positive value
- `dlgReject()`—Closes the dialog; `dlgDialog` returns 0 or a negative value
- `dlgReset()`—Resets dialog elements to their default values
- `dlgRedisplay()`—Redisplays the dialog with all changes

The first two functions are particularly important. Many dialogs have an OK button that closes the dialog and accepts the user's changes. The statement block for this button should execute `dlgAccept`. Without arguments, this causes `dlgDialog` to return 1. But `dlgAccept` can be called with an optional positive integer to be returned instead.

Similarly, a Cancel button should execute `dlgReject` to close the dialog and reject the user's changes. Without arguments, `dlgReject` causes `dlgDialog` to return 0. But this function accepts an optional value (zero or negative integer) to be returned instead.

The following code creates two buttons that close the dialog when pressed. The OK button is the default button and calls `dlgAccept` to close the dialog. The second is the cancel button and calls `dlgReject`.

```
dlgDialog("Dialog with Two Buttons") {
  dlgPushButton("+&OK") {dlgAccept();}
  dlgPushButton("-&Cancel") {dlgReject();}
};
```

Figure 13.5 shows what the resulting dialog looks like. By default, the two buttons are stacked on top of one another.

Figure 13.5: A Dialog with Two Buttons

Radio Buttons and Groups

In the early twentieth century, American radios had multiple buttons representing AM stations. When one button was pushed in, the others would pop out. This behavior continues today in radio buttons, which allow the user to select an option out of a group.

Radio buttons are created by `dlgRadioButton`, whose signature is given as follows:

```
dlgRadioButton(string text, int sel) { optional_block }
```

The first argument, `text`, has the same features available for the push button's text. Preceding the text with `&` sets an accelerator key, `+` makes the button the default button, and `-` makes the button a cancel button.

The second argument identifies which radio button in a group is selected. For example, if there are four buttons in the group and `sel` is initially set to 1, the second button will be selected by default. If the user clicks the fourth button, `sel` will equal 3.

Every radio button must belong to a group widget, which is created by the `dlgGroup` function. The signature of `dlgGroup` is given as follows:

```
dlgGroup(string title) {}
```

The following example shows how groups and radio buttons work together. The group contains four radio buttons and the first is selected by default.

```
int color = 0;
dlgGroup("Favorite Color") {
  dlgRadioButton("&Red", color);
  dlgRadioButton("&Green", color);
  dlgRadioButton("&Yellow", color);
  dlgRadioButton("&Blue", color);
}
```

As different radio buttons are selected, the value of `color` will change accordingly.

13.2.4 Edit Widgets

In many dialogs, the application may require more information than a selection from a choice of options. If the application requires character entry, such as a Name field, the dialog should provide an edit widget. The UL provides four different types of edit widgets: two for text entry and two for number entry.

Text Entry Widgets

There are two different types of text entry widgets. Widgets of the first type may contain only a single line, but each keeps a history of previous entries. Widgets of the second type hold multiple lines, but they don't remember past entries. The functions that create these widgets are given as follows:

- `dlgStringEdit(string &text[, string &history[][, int size]])`
- `dlgTextEdit(string &text)`

In both cases, the first argument is a variable that stores the text entered by the user. This can be initialized to set the default text in the widget.

`dlgStringEdit` creates a text box similar to the command box in an EAGLE editor. It has an arrow on the right that allows the user to view and select previous entries. The user can also access previous entries by pressing the down arrow key.

In code, the widget's past entries are stored in the string array provided as `dlgStringEdit`'s second argument. The last argument identifies the maximum number of strings that should be stored.

The code in Listing 13.2 creates a string edit widget capable of storing the last four entries. Each time the first button is pressed, the dialog prints the edit widget's current and past entries to a file.

Listing 13.2: text_box.ulp

```
output("history.txt") {
  dlgDialog("String Edit Box") {
    string edittext = "Enter text";
    string history[];

    // Create string edit widget
    dlgStringEdit(edittext, history, 4);

    // Create push button to print to file
    dlgPushButton("+&Print history to file") {
      printf("New entry: %s\n", edittext);
      for(int i=0; history[i]; i++)
        printf("Past entry %d: %s\n" , i, history[i]);
    }
    dlgPushButton("-&Close") {dlgAccept();};
  };
};
```

The text box created by `dlgTextEdit` doesn't save previous entries. But unlike the text box created by `dlgStringEdit`, the text box created by `dlgTextEdit` makes it possible to enter text on multiple lines. The following code initializes this widget with a string that occupies three lines.

```
string text = "Line 1\nLine2\nLine3";
dlgTextEdit(text);
```

After the user enters text, line breaks are represented by `\n` characters. Empty lines and whitespace at the end of a line are discarded.

Number Entry Widgets

The preceding text boxes don't make any distinction between characters, but UL provides two text widgets that accept only numeric data. Characters that aren't part of a number are ignored.

The first number entry widget accepts only integers within a given range. This is created by the `dlgIntEdit` function, whose signature is given as follows:

```
dlgIntEdit(int value, int min, int max)
```

The second widget accepts only reals within a given range. It's created by the `dlgRealEdit` functions, whose signatures are given as follows:

```
dlgRealEdit(real value, real min, real max)
```

As an example, suppose an application wants the user to enter a percentage. If the value should be given as an integer, the edit widget can be created with the following code:

```
int percent;
dlgIntEdit(percent, 0, 100);
```

If the value should be given as a real, the widget can be created as follows:

```
real percent;
dlgRealEdit(percent, 0.0, 100.0);
```

13.2.5 List Widgets

In addition to radio buttons, UL provides three other widgets that allow the user to select an option from multiple choices:

- **Combo box**—Only the selected item is displayed.
- **List box**—All items are displayed in a single column.
- **List view**—Items can be displayed and sorted in multiple columns.

Despite the similar descriptions, the widgets are noticeably different. Figure 13.6 presents each of them side-by-side.

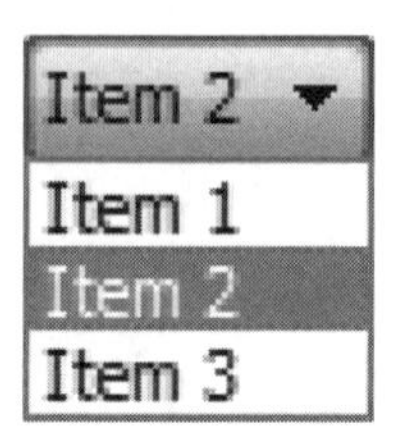

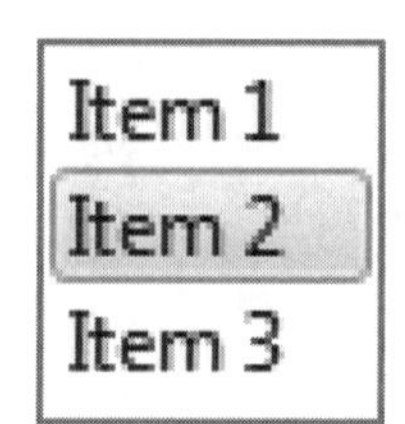

Figure 13.6: List Widgets (Combo Box, List Box, and List View)

This discussion will present all three widgets and show how they can be created and configured in ULPs.

Combo Box

A combo box is a compact widget that makes it possible to select from multiple options. These boxes are created with `dlgComboBox`, whose signature is given as follows:

```
dlgComboBox(string list[], int sel) {optional_block}
```

The first argument provides the list of strings to be displayed in the box. None of the strings in the array may be empty.

The second argument identifies which string is selected in the box. If `sel` is set to an initial value, the corresponding string will be initially selected. `sel`'s value changes as the user selects different options.

The following code creates the combo box shown on the left side of Figure 13.6. `sel` is initially set to 1, so the second list item is selected by default.

```
int sel = 1;
string list[] = {"Item 1", "Item 2", "Item 3"};
dlgComboBox(list, sel);
```

When the user changes the selection, the statements in the optional block execute. These can access the user's selection through the function's second argument.

List Box

A list box is similar to a combo box, but all the options are displayed. This requires more space than the combo box but allows the user to see every list element at once.

List boxes are created with `dlgListBox`, which accepts the same argument as `dlgComboBox`. This is shown by the following signature:

```
dlgListBox(string list[], int sel) {optional_block}
```

These parameters serve the same purposes as those in `dlgComboBox`. The first argument sets the elements in the list, the second argument identifies the current selection, and the optional block is executed when the selection changes.

List View

List views are similar to list boxes, but instead of presenting every option in a single column, list views can display items in multiple columns. Each column has a header that, when clicked, sorts the column's elements.

List views are created by `dlgListView`, whose signature is given as follows:

```
dlgListView(string headers, string list[], int sel [,int sort])
                                             {optional_block}
```

The `list` and `sel` arguments serve the same roles as those in `dlgComboBox` and `dlgListBox`. But the first argument identifies the headers to be displayed in each column. The optional `sort` argument identifies which column should be used to sort the widget's elements.

Tab characters (`\t`) separate headers and list items intended for different columns. For example, the following code creates the list view in Figure 13.6.

```
int sel = 1;
string headers = "Header 1\tHeader 2";
string list[] = {"Item 1\tItem 4",
                 "Item 2\tItem 5",
                 "Item 3\tItem 6"};
dlgListView(headers, list, sel, 1);
```

As shown, the tab character separates each of the list items. The list item preceding `\t` is placed in Column 0 and the item following `\t` is placed in Column 1. It should be clear that adding items to Column 2 is accomplished by adding more tab characters, not by adding more elements to the `list` array.

13.2.6 Checkboxes and Spinboxes

Two last widgets need to be mentioned: checkboxes and spinboxes. A checkbox receives binary information from the user, true or false. A spinbox allows the user to select a numeric value within a range.

Checkboxes

Like a radio button, a checkbox has one of two states: selected or unselected. Unlike a radio button, checkboxes don't have to be placed in groups. In fact, a checkbox can stand alone without any other checkboxes nearby.

EAGLE frequently employs checkboxes to configure aspects of its operation. If you go to Options > Set... in the main menu and select the Misc tab, you'll see that the entire left column is made up of checkboxes.

In a UL program, checkboxes are created with the `dlgCheckBox` function. Its signature is given as follows:

```
dlgCheckBox(string text, int sel) { optional_block }
```

The `text` argument defines the label displayed to the right of the checkbox. The `sel` variable identifies whether the box has been selected (checked). A value of 1 indicates that the box has been checked and a value of 0 indicates that it hasn't. As the widget's state changes, the value of `sel` changes.

The following code creates a checkbox that is initially checked.

```
int checked = 1;
dlgCheckBox("Example Checkbox", checked);
```

Spinboxes

A spinbox allows the user to select an integer between a given minimum and maximum. This widget is created by `dlgSpinBox`, whose signature is given as follows:

```
dlgSpinBox(int val, int min, int max)
```

`val` sets the widget's initial value and changes as the widget's value changes. Its allowable values are between `min` and `max`.

The following code creates a dialog whose spinbox ranges between 0 and 100. The initial value is set to 33.

```
dlgDialog("Spin Box") {
  int percent = 33;
  dlgSpinBox(percent, 0, 100);
};
```

Figure 13.7 shows what the resulting spinbox looks like. These widgets are much more efficient for choosing numbers than combo boxes and list boxes.

Figure 13.7: A Spinbox

13.3 Dialog Layouts

By default, widgets placed in a dialog are stacked on top of one another with little space between them. To customize the arrangement, you need *layout statements*. Layout statements have two important properties:

- A layout statement defines a statement block, but it is not a function. That is, layout statements don't accept any arguments and they don't provide return values.
- When widgets are placed in the block, they're arranged in the manner defined by the statement.

For example, the following code uses `layout_statement` to arrange three push buttons:

```
dlgDialog("Example Layout") {
  layout_statement {
    dlgPushButton(..);
    dlgPushButton(..);
    dlgPushButton(..);
  }
};
```

A layout block can be thought of as an individual widget. With this in mind, it makes sense that one block can be placed inside another. This is demonstrated by the following code, which nests `layout_statement2` inside `layout_statement1`.

```
dlgDialog("Example Nested Layout") {
  layout_statement1 {
    dlgStringEdit(..);
    dlgSpinBox(..);
    layout_statement2 {
      dlgPushButton(..);
      dlgPushButton(..);
      dlgPushButton(..);
    }
  }
};
```

UL's layout statements can be divided into two categories: vertical/horizontal layouts and grid layouts. This section presents the statements in both categories and explains how they arrange widgets.

13.3.1 Horizontal/Vertical Layouts

The simplest way to arrange widgets in a dialog is to place them a row (horizontally) or in a column (vertically). In the first case, the layout statement is `dlgHBoxLayout`. The following dialog uses this statement to place two checkboxes side-by-side.

```
dlgDialog("Horizontal Layout") {
  int checked1, checked2;
  dlgHBoxLayout {
    dlgCheckBox("Checkbox1", checked1);
    dlgCheckBox("Checkbox2", checked2);
  }
};
```

These checkboxes can be positioned vertically by replacing `dlgHBoxLayout` with `dlgVBoxLayout`. Without additional configuration, this vertical arrangement is exactly similar to the default layout.

The `dlgSpacing` function makes it possible to insert space between widgets in a horizontal or vertical arrangment. This accepts an integer that identifies the additional space in pixels. For example, the following code arranges two checkboxes vertically with fifty pixels of space between them.

```
dlgDialog("Vertical Layout") {
  int checked1, checked2;
  dlgHBoxLayout {
    dlgCheckBox("Checkbox1", checked1);
    dlgSpacing(50);
    dlgCheckBox("Checkbox2", checked2);
  }
};
```

It's important to note that one call to `dlgSpacing` doesn't uniformly set the spacing inside the block. This function must be inserted between each pair of widgets that require spacing.

13.3.2 Grid Layouts and Cells

With the right spacing, it's possible to position differently sized widgets in a table with `dlgHBoxLayout` and `dlgVBoxLayout`. But it's easier to use `dlgGridLayout`, which creates a table of widgets and handles the spacing for you.

The `dlgGridLayout` block doesn't accept widgets directly. Instead, each widget must be placed inside a block defined by `dlgCell`. `dlgCell` is a function whose signature is given as follows:

```
dlgCell(int row, int col[, int row2, int col2]) {widget(s)}
```

The first two arguments are required, and identify the row and column where the cell should be placed. If the cell needs to span more than one position, the optional `row2` and `col2` arguments indicate that the cell should occupy the rectangle whose opposite corners are (`row`, `col`) and (`row2`, `col2`).

The block following `dlgCell` contains the widget or widgets to be inserted. By default, widgets in a `dlgCell` are positioned horizontally. This behavior can be changed with `dlgHBoxLayout` and `dlgVBoxLayout`.

The code in Listing 13.3 creates a layout containing four cells. The second and fourth cells both contain two widgets. The second cell arranges them vertically and the fourth cell arranges them horizontally.

Listing 13.3: grid_layout.ulp

```
dlgDialog("Example Grid Layout") {
  int sel;
  dlgGridLayout {

    // Cell in the upper-left
    dlgCell(0, 0) dlgLabel("Cell 0,0");

   // Cell in the upper-right
    dlgCell(0, 1) {
      dlgVBoxLayout {
        dlgPushButton("Cell 0,1");
        dlgPushButton("Cell 0,1");
      }
    }
    dlgCell(1, 0) dlgLabel("Cell 1,0");
    dlgCell(1, 1) {
      dlgCheckBox("Cell 1,1", sel);
      dlgCheckBox("Cell 1,1", sel);
    }
  }
};
```

Figure 13.8 shows what the resulting dialog looks like.

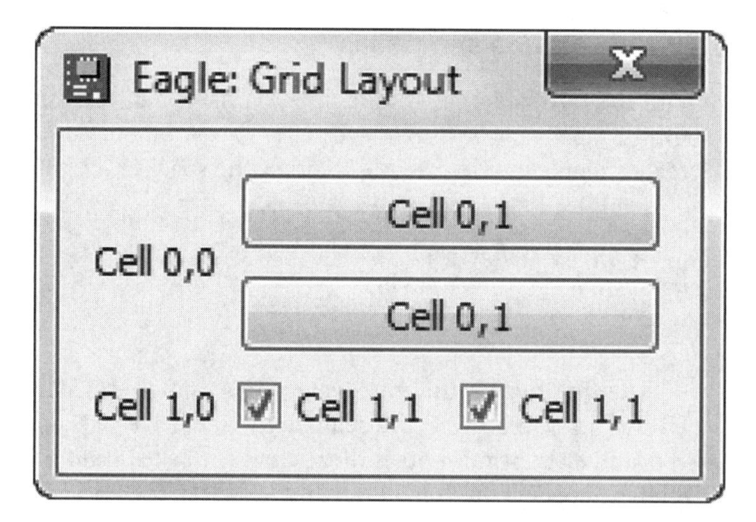

Figure 13.8: An Example Grid Layout

As shown, the grid layout stretches elements in the same column to occupy the available area. The code didn't specify any alignment or spacing. These decisions are all made by the grid layout.

13.4 The Menu Command

Chapter 10, "Editor Commands," discussed many of EAGLE's editor commands but left out one important command: `menu`. Despite its name, this doesn't modify the main menu of any EAGLE window. Instead, it adds an item to the horizontal toolbar of the active editor.

The only required parameter of the `menu` command is the name of a command to execute when the item is pressed. By default, the item takes the command's name, so the following command creates a toolbar item called `auto` that executes `auto` when clicked.

```
menu auto
```

When this executes, EAGLE creates a new toolbar item called `auto` and places it on the far right of the horizontal toolbar. If the user clicks the item, `auto` will execute and bring up the Autorouter Setup dialog.

A different label can be set by separating the label from the command with a colon and surrounding the pair with single quotes. The following command creates an item labeled `Router` that executes `auto` when clicked:

```
menu 'Router : auto'
```

The menu command makes it possible to associate the toolbar item with a menu. When the user clicks the item, the menu will appear and allow the user to select a command to execute. To configure this, the menu options must be separated by | and surrounded by curly brackets.

As an example, the next command creates a toolbar item called Chapter13. Its menu items execute the ULPs presented in Listings 13.2 and 13.3. The first menu item, labeled Text Box, executes 'run text_box.ulp' and the second item, labeled Grid Layout, executes 'run grid_layout.ulp'.

```
menu 'Chapter 13 { Text Box : run text_box.ulp |
                   Grid Layout : run grid_layout.ulp}'
```

In addition to labels and menu items, the menu command can associate toolbar items with images. The image must be specified by surrounding the image file's path with square brackets. The following command associates "C:\images\example.png" with the auto command.

```
menu '[C:\images\example.png] Router : auto'
```

It isn't explicitly stated, but it looks like the PNG format is the only supported format for toolbar images. The image's width can take any value, but for the Windows operating system, the image should be kept to around 18 pixels in height.

13.5 Conclusion

The run command accepts command-line arguments, but when ULPs require complex input, it's more effective to create dialogs. Most of this chapter has focused on creating these dialogs and the last part explained how to add new menus to the horizontal toolbar.

The UL provides four predefined dialogs that can be created quickly. The message box presents text to the user and accepts button presses in response. The directory dialog opens a file explorer and allows the user to select a directory. Similarly, the file open and file save dialogs open explorers that allow the user to open files or store content to files.

To code a custom dialog, it's important to be familiar with the different widgets available. UL provides a wide range of widgets, including labels, text boxes, buttons, and checkboxes. Each of these is created by a function and many widget functions can be followed by optional blocks of statements.

By default, widgets in a dialog are positioned vertically, each widget below the one preceding it. This arrangement can be configured using layout statements. The UL provides two types of layouts: horizontal/vertical layouts and grid layout.

The horizontal and vertical layouts place widgets in a single row or column. The grid layout arranges widgets in a table using cells. In all cases, one layout can be nested inside another for more complex arrangments.

The last part of this chapter explained how to create new items in the horizontal toolbar of the active editor. Many users prefer pressing these toolbar items to typing commands at the command line. Therefore, if you want a feature to be available to the most users, you may want to consider the `menu` command. By combining this command with UL's dialogs, it's possible to add capabilities that don't require the user to ever use the command line.

Chapter 14

Schematic Design for the BeagleBone Black

Chapter 4, "Designing the Femtoduino Schematic," through Chapter 7, "Generating and Submitting Output Files," presented the Femtoduino, the smallest circuit in the Arduino family of circuit boards. Arduino circuits provide many advantages including low cost, ease of programming, and a wealth of online support. For these reasons, they're ideal for students, hobbyists, and entrepreneurs in need of simple, reliable circuits.

But when it comes to serious data processing, circuits like the Femtoduino aren't even close to the state of the art. The Femtoduino's microcontroller, the ATMega328P, has an 8-bit processor that runs at 20MHz and has 32kB of Flash memory. In contrast, the latest microcontroller from NXP is the LCP4300, whose 32-bit processor runs at 204MHz and has 1MB of Flash memory.

For this book's final example, I wanted a circuit that demonstrates the cutting edge of embedded electronics. I considered the Raspberry Pi but decided against it because of the lack of design documentation. I also considered Intel's Minnowboard but decided against it because of the price (nearly $200).

In the end, I chose the BeagleBone Black as this book's example of advanced circuit board design. I made this decision based on five important advantages:

- **Circuit complexity**—Six layers, high-speed ball grid array (BGA) devices.
- **Powerful devices**—The Sitara AM3359 is a 32-bit system on a chip (SoC) that runs at speeds up to 1GHz.
- **Free resources**—Schematics and Gerber files available at http://beagleboard.org.
- **Low cost**—$45 for a complete board.
- **Available support**—The forum at http://beagleboard.org/Community/Forums is very active, with many knowledgeable contributors.

The BeagleBone Black has its drawbacks as well. The circuit board's design files are available to everyone, but some of the devices are difficult, if not impossible, to obtain. For example, the XAM3359AZCZ100 (referred to as the AM3359 throughout this chapter) is an experimental device from Texas Instruments that can't be purchased through regular channels. Also, the company that produces the LPJ0011BBNL Ethernet connector refuses to provide a datasheet or any means of purchasing the component.

The last drawback is printed on the first page of the schematic design: *This schematic is *NOT SUPPORTED* and DOES NOT constitute a reference design.* For this and the preceding reasons, I don't recommend constructing actual boards based on the BeagleBone Black design. But it's an excellent design to learn from.

14.1 Overview of the BeagleBone Black

The majority of this chapter is devoted to schematic design, but first, I'd like to present a high-level overview of the BeagleBone Black, commonly referred to as the BBB. The best way to introduce this board is to contrast it with Arduino circuit boards.

Arduino boards contain microcontrollers, specifically from Atmel. These devices have analog-to-digital converters, general-purpose I/O, and the ability to communicate using different protocols, such as SPI, USB, and Ethernet. In addition, microcontrollers don't need external memory. Because these devices are so versatile and self-contained, Arduino boards are ideal for simple data processing and data conversion tasks.

Unlike Arduino boards, the BBB has a microprocessor instead of a microcontroller. The good news is that the BBB can process data at much higher speed and can perform more complex processing tasks. It even contains an analog-to-digital converter and resources for connecting to various communication protocols, including USB.

The bad news is that the BBB requires external memory devices, such as Flash memory and DDR RAM. This significantly increases the cost of the board and the amount of power it consumes. But there's a benefit to this as well—these external devices provide an immense amount of memory. In the BBB, the DDR RAM stores 512 MB of data and the Flash memory stores 2GB of data. That's much more memory than any microcontroller board I've encountered.

The main reason I prefer programming the BBB over any microcontroller board is that the BBB can support a proper operating system (OS). I like to think of an OS as a butler—I provide broad instructions like "send this data over the USB" and "store this data to a file" and it takes care of the low-level details. In contrast, programming a microcontroller frequently requires understanding the device's registers and the addresses of the memory-mapped peripherals.

In essence, the BBB circuit consists of four elements:

- **The central system on a chip**—The AM3359
- **Data storage**—Flash memory and DDR RAM
- **External connections**—USB, Ethernet, and HDMI
- **Power management**—The TPS65217C

The block diagram in Figure 14.1 depicts these aspects of the circuit.

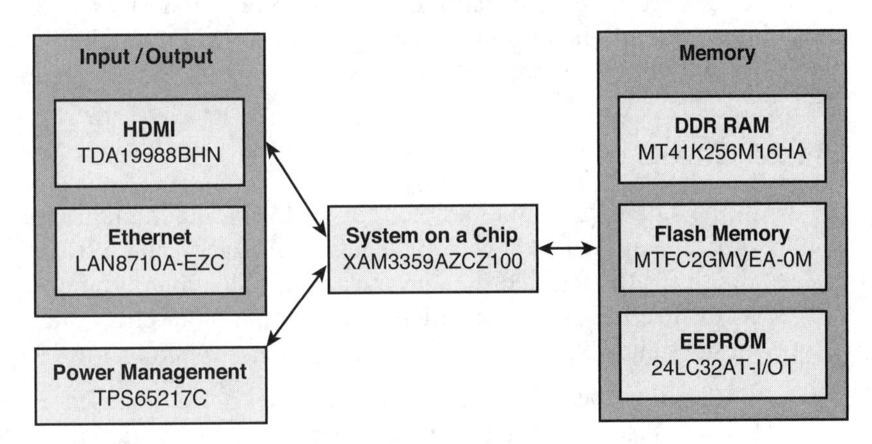

Figure 14.1: Block Diagram of the BeagleBone Black

Though it's conceptually simple, there's nothing simple about the BBB's circuitry. The AM3359 has 324 pins and the circuit board needs 6 layers to route all the connections. The full schematic, provided in the Ch14 folder, consists of 10 sheets. But for the sake of brevity, this chapter presents 6:

1. **AM3359**—Memory connections
2. **AM3359**—Input/output connections
3. **Connections to DDR RAM**
4. **Power management**
5. **Ethernet/USB connections**
6. **HDMI connection**

This chapter doesn't provide directions because I don't expect anyone to design this circuit from scratch. But the BBB serves as a great example of circuit design. As you look over the EAGLE schematic in the Ch15 folder, I hope you'll draw inspiration that you can use in your own circuits.

14.2 Advanced EAGLE Schematics

Chapter 3, "Designing a Simple Circuit," presented an inverting amplifier and Chapter 4 introduced the Femtoduino. In both cases, the schematics were sufficiently simple that the entire design could be drawn on one sheet.

The BBB is significantly more complex and its schematic requires 10 sheets. In addition, the AM3359 has so many pins that it needs to be split into gates. I've mentioned sheets and gates in earlier chapters, and this section presents them in greater depth. In addition, I'll explain how to work with buses, which make it much easier to draw connections between components with large numbers of related pins.

14.2.1 Sheets

In the schematic editor, between the vertical toolbar and the editor area is a thin vertical panel called Sheets. By default, EAGLE creates a single sheet for each design, and this is the selected sheet in the panel. The maximum number of sheets depends on the EAGLE license: the Light version allows one sheet, the Hobbyist and Standard versions allow 99 sheets, and the Professional version allows 999 sheets.

EAGLE doesn't provide toolbar items or main menu items for dealing with sheets. Instead, sheet operations are performed by right-clicking in the Sheets panel. For example, if the schematic has a single sheet and you right-click under Sheet 1, a menu will appear with an option called New. This creates a second sheet, and if you left-click it, the editor area will be empty, reflecting the blank sheet.

It's important to note that sheets don't affect the schematic design in any significant way. That is, when EAGLE generates a board file from a schematic, it doesn't take into account how the circuit has been divided into sheets. It simply combines the sheets' designs into a single schematic and uses it to initialize the board.

But sheets serve a valuable purpose in making circuit designs more readable. Most large-scale circuit designs are divided into sheets according to function: One depicts power circuitry, another depicts memory connections, and so on. This is the methodology employed for the BeagleBone Black, which is divided into 10 sheets according to functions like input/output, DRAM connections, and power management.

EAGLE doesn't allow renaming sheets, but sheets can be removed from the design using the context menu in the Sheets panel. In addition, designers can set a text description for each sheet. This description appears when a user clicks the schematic file in EAGLE's Control Panel.

14.2.2 Gates and the Invoke Tool

In general, every movable element in a schematic represents a different device. But sometimes, a single device will contain multiple elements: A resistor pack contains multiple resistors, an op-amp array contains multiple amplifiers, and so on. In the schematic editor, these individual elements, called gates, can be moved separately. For devices such as the resistor pack and op-amp array, the gates always have the same shape.

Each gate corresponds to a symbol, and Chapter 8, "Creating Libraries and Components," explained how to associate multiple symbols with a device. But there's no requirement that a device's gates have the same symbol, and in many circuits, it's better to have gates with different symbols. This is particularly true when working with devices that are too large to fit on one sheet.

For example, the AM3359 has 324 pins. Rather than create a symbol with 324 pins, I created three: a 117-pin symbol whose pins connect to memory devices, an 86-pin symbol whose pins connect to input/ouput, and a 121-pin symbol whose pins connect to ground and voltage supplies. Each symbol becomes a gate in the BBB schematic. The first gate is in Sheet 1, the second is in Sheet 2, and the third is in Sheet 3.

If a device has multiple gates and you use the Add tool to add it to the design, EAGLE will prepare the first gate for insertion, then the second, third, fourth, and so on. The Invoke tool makes it possible to insert a gate out of sequence. This works in three steps:

1. Select the Invoke item from the vertical toolbar. This is located just above the Text tool.
2. Click a gate in the schematic design or enter the name of the part in the text box above the editor area.
3. Click in the editor area to place the gate.

For example, suppose the part name is U$1 and its three gates are G$1, G$2, and G$3. When the Add tool is activated, G$1 is ready to be inserted into the design. When G$1 is placed, G$2 will be ready for insertion. But let's say you want to insert G$3 before G$2. In this case, select the Invoke toolbar item and click G$1. Figure 14.2 shows what the resulting dialog looks like.

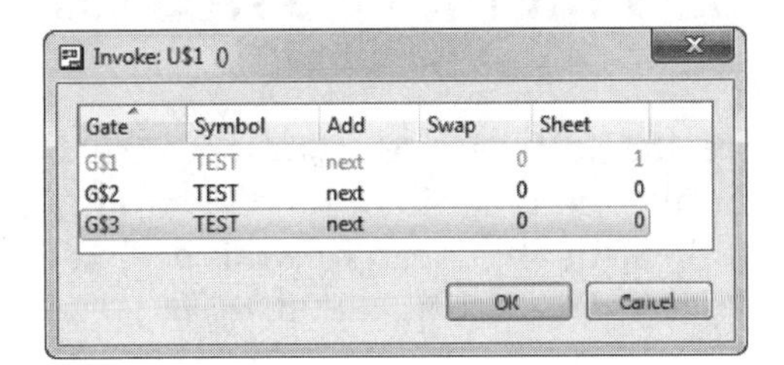

Figure 14.2: The Invoke Dialog

If G$3 is selected from this dialog, clicking OK will prepare G$3 for insertion in the schematic.

An interesting question arises: How do you add a gate to a sheet if no similar gates have been added to that sheet? The answer is to use the text box above the editor area. That is, activate the Invoke tool, enter the part name in the text box (U$1 in the example), and press Enter. The Invoke dialog will appear and allow you to select a gate for insertion.

14.2.3 Buses

For simple designs, components in the schematic editor can be connected by drawing nets between pairs of pins. But for designs containing large components like the AM3359, it's more convenient to groups nets into buses. Buses dramatically simplify the process of designing schematics for circuits containing processors and memory devices.

When a name is assigned to a net, EAGLE understands that every net with the same name is electrically connected. This holds true for buses, but there's an important difference between buses and nets: Bus names follow a specific convention. A bus's name identifies which signals it carries and must take one of three forms:

- *name*—The bus carries one signal.
- *name[low_index..high_index]*—The bus carries multiple signals, from *name[low_index]* to *name[high_index]*.
- *name1,name2,name3*—The bus carries three signals.

These formats can be combined, so a schematic may have a bus called DATA[0..7],ADDR[6..12],RESET. I prefer to use the second naming method exclusively, so my buses usually have names like PORTC[0..7] and ADC[0..2].

A net can be connected to a bus if its name corresponds to one of the bus's signals. For example, a net named RESET can't be connected to a bus named D[0..7], but a net named D3 or D5 can. If a net to be connected doesn't have a name, a dialog will appear and ask which of the bus's signals corresponds to the net.

14.3 AM3359 Memory/JTAG Connections

In essence, the BBB's purpose is to showcase the extraordinary processing capabilities of its central device: the Sitara AM3359 from Texas Instruments. This section presents the first schematic sheet featuring this device, but first, I want to explain why this device deserves so much attention.

14.3.1 AM3359 Architecture

An entire book could be written about the AM3359's features, but I'll focus on two: the ARM-based processor core and the JTAG interface for programming and debugging.

The ARM Cortex-A8 Processor Core

If you've worked with embedded electronics for any length of time, you've probably heard about ARM processors. ARM Holdings plc specializes in designing microprocessors, but they don't manufacturer their own devices. Instead, they license their designs to other companies that construct the actual processors.

This is how the AM3359 was developed. Texas Instruments decided to create a device containing multiple processor designs, called a system on a chip or SoC. Within the SoC, they inserted a processor from ARM, a graphics processor from PowerVR, and many other designs including a Programmable Real-Time Unit and Industrial Communication Subsystem (PRU-ICSS). When discussing SoCs, such processor designs are commonly called cores.

To be specific, the AM3359 contains ARM's Cortex-A8 microprocessor core. The processor defined by this core has the following characteristics:

- 32-bit processing at speeds between 600 MHz and 1 GHz
- Hardware floating-point unit (FPU)
- Support for NEON instructions

The second and third points merit additional attention. Most microcontrollers execute mathematical operations slowly and using only integers. For 16-bit microcontrollers, applications are limited to numbers between –32768 and 32767. Moreover, if you have to perform division, the result will be limited to integer values.

In contrast, the BBB contains powerful resources for number-crunching. Its Cortex-A8 core has a dedicated floating-point unit (FPU), which means it can operate on floating-point values quickly without taxing the rest of the device. In addition, this FPU supports NEON instructions, which process multiple data items at the same time. These groups of multiple data items are called vectors and this FPU is commonly called a vector processor or a Single Instruction, Multiple Data (SIMD) processor.

As mentioned earlier, the primary disadvantage of using the AM3359 over a microcontroller is the need for separate memory devices. The BBB contains three different memory subsystems: one for DDR RAM, one for Flash memory, and one for EEPROM. The first of the BBB's schematics presents the connections from the AM3359 to these three devices. Figure 14.3 displays the AM3359's nets, buses, and symbols related to memory access.

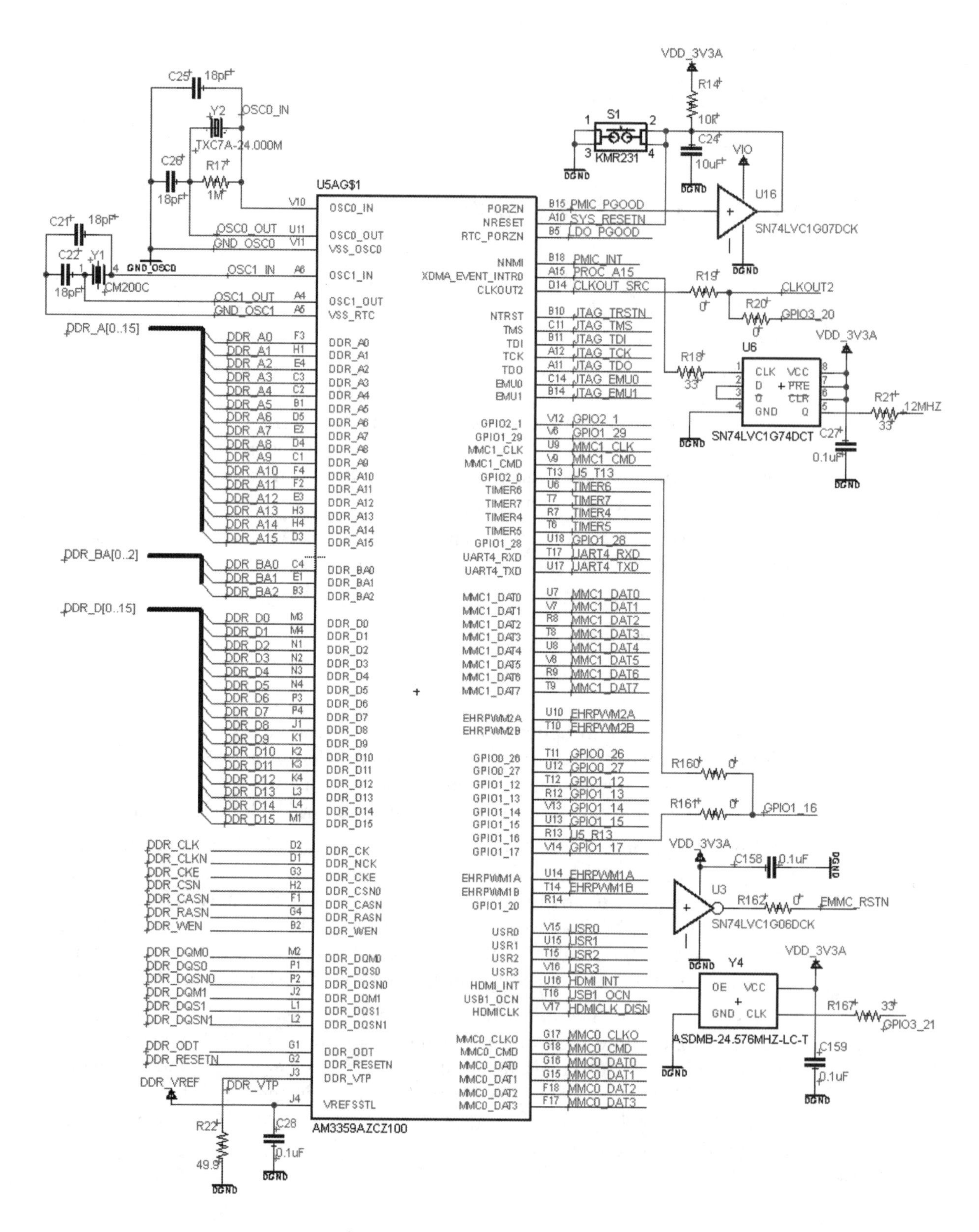

Figure 14.3: Schematic of the AM3359's Memory Connections

Joint Test Action Group (JTAG) Interface

In addition to displaying the AM3359's memory connections, Figure 14.3 also presents its connections to the oscillators (24 MHz for OSC0, 32.768 kHz for OSC1) and JTAG. JTAG is crucial for programmable logic because it serves as a common interface for debugging and programming embedded systems. Devices like the AM3359 support JTAG, and so do most microcontrollers, digital signal processors (DSPs), and FPGAs. If a programmable device has more than 64 pins, you can be confident that it supports JTAG.

JTAG stands for Joint Test Action Group, which is the IEEE group who devised the JTAG standard, IEEE 1149.1. This standard defines the signals and protocols needed to analyze programmable devices. If you have a suitable JTAG adapter, you can connect it to a JTAG-compatible device and debug its programming. JTAG adapters aren't cheap, but they're very useful.

As shown in the schematic, JTAG data transfer is made possible through five signals:

- **TCK**—Test clock
- **TDI**—Test data in
- **TDO**—Test data out
- **TMS**—Test mode select
- **TRST**—Test reset

With each rising edge on TCK, TDI writes 1 bit of data to the device and TDO reads 1 bit of data. These read/write operations access data in registers defined by the JTAG standard. The most important of these JTAG registers is the boundary scan register (BSR).

The incoming bits from TDI form instructions that tell the device what operations it should perform. Every JTAG device is required to support three instructions:

1. **SAMPLE/PRELOAD**—Load test data into the BSR, samples input and output.
2. **EXTEST**—Read output from the device.
3. **BYPASS**—Pass data from TDI to TDO, performing no operation.

The exact operations performed by these instructions are determined by the device's internal state machine. The debugging system can change this state by sending bits on TMS. As with TDI, each bit shift is sampled on rising edges of TCK.

The last of the five signals, TRST, returns the device's state machine to its reset state. This is an optional signal, but when it's present, it's usually active-low.

14.4 AM3359 I/O Connections

The next subcircuit presents the I/O connections of the AM3359, which include resources for data communication and conversion. Figure 14.4 depicts the relevant portion of the BBB's schematic.

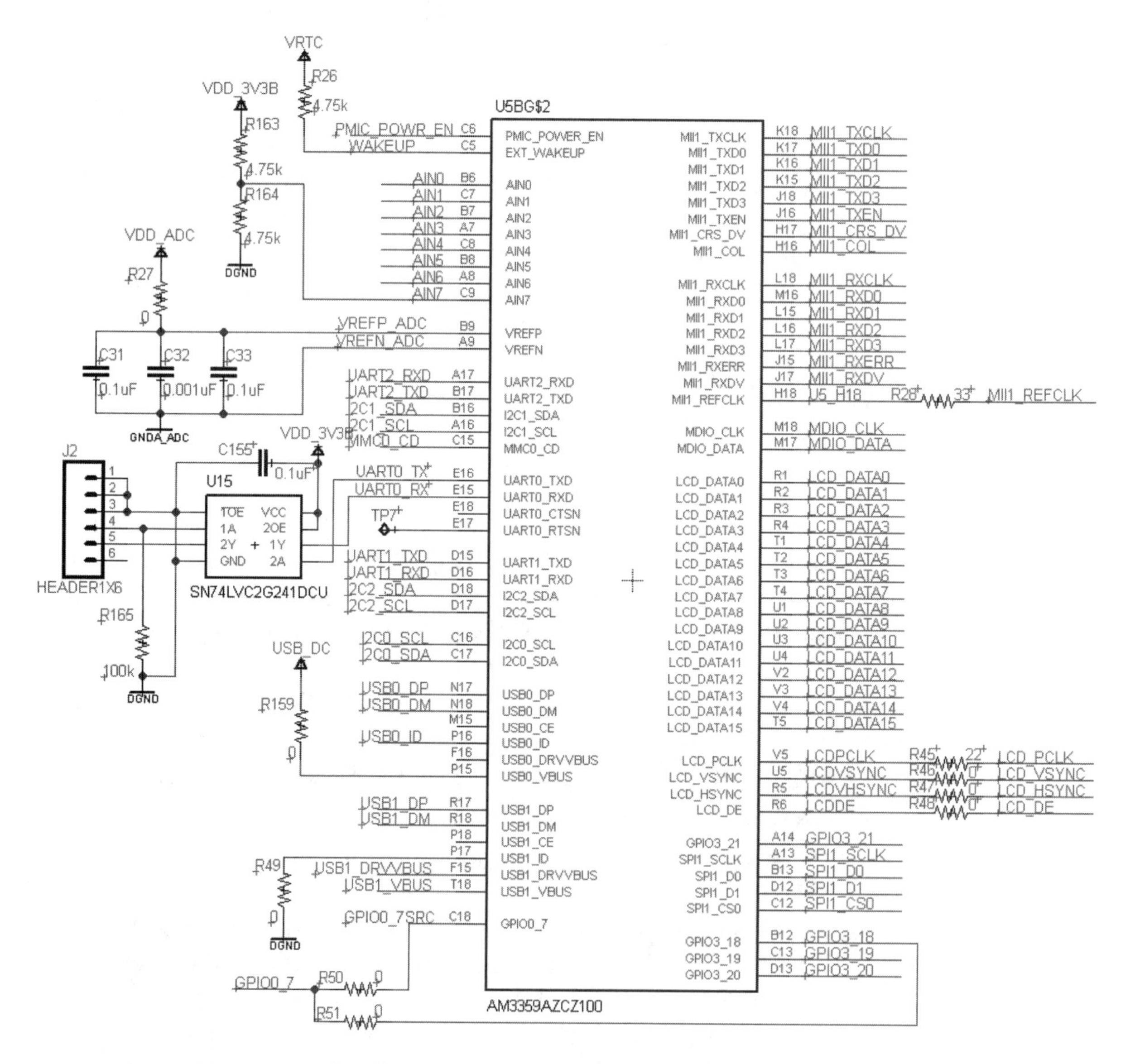

Figure 14.4: Schematic of the AM3359's I/O Connections

These signals can be divided into seven categories:

- **Ethernet**—The signals starting with MII1 transfer data to the LAN8710A, which sends/receives data through the BBB's RJ45 connection.
- **Universal Serial Bus (USB)**—The signals starting with USB0 and USB1 communicate data directly using the BBB's two USB connectors.
- **High-Definition Multimedia Interface (HDMI)**—The signals starting with LCD sends graphic data to the TDA19988 transmitter, which delivers graphic data to an HDMI receiver.
- **UART**—UART0_TX/RX enables serial debugging and UART1_TX/UART1_RX allows modem control.
- **Analog-to-digital conversion**—AIN[7:0] receive analog signals to be converted into digital data.
- **I²C**—The signals starting with I²C transfer data using the I²C protocol.
- **SPI**—The SPI signals transfer audio data to the HDMI transmitter.

Later sections discuss the signals involved in Ethernet, USB, and HDMI communication. This section presents the BBB's capabilities for analog-to-digital conversion, I²C communication, and SPI communication.

14.4.1 Analog-to-Digital Converison

Most single-board computers, such as the Raspberry Pi and PandaBoard, can't read analog signals. But the AM3359 has eight pins (AIN[7:0]) that can be configured to interface a touch screen or convert analog input to digital data. The AM3359's data converter is capable of returning 200,000 samples per second, with each sample containing 12 bits.

The range of analog inputs that can be read is determined by two voltage supplies: the positive reference, VREFP, and the negative reference, VREFN. If the input voltage is greater than VREFP, the converted sample will be 0xFFF. If the input voltage is less than VREFN, the converted sample will be 0x000.

14.4.2 I²C Communication

Of the many protocols available for transferring data between digital devices, I²C (Inter-Integrated Circuit) communication is probably the simplest. It consists of two signals:

- **Serial Data Line (SDA)**—Transfers bits between devices
- **Serial Clock (SCL)**—Data clock (10kHz or 100kHz)

When devices are connected by I^2C, only one can drive the clock signal on SCL. This device is the master and all other devices are slaves, and each slave receives a 7-bit address. In most I^2C data transfers, slaves are memory devices to be read from or written to.

The master communicates by sending messages preceded by START (SDA changes from high to low while SCL is high) and ending with STOP. (SDA changes from low to high while SCL is high.) A message consists of a series of 8-bit sequences, and the first sequence must provide two pieces of information. Its first 7 bits specify the address of the desired recipient and the eighth byte identifies whether the master intends to read data from the slave (1) or write data to it (0).

14.4.3 SPI Communication

Serial Peripheral Interface (SPI) communication is similar to I^2C in that a master drives a clock that enables communication with one or more slaves. But there are three important differences. First, the SPI clock, called SCLK, typically runs at frequencies in the tens or hundreds of megahertz. This is much faster than conventional I^2C clocks.

The second difference between SPI and I^2C is that SPI supports full-duplex communication. That is, instead of sharing a single line, the master and slave can both transmit data at the same time. This is made possible by two signals: MOSI (master output, slave input) and MISO (master input, slave output). While the master transmits data to a slave on MOSI, a slave can transmit data to the master on MISO.

The third difference is that, instead of selecting a slave by address, the master uses one or more signals to identify the recipient. This signal is called slave select (SS), and a master may employ multiple SS signals to communicate with multiple devices. Figure 14.5 shows the signals used by a SPI master to communicate with two slaves.

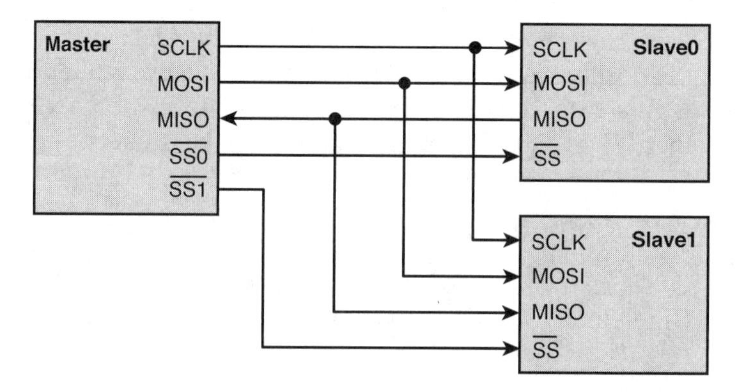

Figure 14.5: SPI Communication Between a Master and Two Slaves

The BBB schematic design doesn't use standard SPI signal names like SCLK, MISO, MOSI, and SS. Instead, its SPI signals have the following names:

- **SPI0_SCLK**—SPI clock
- **SPI0_D0**—Transmits data from master to slave (MOSI)
- **SPI0_D1**—Transmits data from slave to master (MISO)
- **SPI0_CS0**—Chip select, serves the same role as slave select (SS)

The AM3359 uses these signals to transfer audio data to the TDA19988 HDMI transmitter. That is, SPI0_D1 carries audio data from the processor, and the transmitter reads the data on rising edges of SPI0_SCLK.

While I^2C slaves are frequently dumb devices like hard drives, an SPI slave can be any type of device. Because of its simplicity and speed, SPI is frequently employed for programming microprocessors and microcontrollers. This is the case for the Femtoduino board presented in Chapter 4.

14.5 System Memory

The first memory device presented in this chapter is the MT41K256M16, which provides the BBB's system memory. DDR RAM is the common name for this type of device, but the full name is DDR3L SDRAM, which stands for 3rd-Generation Low-Voltage Double-Data Rate Synchronous Dynamic Random Access Memory. As the name implies, this memory loses its data when power is off and all reads/writes are performed on clock edges. In the case of the MT41K256M16, the maximum clock frequency is 800 MHz. Because this device is double-data rate, it performs operations on the clock's rising and falling edges.

The MT41K256M16's memory capacity is 4GB, divided into eight banks of 512MB each. Each bank is further divided into a matrix-like grid containing 32k rows and 1k columns.

To specify a memory address, the AM3359 must provide three pieces of information: the bank address, the column address, and the row address. The bank address is set by BA[2:0], the column address is set by A[14:0], and the row address is set by A[9:0]. During a memory access, the device reads the row address first and the bank address and column address afterward. After the full address is given, the device reads or writes data on the data bus, D[15:0].

In addition to these addressing signals, the MT41K256M16 has many more signals that control the device's state. These are presented in the device's schematic, given in Figure 14.6.

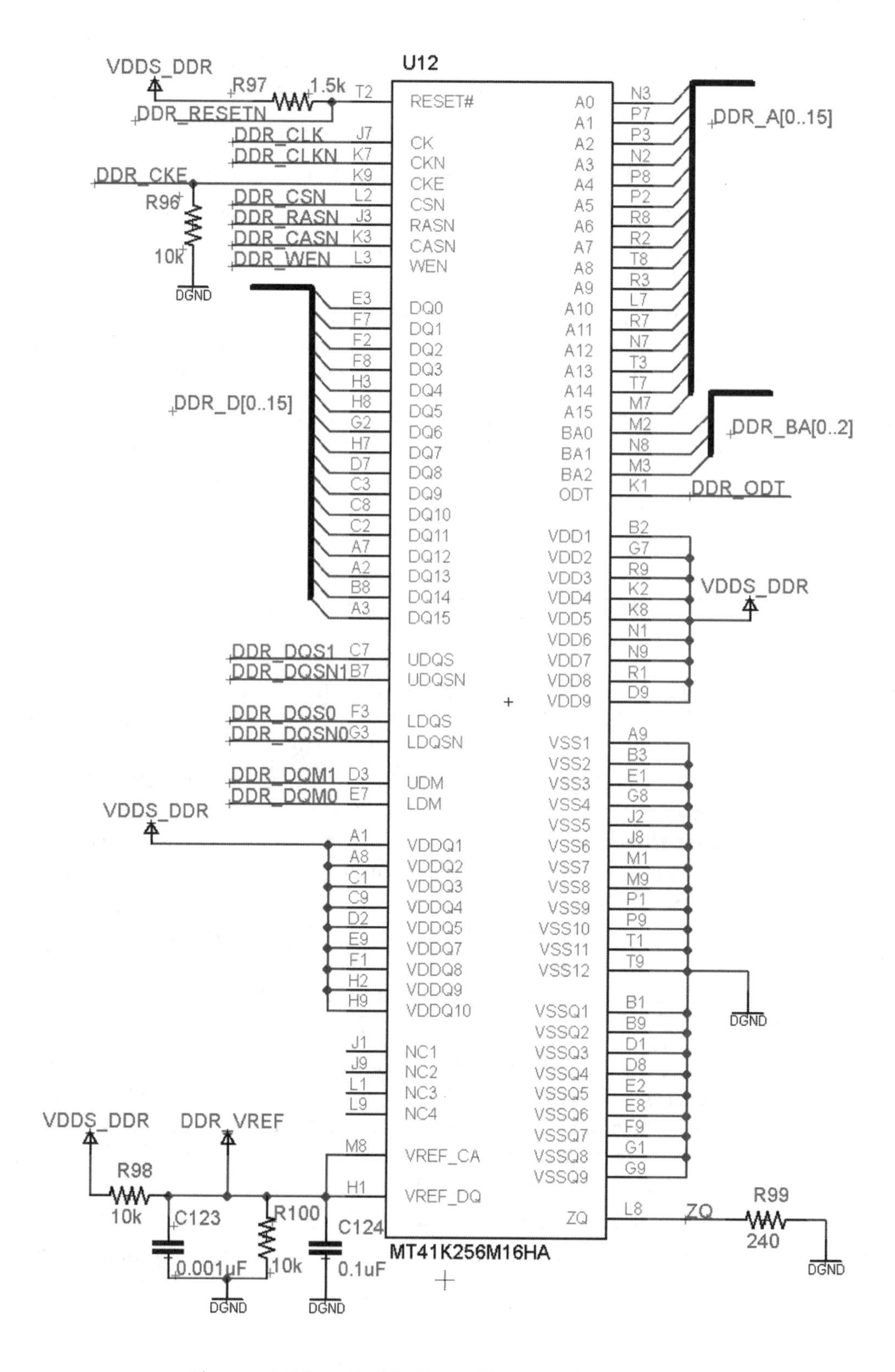

Figure 14.6: Schematic of the System Memory (MT41K256M16HA)

14.6 Power Management

To conserve power, embedded devices support different modes of operation. In active mode, the device's processing and peripheral resources operate at full power. In standby mode, a device may reduce its clock frequency and reduce power to its peripherals. In sleep mode, it may store RAM to main memory and turn off the device's processing resources and peripherals.

Switching between power modes is one of many tasks that fall under the broad heading of power management. Microcontrollers have internal resources for power management but microprocessors need external devices. To meet this need, the BBB relies on the TPS65217C. In addition to supporting power modes, this device makes it possible to power the circuit board using a USB connection or a battery.

The TPS65217C is a complex device with many fascinating capabilities. But in the BeagleBone Black, it serves only five roles:

- **AC input**—Receives transformed power (5V) through a standard power jack
- **USB input**—Receives 5V from the USB's VBUS connection
- **Battery input**—Receives power from a Lithium-ion (Li-ion) or Lithium-polymer (LiPo) battery
- **Battery charging**—Recharges battery from AC input or USB input
- **Voltage regulation/conversion**—Provides power to BBB devices

The last role is the most important. The TPS65217C uses its four LDOs (low drop-out regulators) and three DC-DC converters to provide power to the rest of the BBB. More specifically, it returns the following power connections:

- VDD_MPU—Provides power to the ARM processor in the AM3359
- VDD_CORE—Provides power to the core resources in the AM3359
- VDD_1V8—Provides power to the SRAM and USB connections in the AM3359
- VDDS_DDR—Provides 1.5V to the RAM device
- VIO—Provides power to the input/output devices
- VRTC—Provides power to the real-time clock
- VDD_3V3A—Provides power to the reset circuit and other circuits

Most of the power provided by the TPS65217C goes to the AM3359. This is shown in Figure 14.7, which displays the portion of the schematic containing the TPS65217C.

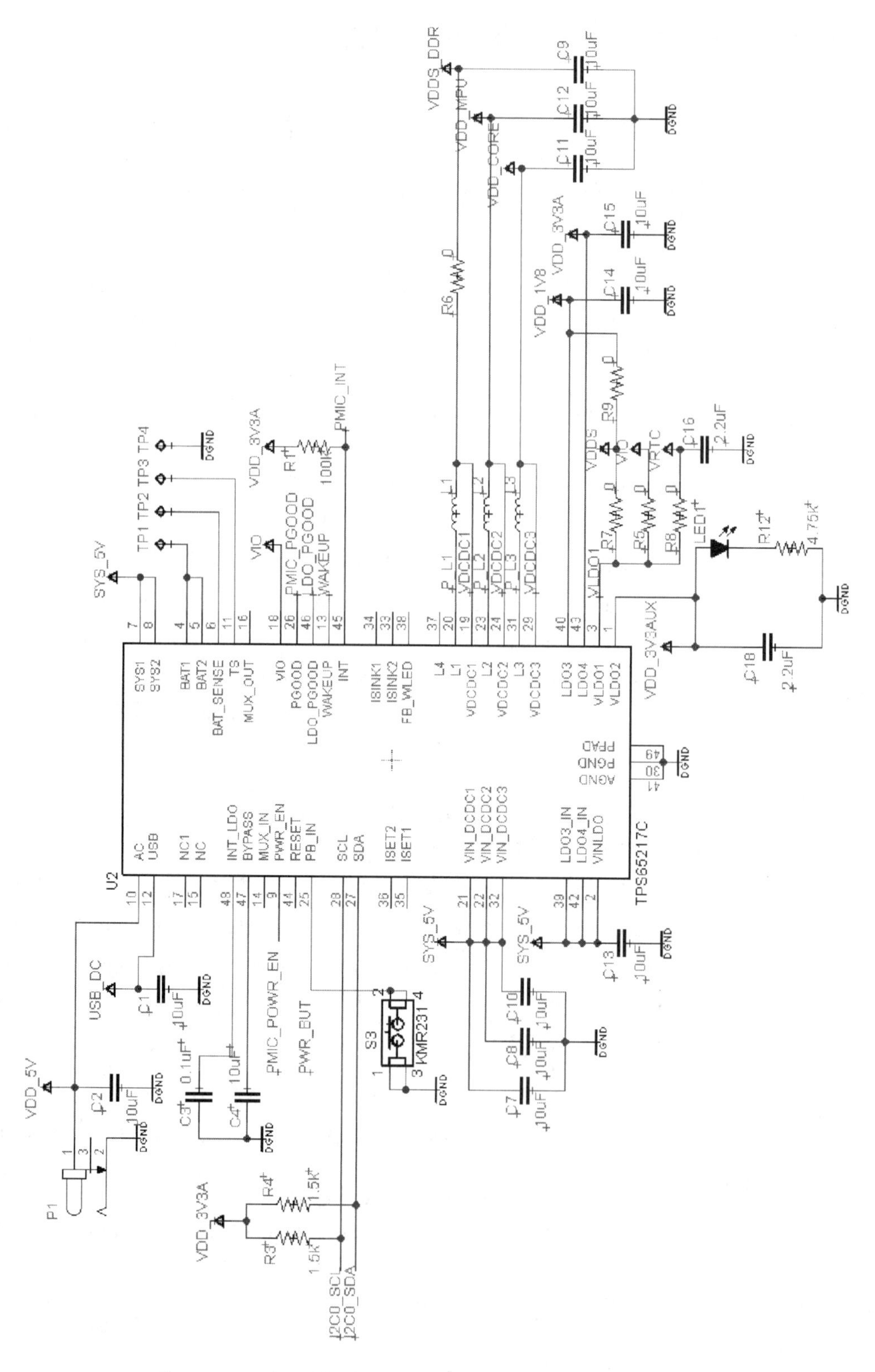

Figure 14.7: Schematic of the Power Management Device (TPS65217C)

14.7 Ethernet and the Universal Serial Bus

Section 14.4 presented the I^2C and SPI protocols, which are fine for simple, low-speed data transfer. This section focuses on two methods that enable complex communication at high speed. These are the Ethernet (IEEE 802.13) and Universal Serial Bus (USB) 2.0 protocols, which are both supported by the BeagleBone Black.

In both cases, high-speed communication is made possible through differential signaling, which transfers data along two conductors carrying opposite voltages. These two conductors form what's called a differential pair. Figure 14.8 shows a transmitter sending seven bits to a receiver across a differential pair.

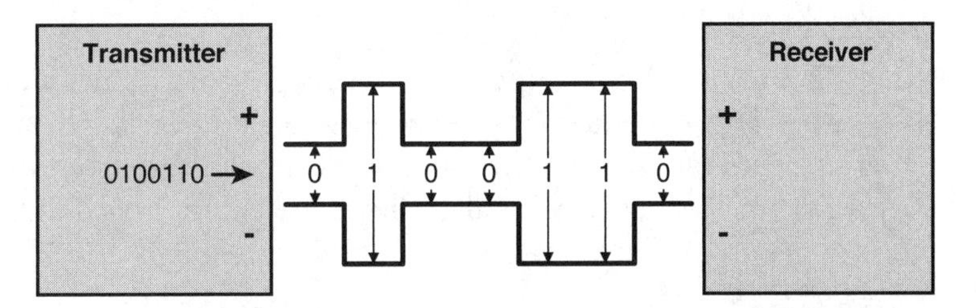

Figure 14.8: Transimitting Seven Bits with Differential Signaling

The negative signal is the inverse reflection of the positive signal, and only the difference between them is important. This ensures that any electromagnetic noise or interference will be removed from both signals. For this reason, differential signaling is less susceptible to crosstalk than single-ended signaling methods, such as SPI, I^2C, and RS-232.

14.7.1 Ethernet

The AM3359 has many pins for input and output, but it doesn't have the processing resources needed to read and write directly to an Ethernet connection. For this reason, the BeagleBone Black contains the LAN7810A. This provides the Media Independent Interface (MII) that the processor needs to send and receive Ethernet traffic.

The AM3359 communicates with the LAN8710A in groups of 4 bits. More precisely, the LAN8710A receives four data signals (MII1_TXD0-MII1_TXD3) from the AM3359 and transmits four signals (MII1_RXD0-MII1_RXD3). The TXD signals are read on the rising edge of TXCLK and the RXD signals are read on the rising edge of MDC, which are both provided by AM3359. For Fast Ethernet (100Mbit/s), both clocks are set to 25MHz.

14.7.2 USB

Since its initial development in the 1990s, the Universal Serial Bus (USB) has established itself as the preeminent protocol for transfering data between computers and peripherals. Like I^2C and SPI, USB communication involves a single master and one or more slaves. However, the USB standard refers to the master as the host and the slaves as devices. The host sends commands to the devices and the devices respond as requested.

Figure 14.9 presents the schematics for the BBB's Ethernet and USB connections. As shown, the circuit has two USB connections: a host port and a device port. If a system connects to the host port, the BBB will assume the role of host and manage data transfer. This allows the BBB to receive data from keyboards, mice, and other peripherals.

The second USB connection, the device port, allows a connected system to take control. This is frequently used to allow a traditional computer to access the BBB as a peripheral. As host, the computer can read files from the BBB and configure its operation. The device port also allows the circuit board to be powered by USB.

Transfer Types

The designers of the USB specification took the term *universal* seriously. The USB protocol is intended to support every possible type of data transfer, from music and video frames to huge data blocks. Different types of data transfer have different requirements, so the USB standard divides transfers into four categories:

- **Control**—The host sends control messages to the device, such as requests for status.
- **Interrupt**—The host regularly checks the device to see if it has data. This is used by PCs when receiving data from keyboards and mice.
- **Bulk**—The host and device transfer data in large blocks. This is used by PCs when reading/writing files to an external USB drive.
- **Isochronous**—The device transfers time-specific data at regular intervals. This is used by PCs when receiving music or video data.

The USB protocol is flexible enough to accommodate different types of data transfer at the same time. But this flexibility makes the USB specification difficult to understand.

Devices and Enumeration

In addition to supporting different types of data transfers, the USB protocol supports many different device types, from cameras and printers to combined camera/printer/keyboards. A device identifies its capabilities through a process called enumeration.

The enumeration process consists of data requests from the host and data responses from the device. The device provides its information using descriptors. For example, the device descriptor identifies device properties such as its USB class and ID numbers. The endpoint descriptor identifies the required data transfer types and packet sizes.

After the host receives these descriptors, it resets the device. Then regular USB communication proceeds normally.

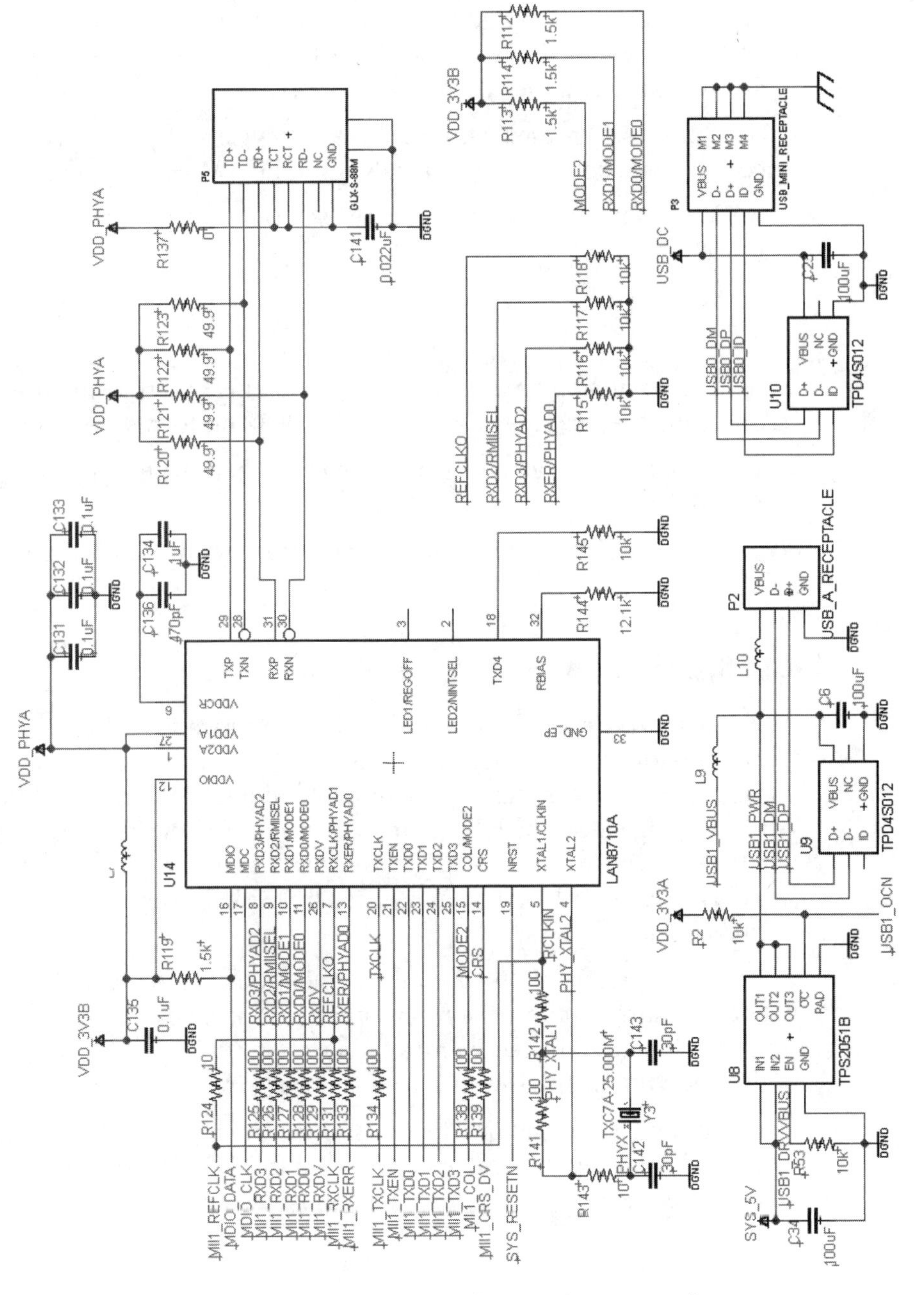

Figure 14.9: Schematic of the Ethernet and USB Connections

14.8 Graphics Display

The BeagleBone Black doesn't have any onboard capabilities for displaying graphics, such as an LCD panel or an OLED display. But it can transfer data to a monitor or television through the High-Definition Multimedia Interface, or HDMI. The AM3359 has 16 pins dedicated to graphic display, LCD_DATA0 to LCD_DATA15, but they can't be directly connected to an HDMI port. For this reason, the BBB circuit contains the TDA19988 device, which serves as the intermediary between the AM3359's display signals and the HDMI port.

Figure 14.10 presents the TDA19988 device and its connections. As shown, it receives three types of input signals from the AM3359:

- Video inputs through the pins VPA0-VPA7, VPB0-VPB7, and VPC0-VPC7
- Audio inputs through pins AP0 and AP1
- Control data inputs through pins ACLK, VSYNC, and HSYNC

This discussion focuses on video processing and transmission. After receiving and converting the incoming data, the TDA19988 performs two important steps:

1. It encrypts the video stream using high-bandwidth digital copy protection, or HDCP.
2. It transmits the stream data using transition-minimized differential signaling, or TDMS.

14.8.1 HDCP Encryption

One major difference between HDMI communication and other types of video transmission is that the transmitting and receiving devices must be authenticated. This is made possible through high-bandwidth digital content protection, or HDCP. Before transmitting video, the HDCP processor in the TDA19988 examines the credentials of the receiving device. To be precise, it will examine the receiver's authentication key, and if the receiver hasn't been licensed by Digital Content Protection LLC, no video will be sent.

If the receiver is licensed and compliant, the TDA19988 will encrypt the video data to ensure that only the receiver can read it. The encryption is performed with a stream cipher, so the receiver uses a series of XOR operations to decode the pixels. This encryption and decryption adds latency to the data communication.

During authentication, the transmitter and receiver communicate using the I^{2}C protocol, which was discussed earlier. On the TDA19988, the I^{2}C pins are DSCL and DSDA. As shown in the schematic, these pins are raised to DVI_+5V. Therefore, the BBB supports digital visual interface (DVI) transmission, but not HDMI. DVI is similar to HDMI in many respects, but DVI doesn't support YCrCb color encoding or encryption.

14.8.2 Transition-Minimized Differential Signaling (TMDS)

To supply pixels to modern high-definition displays, the TDA19988 needs to transmit the encrypted video at very high speed. For this purpose, DVI and HDMI rely on transmission-minimized differential signaling, or TDMS. Like USB and Ethernet signaling, this uses differential pairs to transfer data.

To be precise, the TDA19988 transmits data on three differential pairs, given in the schematic as TX0+/TX0–, TX1+/TX1–, and TX2+/TX2–. These signals have the same clock (CLK+/CLK–), whose frequency is 165 MHz for HDMI 1.0 and 340 MHz for HDMI 1.3.

TDMS encodes data so that every combination of 8 bits is converted to a 10-bit code. The possible values of this code are chosen to reduce the possibility of electromagnetic interference and make it easier for the receiver to recover the original signal.

14.9 Conclusion

A single-board computer (SBC) is a cross between a microcontroller board and a traditional desktop computer. Its small size and (relatively) low power make embedded engineers happy and its powerful processor and memory make programmers happy. Two popular SBCs are the Raspberry Pi from the Raspberry Pi Foundation and the Galileo board from Intel. My favorite SBC is the BeagleBone Black from Texas Instruments, which provides a great deal of processing power in a small form factor.

The source of this power is the AM3359 system on a chip, which contains an ARM processor, a graphics processor, and a handful of other processing resources. This multipurpose device is so large that it can't be fully connected in a single sheet. Instead, the BBB schematic presented in this chapter splits the device into three gates depicted in separate sheets. Each sheet corresponds to a different functional aspect of the device: memory connections, I/O connections, and power input.

The AM3359 isn't a microcontroller, so it needs memory and power management to be provided externally. In the BBB, the AM3359 relies on an MT41K DDR3 RAM device to store data and an MTFC Flash memory device to store the operating system. Power management is the responsibility of the TPS65217C, which makes it possible for the BBB to receive power from a battery, a USB connection, or a regular power adapter.

Given the size and complexity of the BeagleBone Black, many might suppose that designing the circuit is beyond the abilities of students and hobbyists. But I hope this chapter has shown that the BBB isn't difficult to grasp. What's more, EAGLE's schematic editor makes it straightforward to connect the hundreds of required components. The next chapter will progress beyond the schematic editor and show how to design the actual board.

Chapter 15

Board Design for the BeagleBone Black

Now that the schematic for the BeagleBone Black (BBB) is finished, the next step is to design the board. This entails laying out the components and drawing traces between their pads. I performed most of the routing manually, and because of the thousands of pads involved, it's a long and tedious process.

It would take an entire book to present all the details of BBB layout and routing. Instead, this chapter presents three crucial tasks:

- **Configuring the stackup**—Why the BBB uses six copper layers for its board design and the purpose of each layer
- **Creating and routing ball grid array (BGA) components**—The methods used to generate high-density BGA components and route their many pads
- **Routing differential pairs**—The process of routing high-speed signals using differential pairs

In each case, I'll explain the theory behind the task and the manner in which it can be accomplished in EAGLE. The importance of these topics goes beyond just designing the BeagleBone Black. They're necessary for the design of any multilayer board carrying high-frequency signals.

The last part of this chapter looks at the completed BBB design, which can be found in the Ch15 folder. In particular, I'll present the portions of the circuit beneath the AM3359 system on a chip. This will show how high-density routing can be accomplished with EAGLE's board editor.

15.1 Configuring the Stackup

Chapter 2, "An Overview of Circuit Boards and EAGLE Design," discussed the materials that form the structural basis of circuit boards. To review, thin copper sheets are attached to a hard material called core. A double-sided circuit board has one layer of core with copper on its top and bottom sides. As the number of copper layers increases, the board can support an increased number of connections between components.

Multilayer circuit boards are constructed by gluing core layers together with a material called prepreg. The arrangement of copper, core, and prepreg is called the board's *stackup*.

The goal of this section is to present the six-layer stackup used by the BBB. To reach this goal, I'll start by explaining the importance of inserting ground and power planes into a bord's stackup. Then I'll discuss the advantages and disadvantages of four-layer and six-layer stackups and the process of configuring these stackups in EAGLE.

15.1.1 Ground and Power Planes

As presented in Chapter 5, "Layout and Design Rules," the Femtoduino circuit board has two copper layers and both are covered with a copper pour (created with the Polygon tool). These copper pours carry the board's ground signal, and other signal traces are routed inside of them.

Two-layer stackups are fine for low-frequency circuits, but for high-frequency designs, signal traces need to be routed on separate layers. To make this distinction clear, this chapter employs the following terms when discussing stackup layers:

- **Signal layer**—Copper layer whose traces carry signals
- **Ground plane**—Copper layer containing a copper pour connected to ground
- **Power plane**—Copper layer containing a copper pour connected to power

One major reason to use separate ground and power planes is isolation. High-frequency circuits produce electromagnetic radiation that can interfere with nearby circuits. But if a ground or power plane is placed between two signal layers, it blocks this interference.

For this reason, engineers recommend that every signal layer should be adjacent to a ground or power plane. In addition, a signal layer should be kept as close as possible to its adjacent plane. This means that, no matter how many copper layers are present, the stackup should contain at least one ground/power plane for every two signal layers.

15.1.2 Four-Layer Stackup

To simplify soldering, the top and bottom layers of a circuit board are usually signal layers. For this reason, most four-layer boards have signal layers on the top and bottom. The middle layers are a ground plane and power plane in some order. Figure 15.1 shows what this looks like.

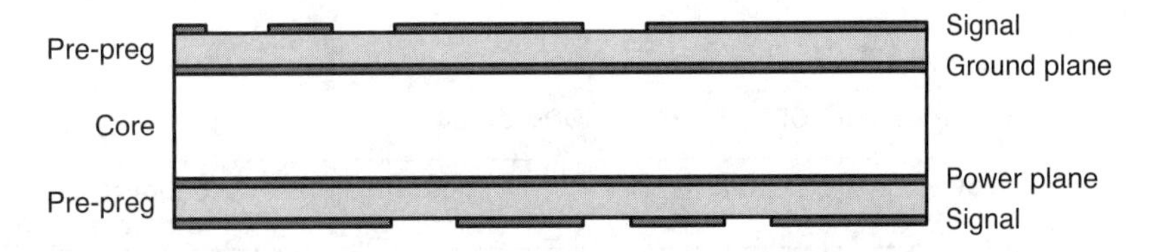

Figure 15.1: Common Four-Layer Stackup

The ground and power planes provide electromagnetic isolation between the signal layers. This can be improved further by routing the traces on the top and bottom layers at right angles to one another. That is, if the majority of the top traces are routed horizontally, the traces on the bottom should be routed vertically.

This stackup reduces crosstalk between the two signal layers, but it doesn't protect external circuits from radiation. This problem can be reduced by keeping the signal layers as close as possible to their adjacent reference plane (ground plane or power plane), and increasing the separation between the ground and power plane. Electromagnetic interference can be reduced further by using a six-layer stackup, discussed next.

15.1.3 The BBB Stackup

There are many ways to arrange six layers in a printed circuit board, but the most popular six-layer stackup consists of the following:

- Signal layers as the top (first layer) and bottom (sixth layer)
- Ground/power planes as the second layer and fifth layer
- Signal layers as the center layers (third and fourth)

This is the stackup used by the BBB. In this case, the second layer is a ground plane and the fifth layer is a power plane. Figure 15.2 shows what this stackup looks like.

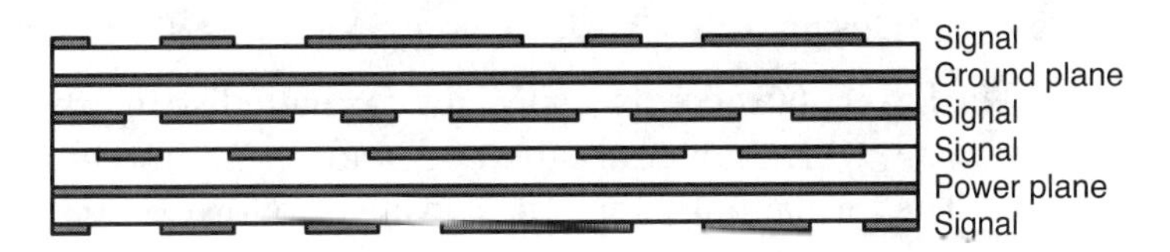

Figure 15.2: Six-Layer Stackup of the BeagleBone Black

The ground and power planes provide excellent shielding for the central signal layers. For this reason, long traces and traces carrying high-speed signals are routed in the central layers.

From top to bottom, the BBB circuit board is 62 mils wide. Table 15.1 presents the thicknesses of the materials in the stackup. Note that the copper, core, and prepreg layers are listed separately.

Table 15.1

Composition of the BeagleBone Black

Layer	Material	Thickness (mil)	Conductivity (mho/cm)
Signal	Copper	2.4	595900
	Prepreg	3.6	0
Ground plane	Copper	1.2	595900
	Core	4.6	0
Signal	Copper	1.2	595900
	Prepreg	36.0	0
Signal	Copper	1.2	595900
	Core	4.6	0
Power plane	Copper	1.2	595900
	Prepreg	3.6	0
Signal	Copper	2.4	595900

The core separating the ground plane from the signal layer beneath it is over six times as wide as the other core/prepreg layers. This increases the isolation between the high-frequency signals in the central layer and the signals on the top layer.

It's interesting to note that the outer copper layers are twice as thick as the inner copper layers. The added thickness increases the amount of current that the traces can carry and reduces the chances that external damage will affect the circuit.

15.1.4 Configuring the BBB Stackup in EAGLE

By default, the board editor assumes that the design will consist of two layers: the top (Layer 1) and the bottom (Layer 16). But depending on the EAGLE version, board designs may use up to 16 layers. These layers are configured using the Layers tab of the DRC (Design Rule Check) dialog, which was discussed in Chapter 5.

The Layers tab accepts a specially formatted string that identifies which layers are present, which materials are used, and the nature of the connections between the layers. The format of this string is given as follows:

- The numbers in the string identify which layers are present.
- If two numbers are separated by an asterisk, the layers will be separated by core. If the numbers are separated by a plus sign, the layers will be separated by prepreg.
- Round parentheses indicate that layers can be connected using buried or through vias. Square parentheses imply blind vias. That is, [x: ... :y] identifies a blind via from Layer 1 (the top) to Layer x and a blind via from Layer y to Layer 16 (the bottom).

As presented in Table 15.1, the BBB has core between Layer 2 and Layer 3 and between Layer 4 and Layer 5. The other layers are separated by prepreg. All of the vias in the design are through vias, which means they run from the top (Layer 1) to the bottom (Layer 16). Therefore, the stackup string for the DRC dialog is (1+2*3+4*5+16). This is shown in Figure 15.3, which depicts the upper half of the Layers tab.

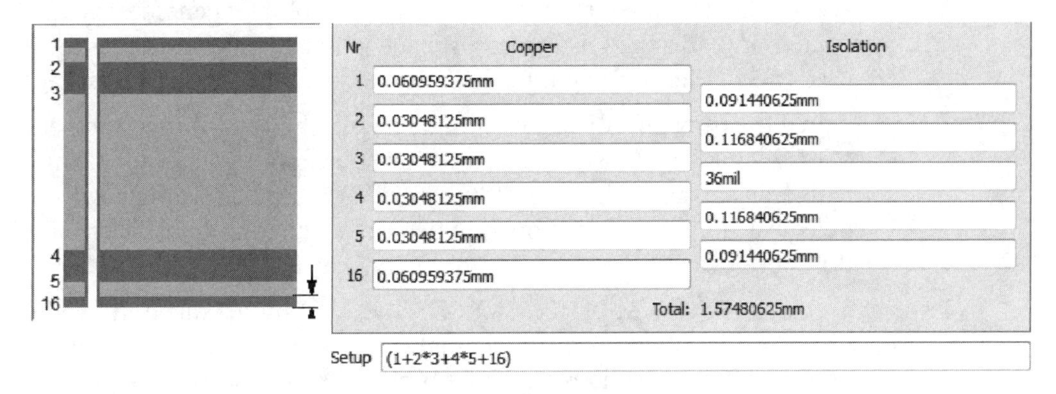

Figure 15.3: Stackup Configuration in EAGLE's DRC Dialog

It's important to configure vias properly in the DRC dialog. If Layer A and Layer B aren't explicitly connected by a via in the Layers tab, the Via dialog won't give you the option of creating a via that runs from Layer A to Layer B.

15.2 Creating and Routing Ball Grid Arrays

One of the many differences between the BBB circuit and the other circuits in this book is the presence of ball grid array (BGA) devices. The BBB has three BGA devices: the AM3359 SoC, the MT41K DRAM, and the MTFC Flash memory. In total, these devices contribute nearly 600 pins to the design.

With so many connections in close proximity, the process of dealing with the components' packages becomes tedious and error-prone. To automate aspects of the design, I've coded two User Language programs (ULPs):

- **make_bga_package.ulp**—Generates a package for a BGA device according to a series of parameters
- **bga_router.ulp**—Connects pads of a BGA device to vias

This section discusses both ULPs and explains what they do. It also shows how I used them in the process of designing the BBB.

15.2.1 Generating BGA Packages

Chapter 8, "Creating Libraries and Components," explains how to create custom components. Each component has three aspects: a symbol, a package, and a device. Section 8.4 demonstrates how the make_symbol ULP generates symbols for integrated circuits and how the make_bga_package ULP generates packages for BGA devices. I used these ULPs to design the BBB circuit board and create its packages.

Of the three BGA packages on the board, the most difficult to generate was the MT41K256M16HA, which serves as the board's system memory. This is because the pads aren't arranged in a simple rectangle or a simple rectangle with a rectangular hole. Instead, they're positioned in two rectangular groups. Figure 15.4 shows what this looks like.

As shown, the MT41K256M16HA has 216 pins arranged in two blocks, each made up of 16 rows of three pads each. make_bga_package can't generate this exact package, but it can create a solid rectangular block with 16 rows of nine pads each. To generate this initial package, I followed three steps:

1. Place the make_bga_package.ulp in EAGLE's top-level ulp folder.
2. Enter `run make_bga_package` in the text box of an EAGLE editor.
3. In the dialog box, enter parameters for the package's dimensions and pins.

Figure 15.5 shows what the dialog looks like with the MT41K256M16HA parameters.

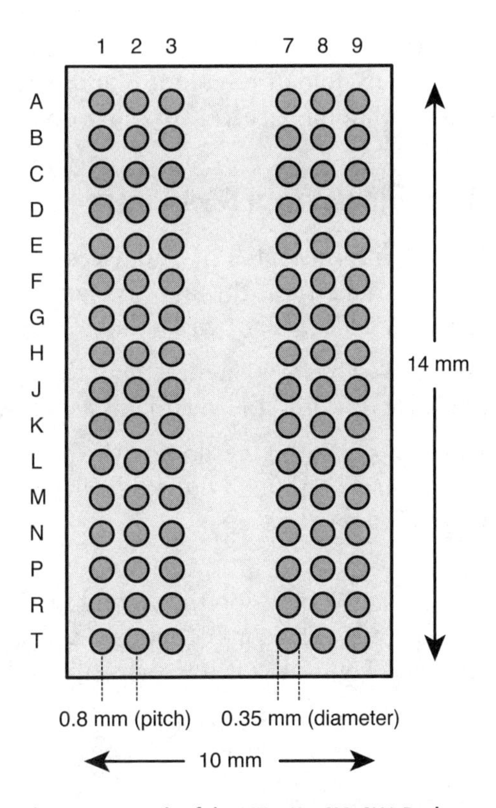

Figure 15.4: Pads of the MT41K256M16HA Package

Eagle: BGA Package Generator

Library name:	example.lbr
Package name:	fbga96
Component width (in mm):	10
Component height (in mm):	14
Radius of BGA leads (in mm):	0.165
Spacing between centers of leads (in mm):	0.8
Total number of leads:	144
Leads in outer rectangle (horizontal):	9
Leads in outer rectangle (vertical):	16
Leads in inner rectangle (horizontal):	0
Leads in inner rectangle (vertical):	0

Generate Package

Cancel

Figure 15.5: The make_bga_package Dialog with MT41K Parameters

When the 144-pin package is open in the editor, the excess pins can be removed by selecting the middle three columns and performing a group delete operation. This leaves 96 pins remaining, each with the proper name and position.

15.2.2 Routing Signals from BGA Pads

When dealing with fine-pitch BGA devices, routing becomes significantly difficult. It isn't feasible to route traces through so many closely spaced pads, so designers connect most pads to vias. There are two methods of doing this:

- **Via-in-pad**—Create a via directly under the pad. This simplifies routing, but solder may flow into the via, making connection difficult.
- **Escape via**—Create a via close to the pad and route a trace from the pad to the via. This may be difficult due to the limited space, but this method doesn't affect solderability.

This discussion focuses on the second method, which is recommended by most fabrication facilities. The connection between a BGA pad and its escape via is referred to as a *dog bone*. Figure 15.6 shows why this is the case.

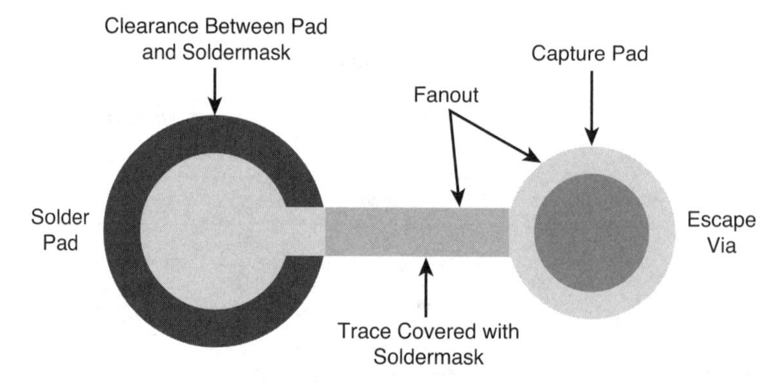

Figure 15.6: A BGA Dog Bone

In this diagram, the pad's position is fixed, but the positions of the trace and via (collectively called the pad's fanout) are chosen by the circuit designer. Vias are usually placed in the center of the four surrounding pads. That is, vias are positioned to the northwest, northeast, southwest, or southeast of the corresponding pad. In many designs, escape vias are always positioned in the same direction relative to their pads.

Another method of arranging fanout is called *quadrant dog bone routing*. In this arrangement, traces are angled according to the pad's quadrant. That is, if a pad is in the upper-right (northeast quadrant), its via will be positioned to the upper-right of the pad. If the pad is in the lower-left (southwest quadrant), the escape via will be positioned to the lower-left of the pad. A key advantage of quadrant dog bone routing is that the central row and central column of the device are available for routing. Figure 15.7 shows what this routing method looks like.

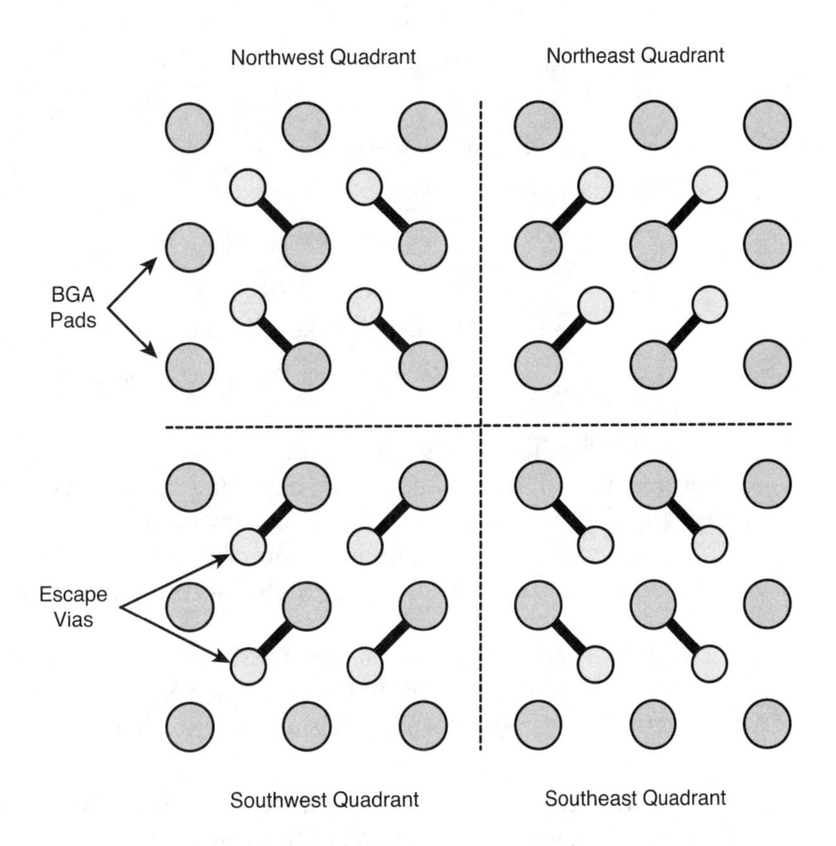

Figure 15.7: Quadrant Dog Bone Routing

Whether the vias are arranged by quadrant or unidirectionally, the process of creating them can be automated with ULPs. In the Ch15 folder, bga_route.ulp creates a dialog that allows the user to select the BGA component to be routed and the size of the traces and vias. Figure 15.8 shows what the dialog looks like.

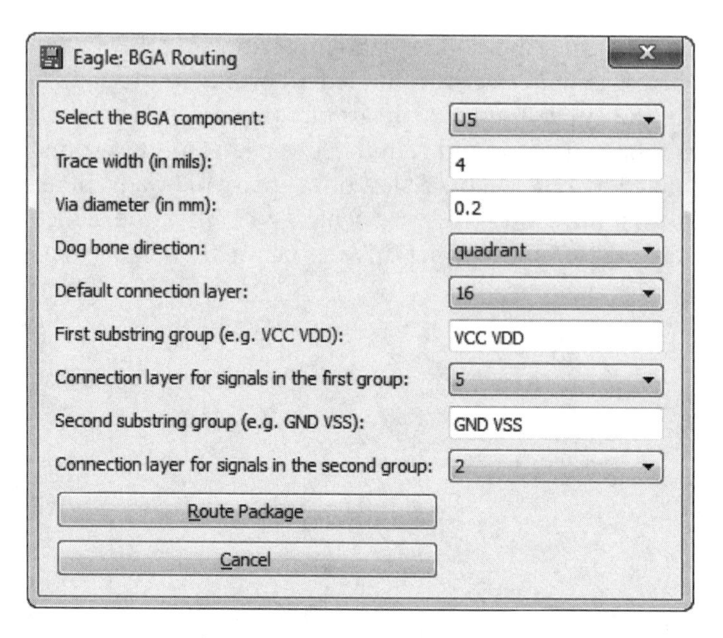

Figure 15.8: The bga_route.ulp Dialog

The fields of this dialog can be split into two groups. The first four entries request basic information about the routing, including the device's name, the via diameter (in mm) and the trace width (in mils). The entry labeled Dog bone direction specifies whether the device's traces should be routed in the same direction (northeast, northwest, southeast, southwest) or routed according to the pad's quadrant.

> **NOTE** The via diameter and trace width are controlled in part by the design rules. Therefore, in addition to using bga_route.ulp, these dimensions need to be set in the Sizes and Restring tabs of the DRC dialog.

The next group of fields specify the layer to which the via should be connected. This is set to Layer 16 by default, but this can be changed with the entry labeled Default connection layer.

In many cases, a designer may want all signals with a specific prefix to be connected to a specific layer. For example, you may want all signals starting with GND or VSS to be connected to the ground plane. You may also want signals starting with VCC or VDD to be connected to the power plane. This matching can be accomplished with the last two dialog entries, and the example in Figure 15.8 shows how this works in practice.

> **NOTE** Two layers can be directly connected by a via if they've been specifically configured in the Layers tab of the DRC dialog.

Before leaving this topic, there's one last point I'd like to mention. Charles Pfeil, Engineering Director at Mentor Graphics, has given a great deal of thought to arranging fanouts. His book, *BGA Breakouts and Routing*, explores the subject in detail and provides a great deal of insight. It can be downloaded for free from Mentor Graphics at http://go.mentor.com/2yss7.

15.3 Trace Length and Meander

In most designs, the length of a routed trace isn't a major concern—so long as a trace connects its start/end points without intersecting other traces, it doesn't matter how long it is. But when a signal has to reach its destination in a particular time, the trace's length must be explicitly set. For example, the control signals entering a high-speed DRAM device need to reach the device at the same time as the incoming data.

In EAGLE, configuring a trace's length is made possible through the Meander tool. This makes it possible to obtain the length of a selected trace. It can also be used to extend a trace to a specific length. This becomes useful when routing differential signals, which is the last topic discussed in this section.

15.3.1 Obtaining the Length of a Trace

In the board editor's vertical toolbar, the Meander tool is located immediately above the Route tool. When this is active, the length of a trace can be found by clicking on the trace in the editor. A small pop-up dialog will appear above the trace, presenting the trace's length. This is shown in Figure 15.9.

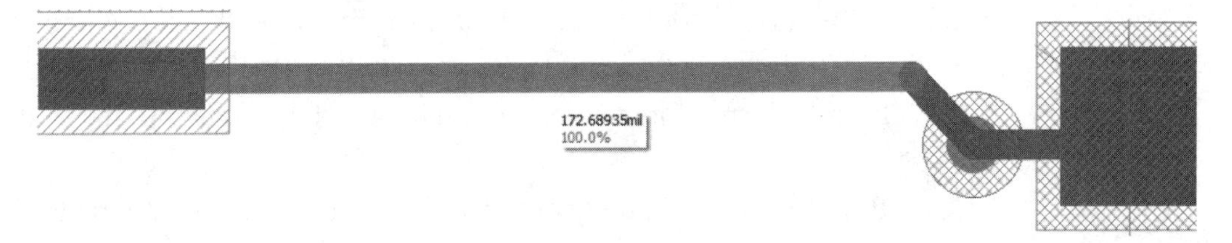

Figure 15.9: Finding a Trace's Length

The length given in the dialog is the length of the entire trace, and not just the length of a particular segment. Therefore, no matter which segment of a trace you select, Meander will always return the same length.

15.3.2 Extending the Length of a Trace

The process of extending a trace to a specific length consists of four steps:

1. Activate the Meander tool.
2. Click the trace in the editor.

3. Enter the desired length in the text box above the editor area and press Return.
4. Move the mouse near the selected point to extend the trace.

Figure 15.10 shows an extended version of the trace depicted in Figure 15.9. I extended its length to 200 mils by activating Meander, clicking the trace, and entering 200 in the text box.

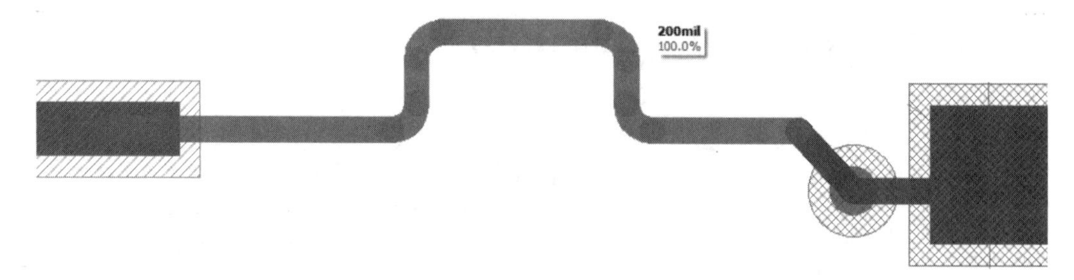

Figure 15.10: Extending Traces with Meander

15.3.3 Routing Traces in a Differential Pair

Chapter 14, "Schematic Design for the BeagleBone Black," introduced the topic of differential signals and explained why the BBB uses them for high-speed data transfer. When routing traces that form a differential pair, it's crucial that the two signals reach their destinations at the same time. This can be accomplished by making sure both traces have the same length.

If two signals have similar names but one ends with _P and the other ends with _N, EAGLE assumes they form a differential pair. When you route either one of them, traces will be formed for both signals at the same time. If you press the Escape key, the symmetric routing will end and you'll be able to route the selected signal normally.

With this information, we can obtain a four-step procedure for routing differential signals:

1. With both signals named correctly (*_N and *_P), activate the Route tool.
2. Click the start of one of the signals and route traces for both toward their destination.
3. Press the Escape key to end simultaneous routing. Connect the signals separately to their end points.
4. If the traces' lengths need to be increased, use the Meander tool.

An example will make this clear. Figure 15.11 shows two differential signals, DPAIR_P and DPAIR_N.

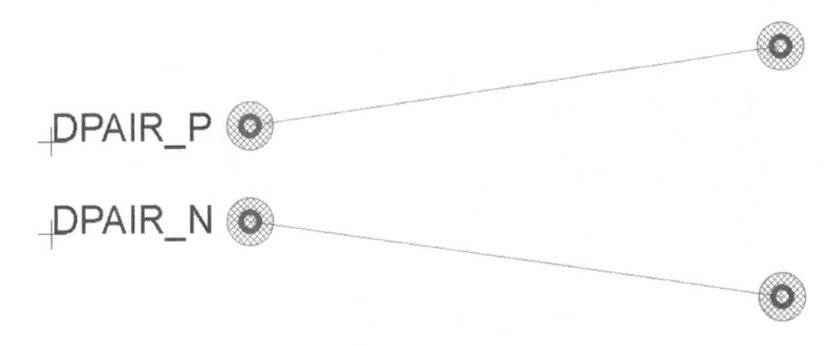

Figure 15.11: Unrouted Differential Pair

When Route is active, both of the differential signals can be routed at the same time. For the most part, the gap between the two traces is kept constant. This is shown in Figure 15.12.

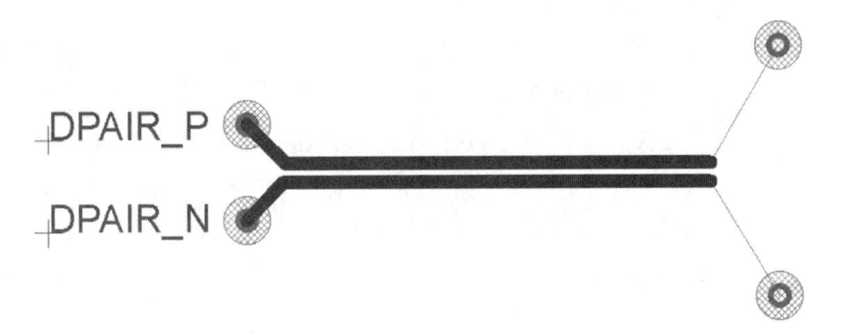

Figure 15.12: Partially Routed Differential Pair

DPAIR_P and DPAIR_N can't be connected simultaneously to their end points. Pressing the Escape key makes it possible to connect each signal individually.

After the connections are made, the length of the traces can be extended as needed. Like the Route tool, the Meander tool operates on both traces at once. Figure 15.13 shows what the traces look like after the Meander tool is used twice.

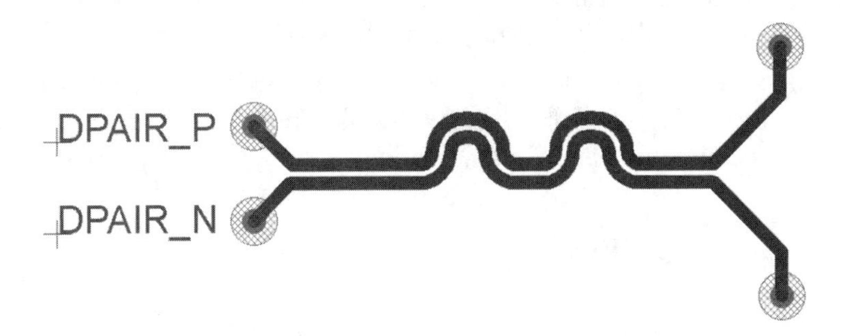

Figure 15.13: Fully Routed Differential Pair Extended by Meander

By default, the gap between the differential traces is a multiple of the clearance parameter set in the design rules. This gap factor is 2.5 by default, but it can be changed by updating the value in the Misc tab of the DRC dialog. This tab also makes it possible to set a maximum value for the difference in length between traces that make up a differential pair.

15.4 The BBB Board Design

I don't expect anyone to design the BBB circuit on their own, but some may find it helpful to see how I created the design. The beagleboneblack.brd file is in the Ch15 directory and the design process required seven steps:

1. I designed the schematic presented in Chapter 14 and used the Generate board tool to create the board design.
2. I downloaded the Allegro board file from the BeagleBone wiki (http://elinux.org) and examined it using the Allegro Free Physical Viewer from Cadence.
3. In EAGLE's board editor, I set the board's perimeter to 3.4" by 2.1".
4. Following the Allegro board file, I did my best to place the component packages on the top and bottom layers of the board.
5. I set the design rules for the board in the editor's DRC dialog. I'll present the exact values shortly.
6. I defined polygons on the second and fifth layers to serve as copper pours. As explained earlier, the board's second layer is the ground plane and the fifth layer is the power plane.
7. I routed the design on all six layers of the circuit board. I started with the top and bottom layers and then routed the signals on the third and fourth layers.

Most of these steps are straightforward and have been discussed earlier in this book. But in this section, I'd like to go into greater detail with regard to the design rules. I'd also like to show how the AM3359 was routed.

The online design files for the BBB are not officially supported. In my design, I did my best to follow the online files, but due to unavailable components, I strongly recommend against using my board design for any purpose other than learning.

15.4.1 Design Rules

Given the high component density of the BBB, the design rules require more precise fabrication than the Femtoduino circuit presented earlier. Table 15.2 lists the design rules I set in the DRC dialog.

Table 15.2
Design Rules for the BeagleBone Black Board Design

DRC Dialog Tab	Field	Value
Layers	Setup string	(1+2*3+4*5+16)
Clearance	Wire-Wire	2 mil
	Wire-Pad	2 mil
	Pad-Pad	2 mil
	Wire-Via	2 mil
	Pad-Via	2 mil
	Via-Via	2 mil
Distance	Copper/dimension	4 mil
	Drill/hole	4 mil
Sizes	Minimum width	4.75 mil
	Minimum drill	4.75 mil
	Minimum micro via	1 mm
	Minimum blind via ratio	0.5
Restring	Top pad - min	2 mil
	Top pad - max	6 mil
	Inner pad - min	2 mil
	Inner pad - max	6 mil
	Bottom pad - min	2 mil
	Bottom pad - max	6 mil
	Outer via - min	2 mil
	Outer via - max	6 mil
	Inner via - min	2 mil
	Inner via - max	6 mil
	Outer micro via - min	2 mil
	Outer micro via - max	6 mil
Shapes	Roundness - min	0 mil
	Roundness - max	0 mil
Supply	Thermal isolation	4 mil
Masks	Stop - min	4 mil
	Stop - max	4 mil
	Cream - min	0 mil
	Cream - max	0 mil
	Limit - min	0 mil

This table doesn't list the thicknesses of the board's layers. This information is presented in Table 15.1.

15.4.2 Routing the AM3359's Signals

The BBB's board design is so complex that displaying its six layers wouldn't be helpful from a learning perspective. Instead, this discussion focuses on the AM3359 and its routing. If you'd like to see the complete design, the beagleboneblack.brd file can be found in the Ch15 directory.

The AM3359 is the largest integrated circuit in the BBB and Chapter 14 explained its processing functions. Its package is a New Fine-Pitch Ball Grid Array (NFBGA) with 324 pins arranged in an 18 x 18 square. Routing its signals was, without question, the most difficult task in the process of designing the BBB circuit.

The AM3359 component is soldered to the top layer of the BBB and Figure 15.14 depicts its package and routing. For the most part, the device relies on escape vias to route signals and the vias are positioned according to the pad's quadrant. For example, if a pad is in the upper left (northwest quadrant), its via will usually be positioned to the upper left of the pad.

One notable exception is that pads located on or near the AM3359's perimeter do not have escape vias. These pads are routed away from the AM3359 on Layer 1.

Inside the package's perimeter, many of the escape vias are not positioned according to their quadrant. If you look closely, you may notice a pattern. The vias are positioned to allow small components to be soldered to the bottom layer beneath the AM3359.

Figure 15.15 shows what the bottom layer looks like. As shown, there are more than 20 components directly beneath the AM3359. On the top layer, the escape vias are arranged to ensure that they don't interfere with the components' placement.

All the devices beneath the AM3359 are capacitors with capacitances of 0.1 microfarads. Most of them are decoupling capacitors and serve to conduct noise and transient signals from the power supply (VDD_CORE) to ground.

Comparing Figure 15.15 to Figure 15.14, it's clear that most of the vias aren't connected to traces on the top or bottom layers. Many of them are directly connected to ground (Layer 2) or power (Layer 5).

In addition, the high-speed and hard-to-route signals are connected to traces on the central signal layers, Layers 3 and 4. To see these traces, I recommend that you open the beagleboneblack.brd design in EAGLE and use the Layer settings tool to display Layers 3 and 4.

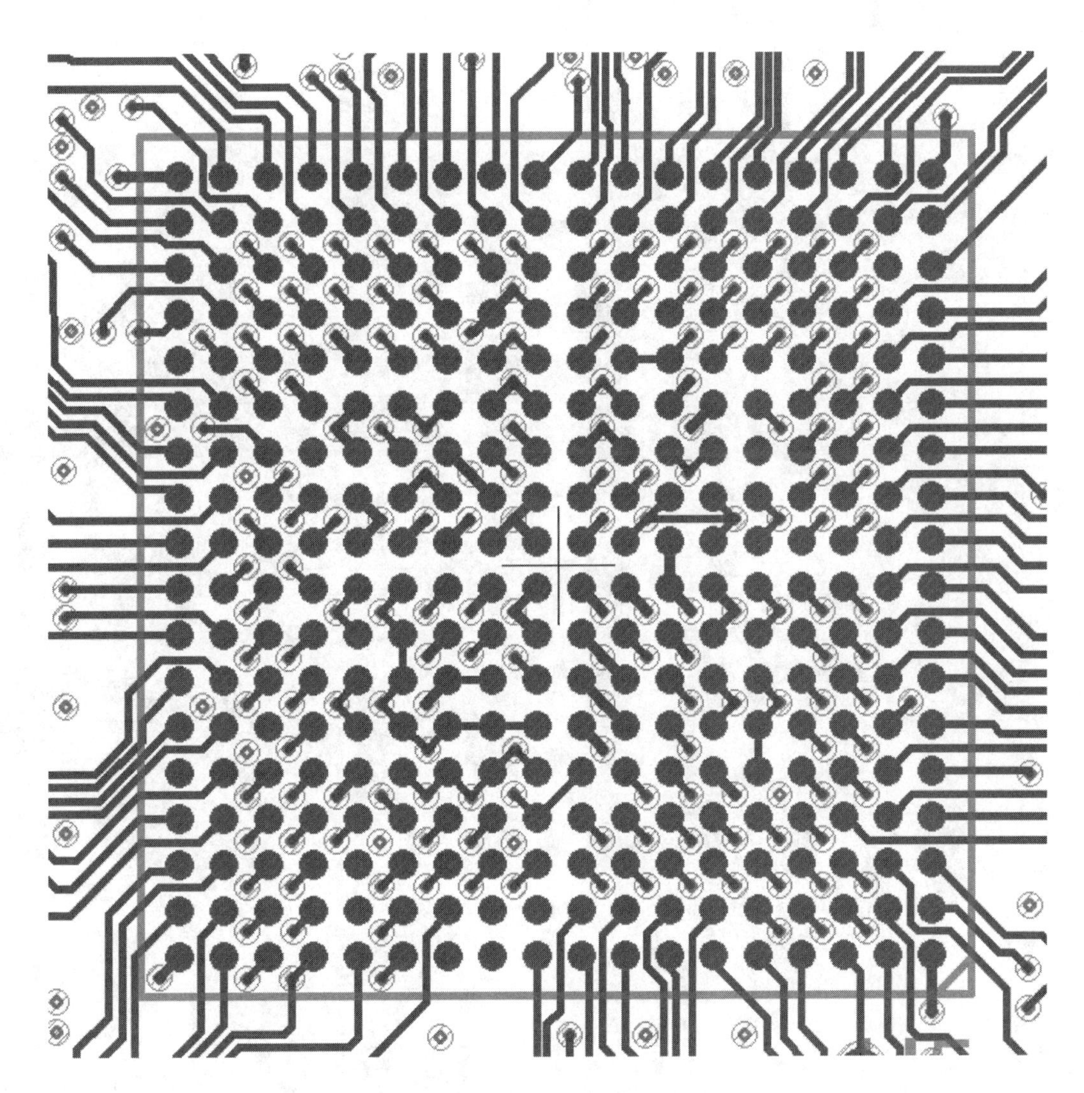

Figure 15.14: AM3359 Routing on the Top Layer

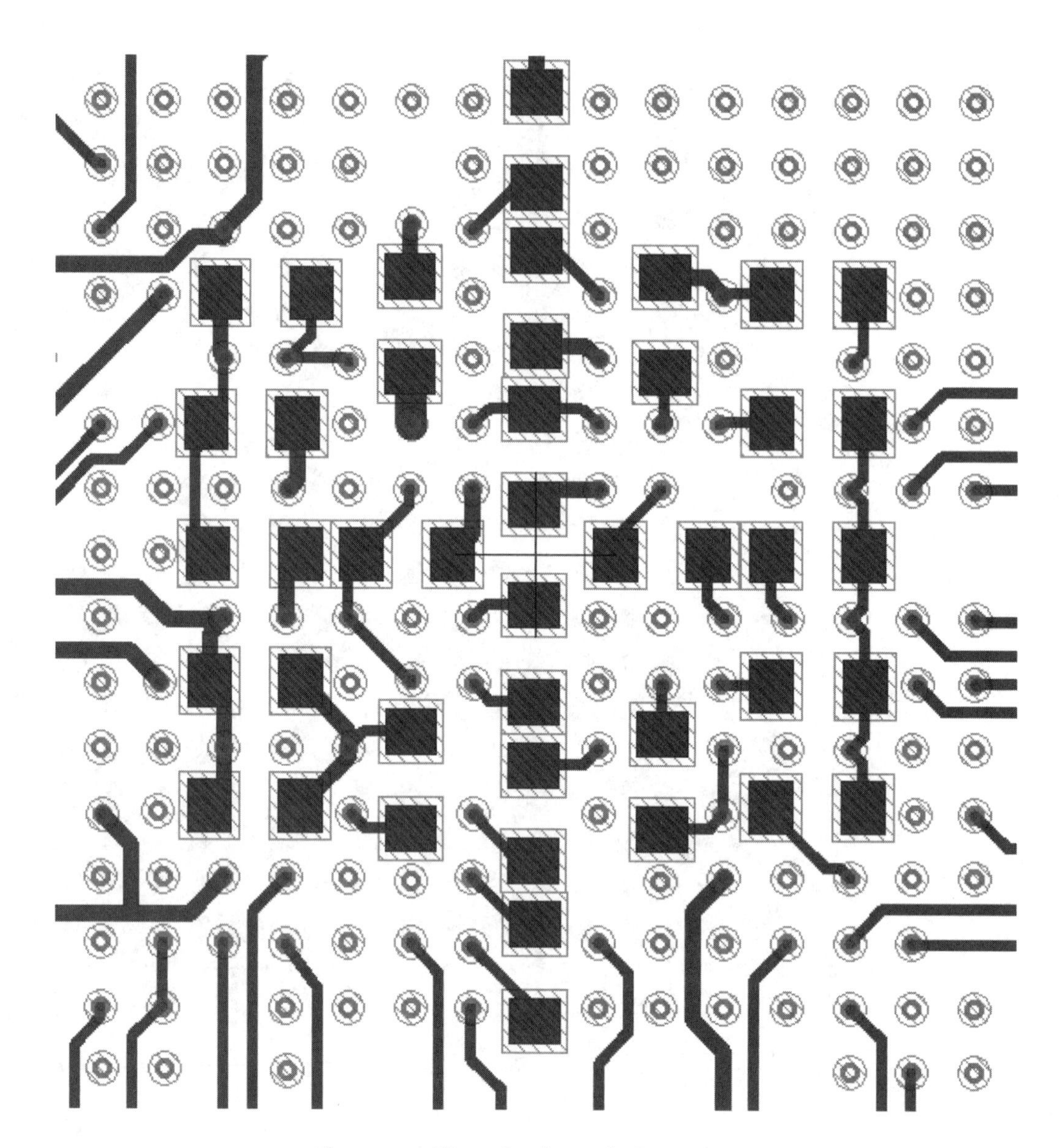

Figure 15.15: AM3359 Routing on the Bottom Layer

15.5 Conclusion

I've heard from many engineers that EAGLE is fine for simple designs like Arduino boards, but isn't suitable for large-scale professional designs. In writing Chapter 14 and this chapter, my goal has been to demonstrate that EAGLE provides all the features needed to design the BeagleBone Black, a single-board computer that any professional would be proud of.

The key to designing large-scale circuits is to take advantage of automation at every opportunity. In the case of EAGLE, this means running ULPs to perform repetitive tasks like drawing symbols and packages. In this chapter, I've explained how I used ULPs to generate the BGA packages for the BBB and create escape vias for the packages in the board design. I hope the message is clear: ULPs save time and reduce the potential for error.

This chapter has also discussed stackups, particularly the six-layer stackup used by the BBB. There are many variables that go into choosing a stackup for a circuit board, and two primary variables are connectivity and electromagnetic interference. If you remember nothing else about mutlilayer stackups, remember that every signal layer should be adjacent to a ground plane or power plane. This is a very useful rule of thumb.

Routing is a difficult task in any nontrivial circuit design, and it's particularly difficult when dealing with high-frequency signals. Trace length becomes a significant concern, and in these instances, the tool to use is the Meander tool. If the signals form a differential pair, they can be routed simultaneously if their names end in _P and _N.

This chapter and the preceding chapter have presented the BeagleBone Black in extraordinary detail. If you'd like to learn more, I recommend the Wiki at http://elinux.org. I'd like to thank BeagleBoard.org for making the design files available. I'd also like to thank you, gentle reader, for continuing so far into this book. I hope you've learned a great deal and I wish you the utmost of success with your circuit board designs.

Appendix A

EAGLE Library Files

EAGLE provides access to electrical components through libraries. Each library corresponds to a *.lbr file in the top-level lbr directory. Most library files contain components for a single manufacturer. For example, the zilog.lbr library file contains components from Zilog, Inc.

Library files are plain text files that can be easily read or written to. Their content is formatted according to XML, the eXtensible Markup Language. XML is popular in the world of software development because it provides a structured, hierchical means of storing data.

For most circuit designers, there's no need to open the lbr directory or look at any of its library files. But if you're interested in accessing components outside of EAGLE, it's important to understand how these XML documents are structured. The goal of this appendix is to present this information so that you can read and modify existing libraries or create new libraries of your own.

A.1 Anatomy of a Library File

If you examine files in the lbr directory, you'll see that most of them have the same overall structure, as shown in Listing A.1.

Listing A.1: Structure of a Generic EAGLE Library

```
<?xml version="1.0" encoding="utf-8"?>
<!DOCTYPE eagle SYSTEM "eagle.dtd">
<eagle version="6.5">
   <drawing>

      <settings>
         <setting.../>
      </settings>

      <grid.../>

      <layers>
         <layer.../>
      </layers>

      <library>

         <description>...</description>

         <packages>...</packages>

         <symbols>...</symbols>

         <devicesets>...</devicesets>

      </library>

   </drawing>
</eagle>
```

Each library file begins with the same three lines: an `<xml>` tag that identifies the XML version, a `DOCTYPE` tag, and an `<eagle>` tag that serves as the root tag of the document.

The `DOCTYPE` tag identifies a file called eagle.dtd as the document type definition (DTD) of EAGLE's specific type of XML. You can find this in the doc folder of EAGLE's installation directory, and it defines the full hierarchy of tags in *.lbr files.

If you're familiar with XML, you may know that XML schemas are more commonly defined using XML schema document (*.xsd) files than document type definitions (*.dtd). Both types of files serve the same purpose: to define the elements and attributes inside XML document types.

For example, the following markup in eagle.dtd defines the content of the `<device>` element:

```
<!ELEMENT device (connects?, technologies?)>
<!ATTLIST device
          name          %String;        ""
          package       %String;        #IMPLIED
          >
```

This states that the <device> element contains two optional subelements, <connects> and <technologies>. Each subelement name is followed by ?, which means that the subelement may occur 0 or 1 times. The <device> element has two attributes, name and package, which are both given as strings.

The top-level element of every library file is <eagle>, and its most important subelement is <drawing>. For the purpose of this appendix, the primary subelement of <drawing> is <library>, which identifies the characteristics of a component library. Most of this appendix focuses on this tag, but the first section discusses the first three tags: <settings>, <grid>, and <layers>.

A.2 Settings, Grid, and Layers

Figure A.1 shows the subelements inside the <drawing> element. The first three, <settings>, <grid>, and <layers>, contain top-level data that applies to the entire library. This section discusses each of them in detail.

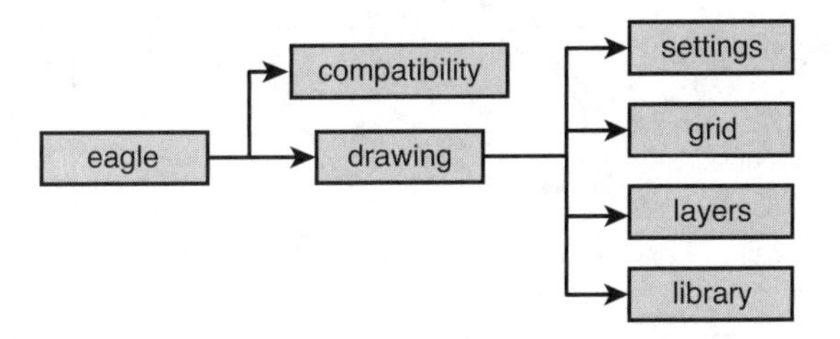

Figure A.1: XML Hierarchy for the Drawing Element

A.2.1 Settings

The <settings> tag is the simplest of the three initial elements. It contains <setting> subelements that configure how text in the library is displayed. These subelements can take one of two attributes:

- alwaysvectorfont—Can be set to yes or no. If set to yes, text will be printed using vector fonts.
- verticaltext—Can be set to up or down. If set to up, vertical text will be printed from bottom to top. If set to down, vertical text will be printed from top to bottom.

As an example, the following XML configures text to be printed with vector fonts, and vertical text will be printed from top to bottom:

```
<settings>
   <setting alwaysvectorfont="yes" />
   <setting verticaltext="down" />
</settings>
```

A.2.2 Grid

The `<grid>` tag identifies how the library's components interact with EAGLE's grid. This sets values for a series of attributes, as shown in Table A.1.

Table A.1

Attributes of the <grid> Tag

Attribute	Value	Default	Description
`distance`	Positive real number	—	Distance between grid markings
`unitdist`	`mil`, `mic`, `inch`, or `mm`	—	Units of grid distance
`unit`	`mil`, `mic`, `inch`, or `mm`	—	Base unit of the grid
`style`	`lines` or `dots`	`lines`	Synchronize data
`multiple`	Integer	`1`	Identifies the grid distance as a multiple of grid units
`display`	`yes` or `no`	`no`	Sets whether the grid should be displayed
`altdistance`	Positive real number	—	Alternative distance between grid markings
`altunitdist`	`mil`, `mic`, `inch`, or `mm`	—	Alternative units of grid distance
`altunit`	`mil`, `mic`, `inch`, or `mm`	—	Alternative base unit

The `unitdist` and `unit` members have similar names but serve different roles. `unitdist` identifies the units used by the primary grid spacing. `unit` identifies the units used for the displayed text, which is usually set with the alternate grid. Chapter 3, "Designing a Simple Circuit," and Chapter 4, "Designing the Femtoduino Schematic," discuss grid settings in greater detail.

A.2.3 Layers

The components in a library must all use the same set of layers to identify which regions they occupy. Layers are identified by number, and the `<layers>` tag creates the relationship between layer numbers and layer names. Table A.2 lists its attributes.

Table A.2

Attributes of the <layers> Tag

Attribute	Value	Default	Description
`number`	Positive integer	—	ID number for the layer
`name`	String	—	Name of the layer
`color`	Positive integer	—	Color of components assigned to the given layer
`fill`	Positive integer	—	Fill style
`visible`	`yes` or `no`	`yes`	Whether the layer is visible in the editor
`active`	`yes` or `no`	`yes`	Whether the layer is displayed in the editor's dialog

A.3 Overview of Library Elements

The `<library>` element contains information about a single EAGLE library and is the main focus of this appendix. It contains four subelements:

- `<description>`—The text describing the library
- `<devicesets>`—The library's device hierarchy
- `<packages>`—The packages corresponding to the devices
- `<symbols>`—Geometric primitives used by the library components

The last three names are plural, and each contains subelements in the singular. That is, a `<devicesets>` element may contain multiple `<deviceset>` subelements, a `<packages>` element may contain multiple `<package>` subelements, and a `<symbols>` element by contain multiple `<symbol>` subelements. Figure A.2 displays these relationships graphically.

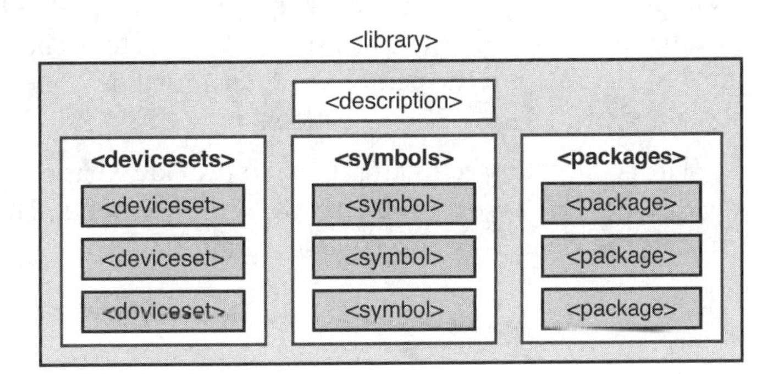

Figure A.2: Elements of an EAGLE Library File

An example will clarify how these work. Figure A.3 displays the library from LSI Computer Systems. On the right, the description is given as `LSI Computer Systems, Inc.` The XML file for this library is lsi-computer-systems.lbr, and if you find the `<library>` element, you'll see that the `<description>` subelement contains the text `LSI Computer Systems, Inc.`

As shown in the figure, the LSI library provides six devices: LS7766DH-TS, LS7766DO-S, LS7766DO-TS, LS7766SH-TS, LS7766SO-S, and LS7766SO-TS. In this case, each device name has two parts: the *deviceset* is the device name before the hyphen and the hyphenated portion is the *package designation.*

NOTE Some libraries do not use hyphens in their package designations. Further, many devices have an additional designation that identifies their technology.

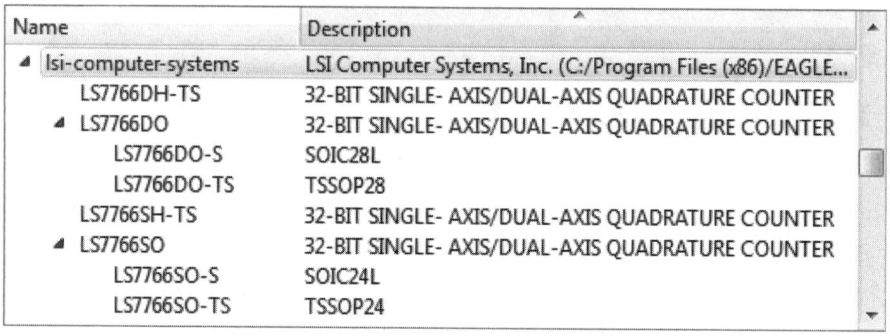

Name	Description
◢ lsi-computer-systems	LSI Computer Systems, Inc. (C:/Program Files (x86)/EAGLE...
LS7766DH-TS	32-BIT SINGLE- AXIS/DUAL-AXIS QUADRATURE COUNTER
◢ LS7766DO	32-BIT SINGLE- AXIS/DUAL-AXIS QUADRATURE COUNTER
LS7766DO-S	SOIC28L
LS7766DO-TS	TSSOP28
LS7766SH-TS	32-BIT SINGLE- AXIS/DUAL-AXIS QUADRATURE COUNTER
◢ LS7766SO	32-BIT SINGLE- AXIS/DUAL-AXIS QUADRATURE COUNTER
LS7766SO-S	SOIC24L
LS7766SO-TS	TSSOP24

Figure A.3: An Example Library in the ADD Dialog

If you look through the XML file, you'll see that the `<devicesets>` element contains four `<deviceset>` elements. Their names are LS7766DH, LS7766DO, LS7766SH, and LS7766SO. In the schematic editor, each deviceset corresponds to a single symbol. This is why the `<symbols>` subelement contains four `<symbol>` elements.

Similarly, the `<packages>` subelement contains 13 `<package>` elements, and six of their names are SOIC24L, SOIC28L, TSSOP24, TSSOP28, TSSOP38, and TSSOP48. It's important to note that the package designations (-S, -TS) have different names than the package names (SOIC24L, TSSOP24, etc.). The relationship between the designations and the names are determined by `<device>` elements, which are contained inside `<deviceset>` elements.

The goal of this section has been to provide a basic understanding of the `<symbols>`, `<packages>`, and `<devicesets>` elements. The remainder of this section presents the `<symbols>` , `<packages>`, and `<devicesets>` in detail.

A.4 Symbols

A symbol defines a shape in the schematic editor. In general, each device in a library is associated with a single symbol. But this isn't always the case. If a device contains an array of elements, each component in the array will be associated with a symbol. For example, if a device contains four operational amplifiers, the device will be associated with four symbols instead of one.

The `<symbols>` element may contain one or more `<symbol>` subelements. The only attribute of `<symbol>` is `name`, which uniquely identifies the symbol. This ID will be used by `<device>` elements to associate the symbol with a device. `<device>` elements will be discussed in a later section.

A symbol's appearance has three aspects: its body, its pins, and text. The body's shape can be defined using lines, rectangles, circles, or polygons. The body's text provides information about the component. For example, if the text is set to >NAME, EAGLE will replace the text with the component's name in the schematic editor.

To specify a shape in XML, the `<symbol>` element can contain a number of possible subelements, such as `<wire>`, `<circle>`, and `<polygon>`. The `<pin>` subelement defines the location of its pins. Table A.3 lists each of these subelements along with their attributes.

Table A.3

Subelements of the <symbol> Element

Subelement	Attributes	Description
`description`	—	Text description of the symbol
`text`	`x, y, size, font, ratio, rot, align, layer`	Text added to the schematic
`wire`	`x1, y1, x2, y2, width, extent, style, curve, cap, layer`	Draw a line from (x1, y1) to (x2, y2)
`rectangle`	`x1, y1, x2, y2, rot, layer`	Draw a rectangle with opposite corners at (x1, y1) and (x2, y2)
`circle`	`x, y, radius, width, layer`	Draw a circle with center (x, y) and radius
`polygon`	Contains one or more `<vertex>` elements Attributes: `width`, `pour`, `spacing`, `rank`, `isolate`, `orphans`, `thermals`, `layer`	Draw a polygon between the given vertices
`pin`	`name, x, y, visible, length, direction, function, swaplevel, rot`	Set an input/output terminal for the schematic

For integrated circuits, symbols are usually defined with four wires that set the rectangular perimeter, text that defines the IC's name and value, and pins along the perimeter. The best way to understand these elements and their attributes is to look an example symbol and compare it to its `<symbol>` element. Figure A.4 presents the symbol for the 74LS113N, a dual J-K negative-edge triggered flip-flop with preset.

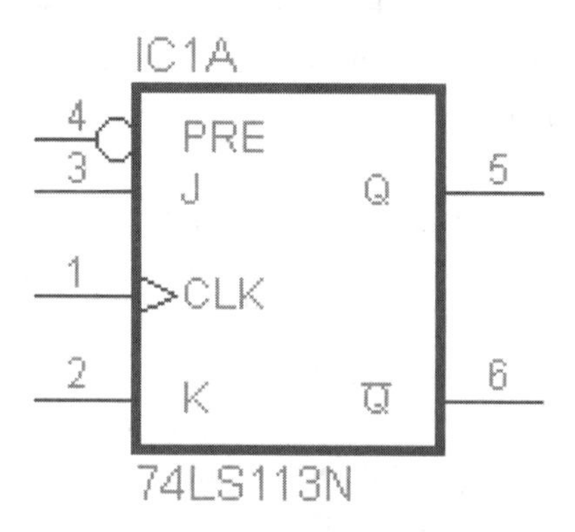

Figure A.4: Symbol of the 74LS113N

The 74xx-us.lbr library contains the definition for the 74LS113N's symbol. The following markup shows how it's defined:

```
<symbol name="74113">
   <wire x1="-7.62" y1="-7.62" x2="7.62" y2="-7.62" width="0.4064"
         layer="94"/>
   <wire x1="7.62" y1="-7.62" x2="7.62" y2="10.16" width="0.4064"
         layer="94"/>
   <wire x1="7.62" y1="10.16" x2="-7.62" y2="10.16" width="0.4064"
         layer="94"/>
   <wire x1="-7.62" y1="10.16" x2="-7.62" y2="-7.62" width="0.4064"
         layer="94"/>
   <text x="-7.62" y="10.795" size="1.778" layer="95">&gt;NAME</text>
   <text x="-7.62" y="-10.16" size="1.778" layer="96">&gt;VALUE</text>

   <pin name="CLK" x="-12.7" y="0" length="middle" direction="in"
        function="clk"/>
   <pin name="K" x="-12.7" y="-5.08" length="middle" direction="in"/>
   <pin name="J" x="-12.7" y="5.08" length="middle" direction="in"/>
   <pin name="PRE" x="-12.7" y="7.62" length="middle" direction="in"
        function="dot"/>
   <pin name="Q" x="12.7" y="5.08" length="middle" direction="out"
        rot="R180"/>
   <pin name="!Q" x="12.7" y="-5.08" length="middle" direction="out"
        rot="R180"/>
</symbol>
```

The `<symbol>` element contains three types of subelements. The symbol's shape is a rectangle formed from four `<wire>` elements. The labels in the schematic are set by the `<text>` elements and the pins of the symbol are configured with the `<pin>` elements. This section examines these `<wire>`, `<text>`, and `<pin>` elements and then presents the `<polygon>` and `<frame>` elements.

A.4.1 Wires

The shape of the 74LS113N is defined using four lines, each represented by a `<wire>` element in the `<symbol>` element. Most of the `<wire>` attributes are easy to understand. The `x1` and `y1` attributes define the location of the line's initial point. The `x2` and `y2` attributes define its ending point. `width` identifies the line's width in pixels.

The `cap` attribute identifies the shape of the line's starting and ending points. If `cap` is set to `round`, the ends of the line are rounded. If set to `flat`, the ends of the line are flattened. Ends are rounded by default.

Lines can be made dashed or dotted by setting the `style` attribute. The default value is `continuous`, which means each line is solid from start to end. For a dashed line, set `style` to `longdash` or `shortdash`. For a line with dashes and dots, set style to `dashdot`.

The curvature of a circular arc is given by 1/r, where r is the circle's radius. By default, each line in the editor is straight, which means its curvature equals 0. In the schematic editor, line curvature is set with the `curve` attribute, which can be set to any real value between –359.0 and 359.0. This makes it possible to create arcs in the editor.

The last attributes of `<wire>` that need to be mentioned are `layer` and `extent`. Each of the shapes in Table A.3 have a `layer` attribute that define which board layer they're intended to be drawn for. But the `extent` attribute of `<wire>` sets two layers—the highest layer for the wire and the lowest. This attribute is set to a string, so if the highest layer is LayerA and the lowest layer is LayerB, `extent` should be set to `"LayerA-LayerB"`.

A.4.2 Text

In Figure A.4, the symbol has two labels: one on the top and one on the bottom. These are set by the symbol's `<text>` element, whose attributes are straightforward. The `x` and `y` attributes set the position of the text and `size` sets its size. The `font` attribute can be set to `vector`, `proportional`, or `fixed`. The `align` attribute can be set to `bottom-left`, `bottom-center`, `bottom-right`, `center-left`, `center`, `center-right`, `top-left`, `top-center`, or `top-right`.

The body text between `<text>` and `</text>` will be displayed in the schematic editor. If this text is set to `>NAME`, the editor will display the symbol's unique identifier, such as U1 or R5. If the body text is `>VALUE`, the editor will display the device name. This is why `IC1A` is drawn at the top of the symbol and `74LS113N` is printed below.

EAGLE refers to these text variables as *placeholders*. Each placeholder starts with the `>` character and there are 11 of them in total:

- `>CONTACT_XREF`—The cross-reference of a contact, possibly for switches. By default, this provides the page number, column, and row.
- `>DRAWING_NAME`—The drawing name.
- `>GATE`—The gate instance.

- `>LAST_DATE_TIME`—Last time/date the design was modified.
- `>NAME`—The name of the package and symbol.
- `>PART`—The name of the symbol.
- `>PLOT_DATE_TIME`—Time/date the plot was created.
- `>SHEET`—Current sheet number for circuit diagrams.
- `>SHEETNR`—Current sheet number for circuit diagrams and symbols.
- `>SHEETS`—Total number of sheets.
- `>VALUE`—Component value and type.

In each case, the editor automatically inserts the value in the drawing. Note that many of the placeholders display similar pieces of information.

A.4.3 Pins

The wires leading into and out of the symbol are set using `<pin>` elements. Each `<pin>` must have a name (`name`) and a location (`x` and `y`). The `length` attribute defines the length of the pin and can be set to `short` (0.1 in), `middle` (0.2 in), `long` (0.3 in), or `point` (0.0 in). The default value is `long`.

The `direction` attribute, which sets the pin's logical direction, is more involved. It can take any of the following values:

- `in`—Input pin
- `out`—Output pin
- `io`—Input/output pin
- `pwr`—Power
- `sup`—Supply
- `oc`—Open-collector/open-drain
- `pas`—Passive
- `hiz`—High-impedance
- `nc`—No connection

Most of the pin directions are easy to understand, but I want to clarify the distinction between a supply pin (`sup`) and a power pin (`pwr`). Schematics contain supply elements like VDD, VCC, and GND. When a wire is connected to a supply, its signal takes the supply's name. This is why you don't have to set names for wires connected to voltage supplies and grounds.

If a component's pin is intended to be connected to a voltage supply or ground, it should be given the `pwr` direction. This tells EAGLE that the pin should be connected to a VCC or a GND component. For example, the V+ pin of an operational amplifier should be set to pwr so that EAGLE will know that it's intended to be connected to a sup pin.

In the symbol depicted in Figure A.4, the four pins on the left receive input and the two pins on the right provide output. This is shown in the <symbol> declaration, in which the first four <pin> elements have their direction attributes set to in and the last two have their direction attribute set to out.

The <pin> element contains three more attributes that need to be mentioned: swaplevel, visible, and rot. The swaplevel is an integer, and if two pins have the same value, their positions can be swapped in the schematic editor. Swapping pins is helpful because it allows circuit designers to draw cleaner schematics. But for the 74113 symbol, each pin has the default swaplevel, which is 0. A pin with swaplevel set to 0 cannot be swapped with another.

The visible attribute can be set to off, pad, pin, or both. This determines how the symbol's pins are labeled. In Figure A.4, each pin has two labels: a number on the outside (1 through 6) and a character sequence on the inside (CLK, J, K, and so on). Each character sequence is set by the name attribute of <pin>, and if visible is set to pin, this is the only label that will be shown in the editor. The numbers correspond to pad labels, and if visible is set to pad, these will be the only labels shown. The pins in the 74113 symbol have the default visibility, both, which means the editor displays pin and pad labels.

Most of the subelements in Table A.3 have an attribute called rot, which sets the element's initial angle. This is given in degrees, so rot can be set to 0 through 360.

A.4.4 Polygons and Vertices

The symbol in Figure A.4 doesn't contain any polygons, but if a device has an irregular shape, its definition should contain a <polygon> element, which is the only element in Table A.3 that has subelements. The points of a <polygon> are defined by <vertex> elements, and each <vertex> has three attributes:

- x—The x coordinate of the vertex point
- y—The y coordinate of the vertex point
- curve—The curvature of the line from this vertex to the next

As discussed earlier, curvature defines a circular arc, and curve should be set to a real value between –359.9 and 359.0. This value equals 1/r, where r is the circle's radius. By default, curve is set to 0.0, which means the line is straight.

In addition to setting <vertex> elements, each <polygon> has attributes that can be configured. We've already seen two of them: width sets the width of the polygon's lines and layer identifies the layer upon which the polygon should be drawn.

When board designers want to fill regions of a board with copper, they use polygons to define the shape. For this reason, <polygon> has a number of attributes that we've never encountered before. These are given as follows:

- rank—Real value that sets priority among overlapped polygons. Lower rank means higher priority. That is, if a polygon has high rank, its shape will be subtracted from polygons with low rank to prevent overlapping.
- isolate—Real value that sets the minimum distance between the polygon and other objects on the board.
- pour—Configures how the polygon should be filled with copper. If set to solid, the copper will fill the entire area. If set to hatch, copper will be patterned in a cross-hatched grid. If set to cutout, copper will be cut away to prevent overlapping.
- spacing—If pour is set to hatch, spacing identifies the spacing of the grid lines used to determine how copper is to be poured.
- orphans—Identifies whether the polygon may contain regions that aren't connected to the polygon's signal, may be set to yes or no. The default value is no.
- thermals—Configures the nature of the connections between the polygon's pads. If set to yes, the pads will be connected using thermal symbols. If set to no, all the pads will be connected to the copper plane.

As an example, the following markup defines a <symbol> whose shape is determined by a <polygon> element.

```
<symbol name="EXAMPLE_POLYGON">
   <polygon width="0.254" layer="94">
      <vertex x="7.62" y="0"/>
      <vertex x="2.54" y="-5.08"/>
      <vertex x="0" y="-2.54"/>
      <vertex x="-2.54" y="-5.08"/>
      <vertex x="-7.62" y="0"/>
      <vertex x="-2.54" y="5.08"/>
      <vertex x="0" y="2.54"/>
      <vertex x="2.54" y="5.08"/>
   </polygon>
</symbol>
```

Figure A.5 shows what the resulting symbol looks like.

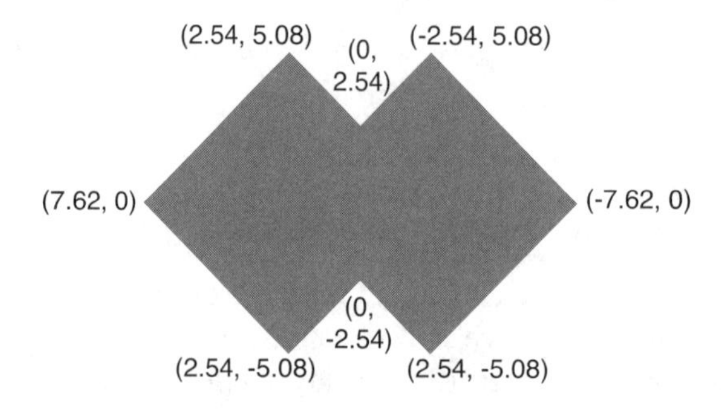

Figure A.5: A Symbol Consisting of a Polygon

The shapes created as polygons, rectangles, and circles are filled in. This contrasts with shapes bordered by wires, which aren't. This is why wires are usually employed to create symbols instead of other shapes.

A.4.5 Frames

Unlike the other elements in Table A.3, `<frame>` elements do not represent shapes of components. Instead, frames are used for labeling purposes. More specifically, a frame creates a rectangle that surrounds the design that provides labels for names of the designer and sheet.

EAGLE provides many example frames in the frames.lbr library. Figure A.6 shows one of the simplest frames, called DINA5_L.

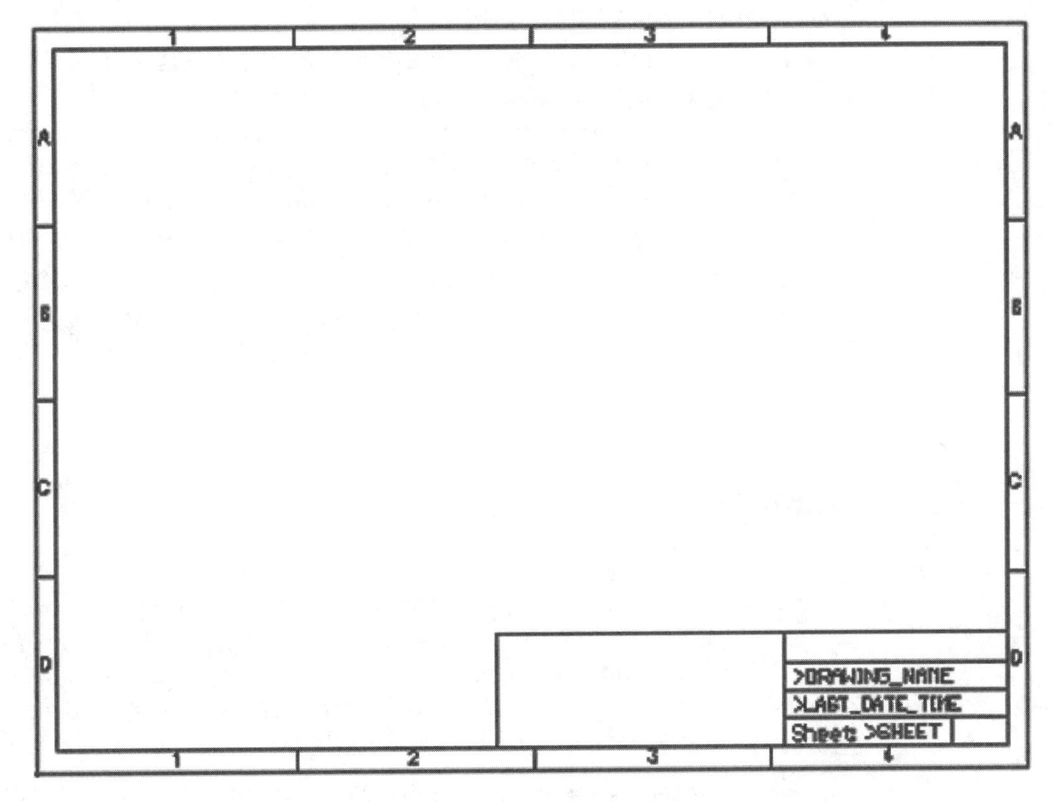

Figure A.6: A Frame with Four Borders

The definition of the DINA5_L component can be found in the frames.lbr library file. The symbol in Figure A.6 is A5L-LOC, whose markup (abridged) is given as follows:

```
<symbol name="A5L-LOC">
   <wire x1="85.09" y1="3.81" x2="85.09" y2="24.13" width="0.1016"
         layer="94"/>
   ...

   <text x="140.97" y="15.24" size="2.54" layer="94"
         font="vector">&gt;DRAWING_NAME</text>
   ...

   <frame x1="0" y1="0" x2="184.15" y2="133.35"
         columns="4" rows="4" layer="94"/>
</symbol>
```

The markup for the frame symbol consists of a series of `<wire>` elements, a series of `<text>` elements, and a `<frame>` element. The `x1` and `y1` attributes set the location of the frame's lower-left corner and the `x2` and `y2` attributes set the location of the frame's upper-right corner.

The border surrounding the frame identifies the sheet's grid. Numerical coordinates identify horizontal locations (columns) and alphabetic coordinates identify vertical positions (rows). In the `<frame>` element, the `rows` attribute identifies the number of rows in the grid and the `columns` attribute identifies the number of columns. In the markup for A5L-LOC, `rows` and `columns` are both set to 4. This explains why, in Figure A.6, the numbers run from 1–4 and the letters run from A–D.

Grid borders are added by default, but they can be disabled by setting attributes of `<frame>`: `border-top`, `border-bottom`, `border-left`, and `border-right`. All these can be set to `yes` or `no`, and if any of them are set to `no`, the corresponding border won't be displayed.

A.5 Packages

Inside `<drawing>`, the `<packages>` element contains one or more `<package>` elements. Just as a `<symbol>` represents a component in the schematic editor, a `<package>` represents a component in the board editor. Therefore, a package's shape should be set equal to the exact shape of the actual device. Any difference between the package's geometry and the device's geometry will produce errors in the circuit board.

Like symbols, each package must have a unique name, and this is specified by the `name` attribute. Also like symbols, the shape of a package shape is defined using a set of primitive shapes. Each primitive shape corresponds to a subelement of `<package>`, and Table A.4 lists them all.

Table A.4

Subelements of the <package> Element

Subelement	Attributes	Description
description	—	Text description.
text	x, y, size, font, ratio, rot, align, layer	Text added to the schematic.
wire	x1, y1, x2, y2, width, extent, style, curve, cap, layer	Draw a line from (x1, y1) to (x2, y2).
rectangle	x1, y1, x2, y2, rot, layer	Draw a rectangle with opposite corners at (x1, y1) and (x2, y2).
circle	x, y, radius, width, layer	Draw a circle with center (x, y) and radius.
polygon	Contains one or more <vertex> elements Attributes: width, pour, spacing, rank, isolate, orphans, thermals, layer	Draw a polygon between the given vertices.
frame	x1, y1, x2, y2, columns, rows, layer, border-left, border-top, border-right, border-bottom	Create a rectangular frame for the symbol.
hole	x, y, drill	Create a hole at the given position.
pad	x, y, name, drill, diameter, shape, stop, thermals, first, rot	Set a through-hole pad at the given location.
smd	x, y, dx, dy, roundness, stop, thermals, cream, rot, layer	Set an SMD pad at the given location.

Most of the entries in this table are similar to the entries in Table A.3. But the last three subelements—<hole>, <pad>, and <smd>—are new. <hole> identifies a circular region to be drilled. It accepts three attributes: x and y define the center of the hole and drill sets the hole's diameter in millimeters. For example, the following element defines a 2.5mm-diameter hole whose center is at (0, 0).

```
<hole x="0" y="0" drill="2.5"/>
```

As discussed earlier, the <symbol> element defines connections using pins. But pins are used only for connections in the schematic editor. The connection points of a <package> element are set using *pads*. Pads come in one of two types: through-hole pads and surface mount pads. These are represented by the last two subelements in Table A.4: <pad> and <smd>.

NOTE EAGLE provides support for through-hole and surface mounted pads only. If you want to design circuits with ball grid array (BGA) components, you need to create pads specifically for BGA devices. This is explained in Chapter 8, "Creating Libraries and Components."

A.5.1 Through-Hole Pads

If a device's terminal wires require drilling through the board, the device's package connections should be defined as through-hole pads. An example will show what these pads look like and how they're defined. Figure A.7 illustrates the package of the AD648N device from Linear Technologies.

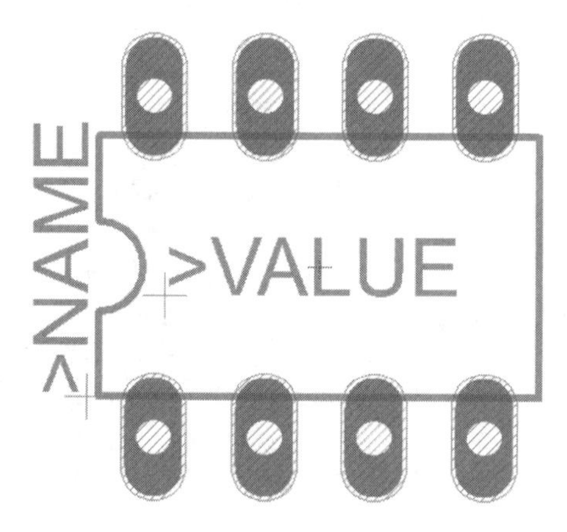

Figure A.7: Package with Eight Through-Hole Pads

The package's name is DIL08, and it uses through-hole pads for each of its eight connections. Inside linear.lbr, the eight <pad> definitions for DIL08 are given as follows:

```
<pad name="1" x="-3.81" y="-3.81" drill="0.8128" diameter="1.27"
     rot="R90"/>
<pad name="2" x="-1.27" y="-3.81" drill="0.8128" diameter="1.27"
     rot="R90"/>
<pad name="7" x="-1.27" y="3.81" drill="0.8128" diameter="1.27"
     rot="R90"/>
<pad name="8" x="-3.81" y="3.81" drill="0.8128" diameter="1.27"
     rot="R90"/>
<pad name="3" x="1.27" y="-3.81" drill="0.8128" diameter="1.27"
     rot="R90"/>
<pad name="4" x="3.81" y="-3.81" drill="0.8128" diameter="1.27"
     rot="R90"/>
<pad name="6" x="1.27"  y="3.81" drill="0.8128" diameter="1.27"
     rot="R90"/>
<pad name="5" x="3.81"  y="3.81" drill="0.8128" diameter="1.27"
     rot="R90"/>
```

The first four attributes are required in every <pad> element. name provides a unique identifier for the pad, x and y set the location of the hole to be drilled, and drill sets the minimum diameter of the drilled hole.

The actual diameter of the hole is specified by the diameter attribute. The default value is 0, which means the diameter will be computed by the design rules. In the example package, the minimum diameter is 0.8128mm and the actual diameter is 1.27mm.

The pads in Figure A.7 are round, but each pad's shape can be configured with the shape attribute. Its default value is round, but other values include square, octagon, long, and offset.

A.5.2 Surface Mount Pads

Surface mount devices (SMD) are soldered onto the surface of the circuit board and do not require holes. Most SMD pads are rectangular and face outward from the device. This is depicted in Figure A.8, which shows the package of the AD648D device from Linear Technologies.

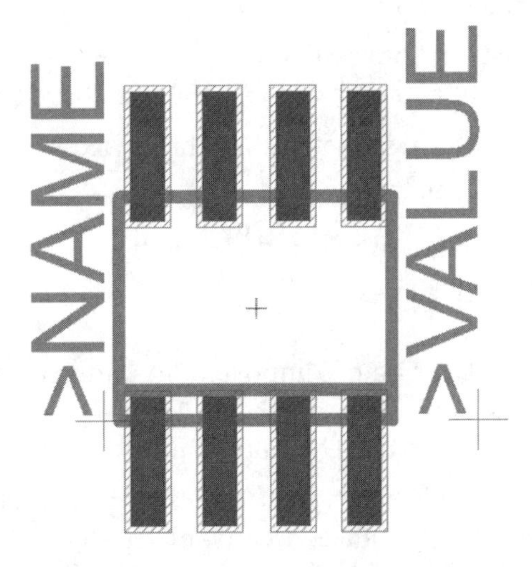

Figure A.8: Package with Eight SMD Pads

The package's name is SO08, and it uses SMD pads for each of its eight connections. Inside linear.lbr, the eight <smd> definitions for SO08 are given as follows:

```
<smd name="2" x="-0.635" y="-2.6" dx="0.6" dy="2.2" layer="1"/>
<smd name="7" x="-0.635" y="2.6"  dx="0.6" dy="2.2" layer="1"/>
<smd name="1" x="-1.905" y="-2.6" dx="0.6" dy="2.2" layer="1"/>
<smd name="3" x="0.635"  y="-2.6" dx="0.6" dy="2.2" layer="1"/>
<smd name="4" x="1.905"  y="-2.6" dx="0.6" dy="2.2" layer="1"/>
<smd name="8" x="-1.905" y="2.6"  dx="0.6" dy="2.2" layer="1"/>
<smd name="6" x="0.635"  y="2.6"  dx="0.6" dy="2.2" layer="1"/>
<smd name="5" x="1.905"  y="2.6"  dx="0.6" dy="2.2" layer="1"/>
```

As with the <pad> element, the x and y attributes set the pad's location. Similarly, dx and dy set the pad's dimensions in the x and y directions. Each pad is rectangular by default, but the shape can be changed with the roundness attribute, which accepts an integer value. The default value of roundness is 0, but increasing this value increases the roundness of the SMD pads.

In Figure A.8, SMD pads are positioned on the top and bottom of the device. But if a device has SMD pads on every side, it's important to rotate pads with the rot attribute. Also, if any pad doesn't need thermal connections to copper, set thermals to no.

A.6 Devicesets

In general, each EAGLE library is a collection of devices. A deviceset is a group of devices with the same architecture but different packages and technologies. For example, in Figure A.3, LS7766DO is the deviceset and the devices are LS7766DO-S and LS7766DO-TS.

In EAGLE library files, the <devicesets> element contains multiple <deviceset> subelements, and each <deviceset> can have three attributes:

- name—String identifier for the deviceset, such as LS7766DO
- prefix—String identifier for the device type, such as R, L, C, or IC
- uservalue—Boolean that specifies whether the user can assign a value to the device (yes or no)

The last attribute is important to understand. Some devices, such as resistors, capacitors, and inductors, accept a user-specified value. If this is the case, uservalue should be set to yes. The value can be displayed through the >VALUE placeholder.

For example, when you add resistors to a schematic, you assign a value to R, which identifies the resistance. If you look through resistor.dtd, you'll see that the resistor deviceset has uservalue set to yes and prefix set to R.

In addition to these attributes, the <deviceset> element has three subelements: <description>, <gates>, and <devices>. This is shown in Figure A.9.

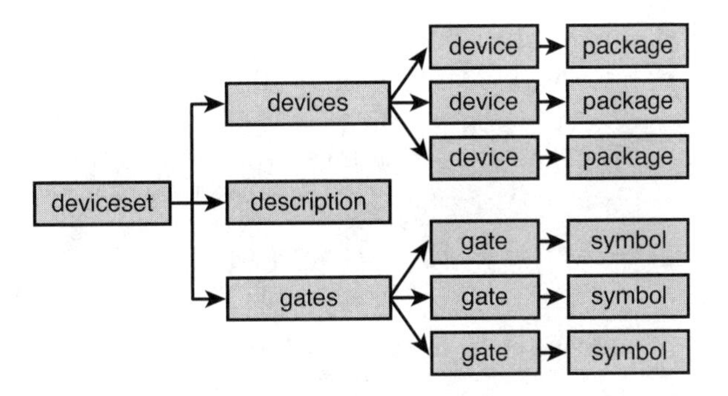

Figure A.9: Hierarchy of the DeviceSet Element

The `<description>` element provides a text description of the deviceset to be presented in the ADD dialog. The remainder of this section will focus on the `<gates>` and `<devices>` elements.

A.6.1 Gates

In EAGLE, a gate is a portion of a device that can be moved separately in the schematic editor. For example, Intersil's CA3081 device consists of seven transistors, and if you add this device to your schematic, you can move each transistor separately. Therefore, each of these transistors is referred to as a gate.

The `<gates>` element can contain multiple `<gate>` subelements, and each `<gate>` represents a gate in the schematic. `<gate>` doesn't contain any subelements of its own but has a number of attributes. Table A.5 lists each attribute along with its value type and default value.

Table A.5

Attributes of the <gate> Tag

Attribute	Value	Default	Description
`name`	string	–	Identifier of the gate
`symbol`	string	–	Identifier of the symbol element
`x`	real value	–	Initial x coordinate
`y`	real value	–	Initial y coordinate
`addlevel`	`must, can, next, request, always`	`next`	The gate's behavior when added to a schematic
`swaplevel`	integer	`0`	Identifies whether the gate can be swapped with another

`x` and `y` set the gate's location, and `symbol` identifies its physical geometry. This must be set equal to the name of a `<symbol>` element, which was discussed earlier.

An example will demonstrate how these attributes are configured. The CA3081 device consists of seven transistors that can be moved separately. This device can be found in the transistor.lbr library. The following markup shows how these seven gates, named A through G, are configured:

```
<gates>
   <gate name="A" symbol="NPN-PAD"
         x="-17.78" y="0" addlevel="always"/>

   <gate name="B" symbol="NPN-C"
         x="-10.16" y="0" addlevel="always" swaplevel="1"/>

   <gate name="C" symbol="NPN-C"
         x="-2.54"  y="0" addlevel="always" swaplevel="1"/>

   <gate name="D" symbol="NPN-C"
         x="5.08"   y="0" addlevel="always" swaplevel="1"/>

   <gate name="E" symbol="NPN-C"
         x="12.7"   y="0" addlevel="always" swaplevel="1"/>

   <gate name="F" symbol="NPN-C"
         x="20.32"  y="0" addlevel="always" swaplevel="1"/>

   <gate name="G" symbol="NPN-C"
         x="27.94"  y="0" addlevel="always" swaplevel="1"/>
</gates>
```

The x and y attributes specify where each gate should be placed relative to the others. For the gates in CA3081, the x attribute changes but the y attribute doesn't. This should make sense because the gates are initially arranged in a horizontal row.

The addlevel attribute determines how the gates behave when the device is added to the schematic. If addlevel is set to always, as in the case of CA3081, all the gates are added to the schematic at once. If addlevel is set to next, the user adds one gate at a time to the schematic, and as each is added, the next gate is attached to the mouse. If this attribute is set to request, EAGLE will ask whether further gates should be inserted in the schematic. The default value is next.

The swaplevel attribute identifies which gates of a device can be swapped with others. That is, if two gates have the same swaplevel, they can be interchanged. In the case of CA3081, Gates B through G can be swapped because they all have a swaplevel of 1. But no gate can be interchanged with Gate A because it has the default swaplevel of 0. Any gate with a swaplevel of 0 cannot be swapped with another.

A.6.2 Devices

The <devices> element may contain one or more <device> subelements. These elements create the correspondence between packages and gates, and identify how pins are connected to pads. Each <device> element has two attributes and two subelements.

The two attributes of <device> are name and package. It's important to note that the name attribute *is not* the full name given in the ADD dialog. Instead, this provides the package designation that will be appended to the name of the deviceset.

For example, as shown in Figure A.3, the LS7766DO-S device is one of two devices that belong to the LS7766DO deviceset. The <device> element corresponding to LS7766DO-S is given as follows:

```
<device name="-S" package="SOIC28L">
```

The name attribute specifies that -S should be appended to the name of the deviceset to specify the package. The value of the package attribute, SOIC28L, identifies the name of the device's <package> element.

The two subelements of <device> are <connects> and <technologies>. The <connects> element may contain one or more <connect> subelements, which match pins to pads. The <technologies> element identifies further subdivisions of the deviceset that relate to the technology used to create the component.

Connects

So far, we've discussed how <pin> elements are defined as part of <symbol> elements and how <pad> elements are defined as part of <package> elements. The <connect> element creates an association between the two, thereby making it possible for the router to translate schematic connections to board connections. The <connect> element has four attributes:

- pad—The name of the <pad> element in the device.
- pin—The name of the <pin> that corresponds to the <pad>.
- gate—The name of the <gate> element containing the pin.
- route—Configuration data for the router, may be set to all or any. The default value is all.

Let's return to the CA3081 device, which contains seven transistors represented by gates A–F. Gate A has three pins named E, B, and C, and the other gates have two pins: B and C. The pads of the DIL16 package are numbered from 1–16. To connect these pins and pads, the <connects> element for the device is defined as follows:

```
<connects>
   <connect gate="A" pin="B" pad="16"/>
   <connect gate="A" pin="C" pad="1"/>
   <connect gate="A" pin="E" pad="15"/>
   <connect gate="B" pin="B" pad="3"/>
   <connect gate="B" pin="C" pad="2"/>
   <connect gate="C" pin="B" pad="13"/>
   <connect gate="C" pin="C" pad="14"/>
   <connect gate="D" pin="B" pad="11"/>
   <connect gate="D" pin="C" pad="12"/>
   <connect gate="E" pin="B" pad="6"/>
   <connect gate="E" pin="C" pad="4"/>
   <connect gate="F" pin="B" pad="10"/>
   <connect gate="F" pin="C" pad="9"/>
   <connect gate="G" pin="B" pad="8"/>
   <connect gate="G" pin="C" pad="7"/>
</connects>
```

Note that each pin of each gate needs a pad connection. Otherwise, an error condition may arise during rule checking or board layout.

Technologies

Different devices in a deviceset may have different packages. Devices may also be distinguished by technology. This can be confusing, so I'll start with an example.

Figure A.10 presents the six devices that make up Atmel's AT8*C51SND1C deviceset. As in earlier examples, the package is denoted by the suffix, which starts with a hyphen.

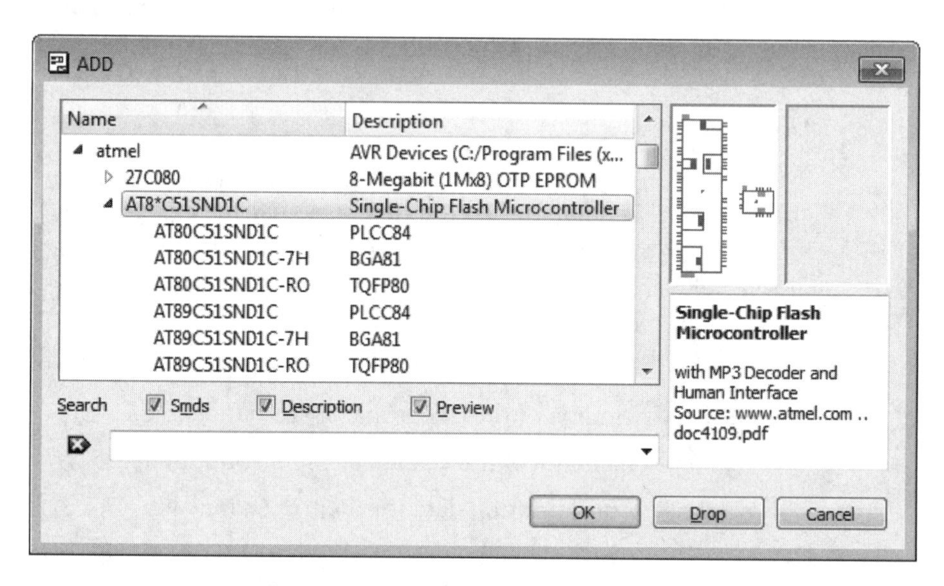

Figure A.10: Seven Gates in the CA3081 Device

In addition to having different packages, these six devices can have different technologies. According to Atmel's data sheets, if the * is set to 9, the microcontroller can be programmed with 64 kB of flash memory. If * is set to 0, the device has no ROM.

In EAGLE, the * in the deviceset name implies that there are multiple technologies available for the device. For each technology, the `<technologies>` element under `<device>` contains a `<technology>` subelement. The only attribute of `<technology>` is `name`, which replaces the * in the deviceset's name.

The following markup presents the `<devices>` element of the AT8*C51SND1C deviceset in the linear.lbr library.

```
<devices>
  <device name="" package="PLCC84">
    <connects>
      ...
    </connects>

    <technologies>
      <technology name="0">
        <attribute name="MF" value="" constant="no"/>
        ...
      </technology>
      <technology name="9">
        <attribute name="MF" value="" constant="no"/>
        ...
      </technology>
    </technologies>
  </device>
  <device name="-RO" package="TQFP80">
    <connects>
    ...
    </connects>

    <technologies>
      <technology name="0">
        <attribute name="MF" value="" constant="no"/>
        ...
      </technology>
      <technology name="9">
        <attribute name="MF" value="" constant="no"/>
      </technology>
    </technologies>
  </device>

  <device name="-7H" package="BGA81">
    <connects>
    ...
    </connects>

    <technologies>
      <technology name="0">
        <attribute name="MF" value="" constant="no"/>
        ...
      </technology>
      <technology name="9">
        <attribute name="MF" value="" constant="no"/>
        ...
      </technology>
    </technologies>
  </device>
</devices>
```

As shown, the `<devices>` element contains three `<device>` elements. In the schematic, they all have the same symbol. But in the board editor, they have different packages. For example, devices in the last type are associated with the BGA81 package.

Each `<device>` element has two `<technology>` elements, and in each case, the technology names are 0 and 9. Because there are three `<device>` elements with two `<technology>` elements each, there are a total of six devices.

It's important to see how each device gets its name. The base name is set by the deviceset, which in this case is AT8*C51SND1C. The `name` attribute of the `<package>` element is appended to the end of the base name, so devices in the last device type are called AT8*C51SND1C-7H. The * is replaced by the `name` attribute of the `<technology>` elements, so the two devices of the last type are AT80C51SND1C-7H and AT89C51SND1C-7H.

There's one more point I'd like to mention about technologies. Each `<technology>` element may contain zero or more `<attribute>` subelements. Each `<attribute>` contains `name` and `value` attributes that identify a property associated with the technology. The `display` attribute identifies whether the property will be displayed and may be set to `name`, `value`, `both`, or `off`. By default, only the value is displayed.

If an aspect of the property is going to be displayed, the text location is set by the `x` and `y` attributes, and the font is set with the `font` attribute. `<attribute>` also has an attribute called `constant`, which is set to `no` by default.

A.7 Conclusion

For most designers, there's no need to examine library files. But if you'd like to read or edit libraries from an external tool, you'll need a basic understanding of how these files are formatted. There's another reason to learn the format of *.lbr files—understanding the XML hierarchy clarifies the relationships between symbols, gates, packages, devicesets, and devices.

The first part of this appendix discussed the basic structure of a library file. The general format is based on the eXtensible Markup Language, or XML. The root element is `<eagle>`, whose `<drawing>` element contains the library's data. At the start, the `<settings>`, `<grid>`, and `<layers>` elements provide top-level information about the library.

The most important element is `<library>`, which defines the library's components. A component's definition has three parts. The `<symbol>` element sets the component's appearance in the schematic editor and specifies its pins. The `<package>` element defines the component's appearance in the board editor and identifies its pads. The `<device>` element, contained within a `<deviceset>`, relates a symbol (or a group of symbols) to a package.

A device's full name depends on a number of factors. The base name is given by the `<deviceset>` element, which may contain wildcards. If a base name contains `*`, the `*` will be replaced with the name attribute of a `<technology>` element. If the `<device>` element has a `name` attribute, that text will be appended to the base name. But if the base name contains a `?`, the package identifier will replace the `?` instead.

Appendix B

The Gerber File Format

EAGLE's final goal is to produce a set of Gerber and Excellon files that fully describe the desired circuit. When these files are generated, they can be sent to a fabrication facility that will construct the board.

Gerber files are popular throughout circuit design, but most designers don't read them directly. Instead, they examine the circuit using a Gerber viewer. Chapter 7, "Generating and Submitting Output Files," explains how EAGLE's CAM processor generates Gerber files and how the gerbv application makes it possible to visualize the board design.

This appendix goes deeper than Chapter 7 and takes a close look at the Gerber format itself. This format is complex and most designers don't need to understand it. But anyone who works in PCB fabrication or automatic circuit design should be familiar with it.

The Gerber file format is defined in the RS-274X standard, also called the Extended Gerber Format Specification. The original Gerber format was developed by Gerber Systems Corporation, and after a long series of mergers and acquisitions, the current owner is a company called Ucamco. Ucamco provides the specification for free, and you can obtain a copy at its download site at http://www.ucamco.com/downloads.php.

B.1 Introducing the Gerber Format

A Gerber file defines its geometry of one layer of a circuit board. It consists of a series of statements that tell the fabrication system about the layer. Each statement keeps to the same basic syntax, given as follows.

```
<code><general-data><specific-data>*
```

The initial code identifies the statement's type. For example, AM identifies an Aperture Macro statement and SF identifies a Scale Factor statement. The text following the code provides further information that proceeds from the general to the specific.

In most Gerber files, the statements can be divided into three stages:

- **Global settings**—Define properties pertaining to the entire board layer.
- **Aperture definitions**—Define the geometry of the shapes that will be placed on the layer.
- **Shape drawing**—Position aperture-defined shapes on the layer.

A brief example will clarify this three-part structure. The Gerber file depicted in Figure B.1 defines two apertures, a circle and a rectangle, and draws one shape for each.

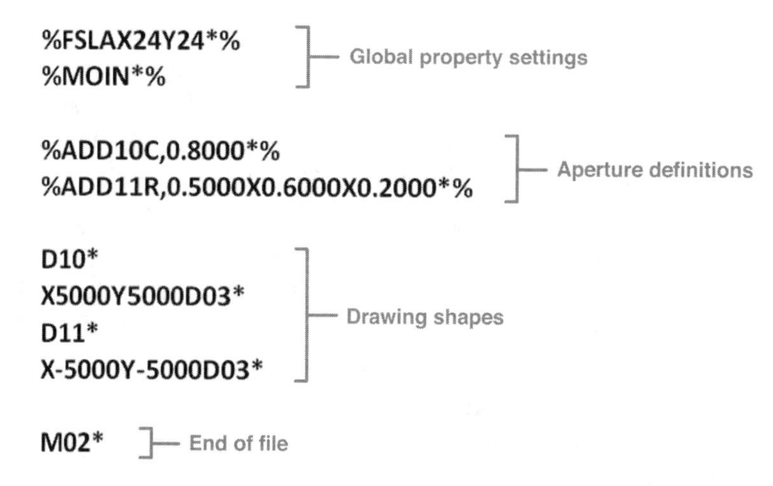

Figure B.1: Contents of a Simple Gerber File

In this example, the global property settings state that all measurements are in inches and that each coordinate should be interpreted with two digits before the decimal point and four digits afterward. In addition, all coordinates are referenced to the origin (0, 0) and the leading zeros may be omitted from the coordinate values.

The first aperture definition associates the code D10 with a circle whose diameter is 0.8". The second definition associates the code D11 with a rectangle of width 0.5" and height 0.6". This rectangle has a circular hole in its center with a diameter of 0.2".

After the apertures are defined, the D10 aperture (circle) is drawn at (0.5, 0.5) and the D11 aperture (rectangle) is drawn at (–0.5, –0.5). Figure B.2 shows what the layer corresponding to the Gerber file looks like.

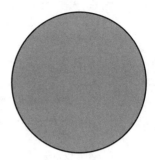

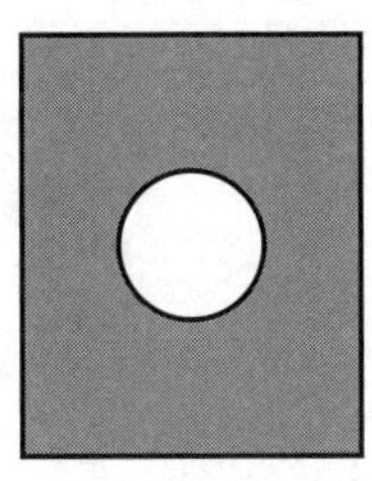

Figure B.2: Shapes in the Example Gerber FIle

For PCBs, these shapes represent material on the board. The shapes may correspond to copper traces, silk-screen ink, soldermask, or another material.

The following sections explain Gerber statements in detail, but this chapter doesn't cover the entire RS-274X standard. In particular, I'm not going to discuss statements involving layers and layer polarity. In my experience, these settings aren't necessary for PCB designs.

B.2 Setting Global Properties

A global property statement defines characteristics that apply throughout the file. These properties are generally defined once in a Gerber file and usually toward the beginning.

As shown in Figure B.1, each global property statement starts and ends with a `%` sign. For example, if a property's code is PR, its statement in a Gerber file might look like `%PRxyz123*%`.

The RS-274X specification defines six global properties. Table B.1 lists them along with their descriptions and basic syntax.

Table B.1
Global Properties

<table>
<tr><th>Parameter</th><th>Description</th><th>Syntax</th></tr>
<tr><td>FS</td><td>Format specification—determines the format of coordinate values</td><td>%FS<L|T><A|I>X[Nn][Gn]
Y[Dn][Mn]*%</td></tr>
<tr><td>MO</td><td>Mode—selects the units of the design (inches or millimeters)</td><td>%MO<IN|MM>*%</td></tr>
<tr><td>AS</td><td>Axis Select</td><td>%AS<string>*%</td></tr>
<tr><td>OF</td><td>Offset</td><td>%OF<C|D>*%</td></tr>
<tr><td>SF</td><td>Scale factor</td><td>%SF<D-code number>
<aperture-type>,
<modifier>[X<modifier>]*%</td></tr>
<tr><td>MI</td><td>Mirror Image</td><td>%MI<aperture macro name>*
<primitive number>,
<modifier$1>,
<modifier$2>, [<...>]*
[<primitive number>
[<modifiers>]]*...*%</td></tr>
</table>

The last four types are officially deprecated by the RS-274X standard. But I've included them in this discussion because you may still encounter them in old Gerber files.

B.2.1 Format Statement (FS)

The first line in many Gerber files is a Format Specification statement, given by the code FS. This is the only required statement in the standard and it identifies the format of the file's coordinates. An FS statement starts with `%FS` and its overall structure is given as follows:

```
%FS<general_config><x_coord_config><y_coord_config>*%
```

The `<general_config>` portion of the statement identifies properties that relate to all coordinates. It consists of two characters:

- `L` to omit leading zeros or `T` to omit trailing zeros
- `A` for absolute coordinates or `I` for incremental coordinates

Each coordinate value in a Gerber file has a fixed number of digits for its integer and decimal parts, so zeros appear frequently. For example, if a value is less than one, each of the integer places will be zero. To reduce wasted space, the FS statement makes it possible to omit leading zeros (with L) or trailing zeros (with T). Only option can be selected.

Each pair of x-y coordinates in a Gerber file specifies a location on the board. This location may be relative to a fixed origin or relative to the preceding pair of coordinates. In the first case, the coordinates are referred to as absolute and this is identified in the FS statement by A.

In the second case, the coordinates are referred to as incremental. This is identified in the FS statement by I. The use of incremental coordinates can produce significant error due to continued operations involving values with minor errors. For this reason, the RS-274X standard recommends the use of absolute coordinates.

The last part of the FS statement describes how the x and y coordinates are formatted. Coordinates are given as fixed-point values, with a maximum of seven places to the left of the decimal point (integer places) and seven places to the right (decimal places). This portion of the FS statement has the following structure:

```
X<int_places_x><dec_places_x>Y<int_places_y><dec_places_y>
```

For example, if the x and y coordinates have four integer places and five decimal places, the format is given as `X45Y45`. The coordinate format is usually the same in both directions.

The FS statement is complex but important to understand. To make matters clearer, here are two full examples of its usage:

1. `%FSLAX34Y34*%`—The coordinates in the Gerber file identify absolute location (`A`) and may have their leading zeros removed (`L`). The x and y coordinates are given with three integer places and four decimal places.
2. `%FSITX27Y16*%`—The coordinates in the Gerber file identify relative (incremental) location (`I`) and may have their trailing zeros removed (`T`). The x coordinates are given with two integer places and seven decimal places. The y coordinates are given with one integer place and six decimal places.

B.2.2 Mode (MO)

Coordinates can be given in inches or millimeters. This distinction is determined by the Mode statement, which is one of the simplest statements in the RS-274X specification. It can be given in one of two ways:

- `%MOIN*%`—Coordinates are given in inches.
- `%MOMM*%`—Coordinates are given in millimeters.

B.2.3 Axis Select (AS) (Deprecated)

The RS-274X standard uses special axes for its coordinates and they're denoted A and B. The Axis Select statement (AS) sets the relationship between the A-B axes and the traditional x-y axes. Like the Mode statement, the AS statement can be given in one of two ways:

- %ASAXBY%—The A axis is the x axis; the B axis is the y axis.
- %ASAYBX%—The A axis is the y axis; the B axis is the x axis.

B.2.4 Offset (OF) (Deprecated)

Some Gerber files add a pair of offset values to each pair of coordinates in the file. The Offset statement (OF) defines these values, which may be positive or negative. Its format is given as follows:

```
%OFA<A-offset>B<B-offset>*%
```

For example, the following OF statement states that (–6, 3) should be added to each coordinate in the file.

```
%OFA-6B3*%
```

The values in an OF statement are always absolute and use the dimensions set by the MO statement. The OF statement is uncommon and will be omitted from future versions of RS-274X standard.

B.2.5 Scale Factor (SF) (Deprecated)

Just as the offset statement adds values to coordinates, the scale factor statement multiples coordinates by a value, magnifying or shrinking the original image. The format of the SF statement is given as follows:

```
%SFA<A-scale>B<B-scale>*%
```

In this statement, A-scale sets the scaling of the A-axis and B-scale sets the scaling of the B-axis. For example, %SFA.1B10*% shrinks the image in the A direction by a factor of 10 and magnifies the image in the B direction by a factor of 10.

B.2.6 Mirror Image (MI) (Deprecated)

The Mirror Image statement (MI) is similar to the Scale Factor statement, but instead of multiplying coordinates by scaling factors, it multiplies one or both coordinates by –1, effectively reflecting the point across one or both axes. For example, if the point (5, 2) is mirrored in the B-axis, the resulting point is (–5, 2).

The format of the MI statement is given as %MIA<0 or 1>B<0 or 1>*%, where 1 signifies that the axis should be mirrored and 0 signifies that it shouldn't. For example, %MIA1B1*% mirrors coordinates in both axes. %MIA1B0*% mirrors coordinates in the A-axis and can be shortened to %MIA1*%.

This statement can throw off an entire design. I've never encountered it in a Gerber file and the RS-274X standard strongly advises against using it. Its recommendation is "Avoid it like the plague."

B.3 Aperture Definitions

After global parameters are defined, Gerber files provide statements that define the shapes to be patterned on the board's layer. The RS-274X standard refers to these shapes as *apertures*. This is because early fabrication systems shined light through specially shaped holes called apertures.

In RS-274X, apertures come in two broad categories: standard apertures and special apertures. A standard aperture is a circle, rectangle, obround, or polygon. Special apertures can be created with the Aperture Macro (AM) command, which will be discussed later in this chapter. In both cases, the shape can be closed or open. A closed shape is completely filled in, whereas an open shape has a hole in its center.

Apertures are defined with the Aperture statement (AD), which is one of the most complex statements in the RS-274X standard. Its overall structure is given as follows:

```
%AD<D-code><aperture_definition>*%
```

The `<D-code>` serves as a unique identifier for the shape. The entire purpose of the AD command is to associate this identifier with a shape. For example, the following command associates a D-code of D23 with a circle whose diameter is 1.5:

```
%ADD23C,1.5*%
```

As a second example, the following command associates D86 with a rectangle of dimensions 0.8-by-1.2:

```
%ADD86R,0.8X1.2%
```

The RS-274X standard reserves the use of the D-codes for 0 through 9, and the largest possible D-code is D-999. Therefore, circuit designers are limited to D-10 through D-999.

Following the `<D-code>`, the `<aperture_definition>` provides two pieces of information. It identifies the shape's type, which can be a circle, rectangle, obround, polygon, or custom type. It also sets the shape's geometric properties, such as the radius of a circle or the length of a rectangle's side.

Table B.2 lists the standard apertures that can be used in an AD statement. In the second column, the Type ID identifies the letter used to identify the type of shape.

Table B.2

Standard Aperture Shapes

Shape	Type ID	Geometric Data
Circle	C	Diameter Hole geometry
Rectangle	R	Dimensions (x-y) Hole geometry

Obround	O	Dimensions (x-y) Hole geometry
Polygon	P	Diameter Number of sides Rotation angle Hole geometry

Each shape can be filled in or have a hole through its center. An aperture's hole may be circular or rectangular.

B.3.1 Circle

The circle is the simplest aperture to understand. Its definition starts with `C` and its properties include its diameter and the geometry of the internal hole. More precisely, a circle's definition is given as follows:

```
C,<diameter>X<hole geometry>
```

The diameter is given in the current units. As an example, a filled circle with a diameter of 0.05 would be defined as `C,0.05`. To associate this circle with a D-code of D22, the AD statement would be given as follows:

```
%ADD22C,0.05*%
```

If the circle has a circular hole through its center, the circle's diameter is followed by an X and the hole's diameter. If a circle with diameter 0.75 has a hole with diameter 0.25, it could be defined with `C,0.75X0.25`.

If the hole is rectangular, two dimensions must be provided: the rectangle's length in the x and y directions. Both dimensions are preceded by `X`, so if a circle with diameter 0.8 has a 0.4-by-0.3 rectangular hole, it's definition could be `C,0.8X.4X0.3`.

B.3.2 Rectangle

Rectangular shapes are defined similarly to circles, but `R` is used in place of `C` and the rectangle's x and y dimensions are used in place of the diameter. The two dimensions are separated by an X. For example, a solid rectangle with dimensions 1.5-by-1.2 could be defined as `R,1.5X1.2`.

The hole geometry for a rectangle is set with the same parameters as for a circle. That is, circular holes are identified with `X` followed by the hole's diameter. Rectangular holes are identified with `X`, the hole's x-dimension, `X`, and the hole's y-dimension. Therefore, the rectangle `R,1.5X1.2X0.5` has a circular hole with the diameter equal to 0.5. The rectangle `R,1.5X1.2X0.6X0.4` has a rectangular hole with dimensions 0.6-by-0.4.

B.3.3 Obround

The names and shapes are similar, but obrounds are not ovals. An obround is a rectangle whose smaller pair of sides are rounded into semicircles. Shape definitions for obrounds are nearly exactly similar to those for rectangles. The only difference is that the shape is identified by `O` instead of `R`.

The following examples show how obrounds are defined:

- `O,1.5X1.2`—Filled obround with dimensions 1.5-by-1.2
- `O,1.5X1.2X0.5`—1.5-by-1.2 obround with a circular hole with 0.5 diameter
- `O,1.5X1.2X0.5X0.5`—1.5-by-1.2 obround with a rectangular hole with dimensions 0.5-by-0.5.

B.3.4 Polygon

In general, a polygon is any closed shape bounded by connected edges. But in RS-274X, a polygon is limited to 3-12 edges, all with the same length. In traditional geometry, this is called a regular polygon.

A polygon's definition starts with `P,` and is followed by the shape's diameter and its number of sides. A polygon's diameter is twice the distance from the center to one of its points. Therefore, a six-sided polygon with a diameter of 1.5 is defined as `P,1.5X6`. A 10-sided polygon with a diameter of 2.5 is defined as `P,2.5X10`.

A polygon can be rotated counterclockwise by following the number of sides with an `X` and the rotation angle. The angle must be given as an integer number of degrees, so `P,0.25X5X30` rotates the five-sided polygon counterclockwise by 30 degrees. If the angle has a negative value, the rotation is clockwise. Therefore, `P,0.25X8X-30` defines an eight-sided polygon rotated 30 degrees in a clockwise orientation.

In addition to diameter, number of sides, and rotation angle, a polygon definition may define hole geometry. These parameters have the same form as that used by obrounds, rectangles, and circles: A circular hole requires a diameter and a rectangular hole requires an x-dimension and a y-dimension. Here are two examples:

- `P,2.75X3X10X.25`—A three-sided polygon (triangle) with a diameter of 2.75, a rotation angle of 10 degrees in the counterclockwise direction, and a circular hole of diameter 0.25.
- `P,1.5X11X-45X.3X.2`—An 11-sided polygon with a diameter of 1.5, a rotation angle of 45 degrees in the clockwise direction, and a rectangular hole whose x-dimension is 0.3 and whose y-dimension is 0.2.

Regarding polygons, there are two last points to keep in mind. First, if a polygon has a hole, the polygon's rotation angle must be included. Second, the outer shape of a polygon can be rotated but its hole can't be rotated.

B.4 Drawing Shapes

After the apertures are defined, the next step is to draw shapes. The simplest method involves selecting the aperture by its D-code and then setting the location of the shape defined by the aperture. For example, the following statements select the aperture D82 and draw the corresponding shape at (0.5, 0.5).

```
D82*
X005000Y005000D03*
```

Most of this should be understandable. The first statement selects an aperture and the second sets a coordinate pair for the shape's location. The structure of this location statement is given as follows:

```
X<X-coordinate>Y<Y-coordinate><D-code>
```

The `<D-code>` at the end deserves an explanation. This is not a D-code associated with an aperture. Instead, it tells the machine whether to perform exposure when the location is reached. I'll discuss this next.

B.4.1 Exposure and Motion

As explained in Chapter 2, "An Overview of Circuit Boards and EAGLE Design," PCB fabrication uses photolithography to pattern copper on a circuit board. When a photoplotter exposes part of a circuit board to UV light, a chemical reaction occurs that results in the presence or absence of copper. Put simply, exposure produces shapes.

In the preceding example (and in Figure B.1), the location is followed by D03. This is one of three exposure commands and it tells the photoplotter to turn the light on and off when the given location is reached. This operation, called flashing, creates a shape corresponding to the current aperture. Consider the following statements.

```
D82*
X005000Y005000D03*
X010000Y010000D03*
```

The second statement tells the photoplotter to move to (0.5, 0.5) and turn the lamp on and off. The last statement tells the photoplotter to move to (1.0, 1.0) and turn the lamp on and off. The result is two separate shapes, both defined by the D82 aperture.

Now suppose you want the photoplotter to keep the light on as it moves from one point to the next. The result will be a line between the two points. To tell the photoplotter to keep the light on, the second location should be followed by D01. The following lines show how this might work:

```
D82*
X005000Y005000D02*
X010000Y010000D01*
```

Figure B.3 shows the results produced by the preceding sets of statements. In this figure, D82 corresponds to a circular aperture.

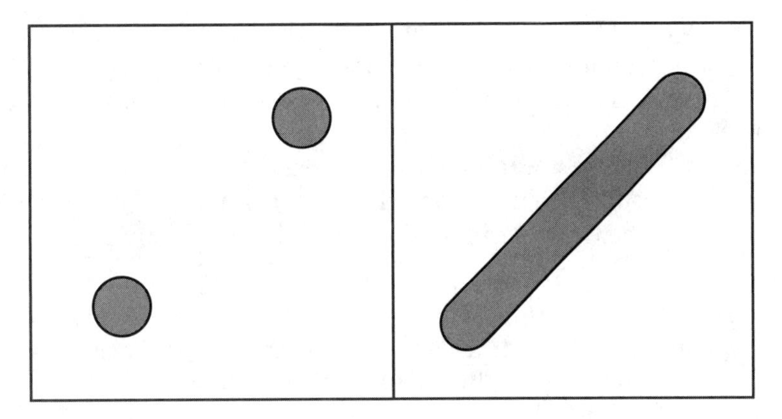

Figure B.3: Shapes Drawn with Flashing (left) and Continuous Exposure (right)

The three D-codes, D01 through D03, are called operational codes. Table B.3 lists each of them and the purposes they serve.

Table B.3

Operational Codes

Code	ID
D01	Move to a location with exposure (lights on)
D02	Move to a location without exposure (lights off)
D03	Move to a location without exposure, then expose (flash)

Another example will help clarify how these codes work. Suppose you want to make D12 the current aperture and draw two lines. The first line runs from (a, b) to (c, d) and the second runs from (f, g) to (h, i). You could use statements such as the following:

`D12*`	Set the aperture
`XaYbD02*`	Set the initial point
`XcYdD01*`	Move to the next point with exposure
`XfYgD02*`	Move to the next point without exposure
`XhYiD01*`	Move to the next point with exposure

D-codes remain active until specifically changed. Therefore, if one location statement ends with D01 and the next statement doesn't have a D-code, the second statement is assumed to end with D01.

B.4.2 Interpolation and G-Codes

By default, when two locations are defined with exposure set to on, the result is a straight line. This is called linear interpolation. But the RS-274X standard also supports drawing circular arcs between points. This is called circular interpolation.

> **NOTE** Linear interpolation is supported if the current aperture is a circle or rectangle. Circular interpolation is supported if the current aperture is a circle. Other shapes cannot be used with interpolation.

Just as aperture selection is controlled by setting D-codes, interpolation is controlled through G-codes. The G stands for general, and Table B.4 lists all the G-codes supported by the RS-274X standard.

Table B.4

General Codes

Code	ID
G01	Linear interpolation
G02	Clockwise circular interpolation
G03	Counterclockwise circular interpolation
G04	Comment
G36	Start outline fill
G37	End outline fill
G54	Select aperture (unnecessary)
G74	Single-quadrant mode for circular interpolation
G75	Multiquadrant mode for circular interpolation
G90	Specify absolute format
G91	Specify incremental format

The G-54 code is unnecessary, but I still encounter statements such as `G54D13*`, which makes the D-13 aperture the active aperture. Modern Gerber files accomplish the same result with the simpler statement `D13*`.

The first three G-codes (G01-G03) control how lines are drawn between points. If G01 is active, every line drawn between points will be straight. The general syntax for this is given as follows:

```
G01X<X-coordinate>Y<Y-coordinate><D-code>
```

At the end of the statement, <D-code> can be set to D01 (exposure on) or D02 (exposure off). It can't be set to D03 because it's incompatible with interpolation.

Circular interpolation is more complicated. This is because there are many ways to draw a circular arc between two points in a plane. To specify which arc should be drawn between the start and end points, the interpolation statement needs to define the location of a third point: the center of the circle containing the arc. The statement doesn't give the center's exact (absolute) location, but instead uses an offset from the starting point.

More precisely, the interpolation statement sets the center position using two values, denoted I and J:

- **I-offset**—The distance from the start point to the center in the x-direction
- **J-offset**—The distance from the start point to the center in the y-direction

These offset values are given without sign. Figure B.4 shows how the offsets determine a circle joining a starting and ending point.

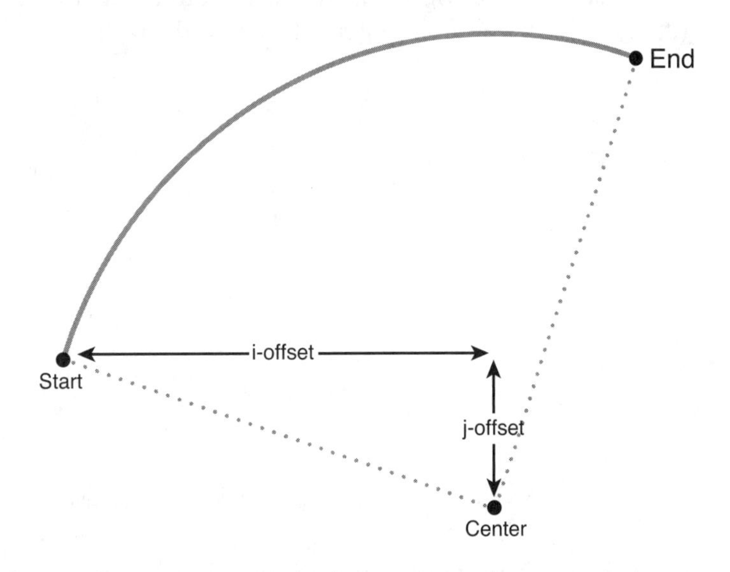

Figure B.4: Determining an Arc from Two Offsets

Arcs can be classified according to their orientation and angular measure:

- Counterclockwise, angle less than 90 degrees
- Clockwise, angle less than 90 degrees
- Counterclockwise, angle greater than 90 degrees
- Clockwise, angle greater than 90 degrees

To specify which arc should be drawn between two points, you need the right G-codes. As given in Table B.5, G02 sets the arc's orientation to be clockwise and G03 sets it to be counterclockwise. The general statement with these codes is given as follows:

```
<G-code>X<x-coordinate>Y<y-coordinate>
                    I<i-offset>J<j-offset><D-code>*
```

In the RS-274X standard, arcs with angles less than 90 degrees are referred to as single-quadrant arcs. These arcs are specified with the `G74*` statement. Arcs with more than 90 degrees (but less than 360) are multiquadrant arcs and are specified with `G75*`.

The RS-274X standard strongly recommends using single-quadrant arcs instead of multiquadrant arcs. This is the default behavior, but some printers assume that multiquadrant arcs should be drawn by default. Therefore, the RS-274X standard recommends that all circular interpolation statements start with G74.

As an example, Figure B.5 depicts a counterclockwise, single-quadrant arc from (2, 5) to (3, 8). The arc's center is (1, 7), so the i-offset is 1 and the j-offset is 2.

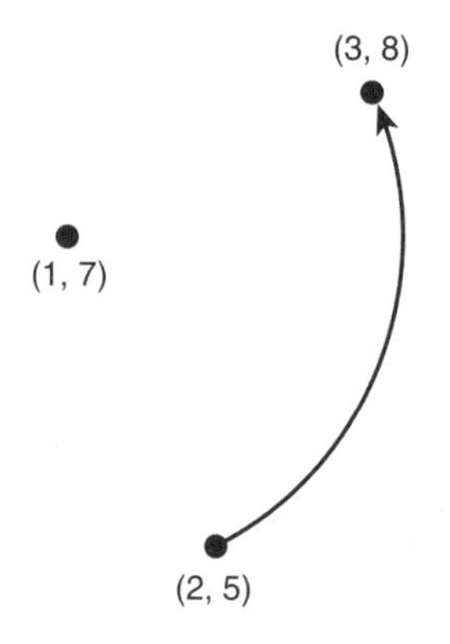

Figure B.5: A Clockwise, Single-Quadrant Arc

The following statements show how to draw this arc in a Gerber file.

```
G74*
X020000Y050000D02*
G03X030000Y080000I010000J020000D01*
```

After the layer's shapes are drawn, the statement M02* tells the machine that the end of the file has been reached. I'll discuss M-codes in a later section, but first I want to explain how to define custom apertures.

B.5 Custom Apertures

Most PCB designs can be drawn using circles, rectangles, obrounds, and polygons. But if these shapes aren't sufficient, the Aperture Macro (AM) statement makes it possible to create new ones. The purpose of the AM statement is to define a custom aperture and associate it with a name. Its overall structure is given as follows:

```
%AM<name>*<shape-code>,<shape-parameter>,<shape-parameter>,...*
          ...
          *<shape-code>,<shape-parameter>,<shape-parameter>,...*%
```

After an AM statement, the AD statement can be used to associate the name with a D-code. For example, if the <name> portion of an AM statement is set to OCTAGON, the following statement associates OCTAGON's shape with the D-code D-49:

```
%ADD49OCTAGON*%
```

A custom aperture is made up of basic shapes called primitives. Each primitive has a code (<shape-code>) and a series of parameters that configures its geometry (<shape-parameters>). Table B.5 lists the seven different primitives available and their parameters.

Table B.5

Aperture Macro Primitives

Shape	ID	Parameter
Circle	1	Exposure Diameter Center coordinates
Rectangle (defined by start/end)	2/20	Exposure Height Starting-point coordinates (x-y) Ending-point coordinates (x-y) Rotation angle
Rectangle (defined by center)	21	Exposure Width Height Center coordinates (x-y) Rotation angle
Rectangle (defined by left corner)	22	Exposure Width Height Lower-left coordinates (x-y) Rotation
Outline	4	(see discussion)

Polygon	5	Exposure Number of sides Center coordinates (x-y) Outer diameter Rotation angle
Moiré	6	Center coordinates (x-y) Outer diameter Ring thickness Space between rings Maximum number of rings Crosshair thickness Crosshair length Rotation angle
Thermal	7	Center coordinates (x-y) Outer diameter Inner diameter Gap thickness Rotation angle

The entries in this table look similar to those in Table B.2, but there are three significant differences between standard apertures and primitives:

- Standard apertures are identified with letters (C, R, O, P) and primitives are identified by number.
- Standard apertures aren't associated with coordinates. But primitives need coordinates so that they can be positioned relative to the other primitives in the custom aperture.
- The first five primitives in Table B.3 have a property called exposure. No standard aperture has this property.

An earlier section explained how to control exposure with D-codes. Exposure for primitives is different. For primitives, exposure identifies whether the primitive is drawn as a solid or as a hole. This value can be set to 0, 1, or 2.

- If exposure is set to 0, the primitive is drawn as a transparent hole. The shape is subtracted from the aperture, revealing any shapes underneath.
- If exposure is set to 1, the primitive is drawn as a solid. The shape is added to the aperture.
- If exposure is set to 2, the exposure toggles between 1 and 0.

The rest of this discussion looks more closely at the individual primitives and how they're defined.

B.5.1 Circle

The circle is the simplest of the primitives. It's defined by its diameter and the coordinates of its center. For example, if a solid circle (ID = 1, exposure = 1) with a diameter of 1.5 is located at (0.5, 0.5), its primitive definition could be given as follows:

```
1,1,1.5,0.5,0.5
```

The following statement creates a macro containing only this circle and gives it the name `SIMPLE`:

```
%AMSIMPLE*1,1,1.5,0.5,0.5*%
```

B.5.2 Rectangle

The AM statement provides three different ways to form rectangles. Each has a different ID and a different method of defining its geometry:

- If the ID is 2 or 20, the rectangle is determined by its height and starting/ending points.
- If the ID is 21, the rectangle is determined by its center.
- If the ID is 22, the rectangle is determined by its lower-left corrner.

The last argument of each rectangle designation sets an optional rotation about the origin. The angle must be given as an integer number of degrees. A counterclockwise rotation has a positive angle and a clockwise rotation has a negative angle.

Figure B.6 presents a rectangle with a height of 3 and a width of 8. The center is located at (0, 0) and the lower-left corner is at (–4, –1.5).

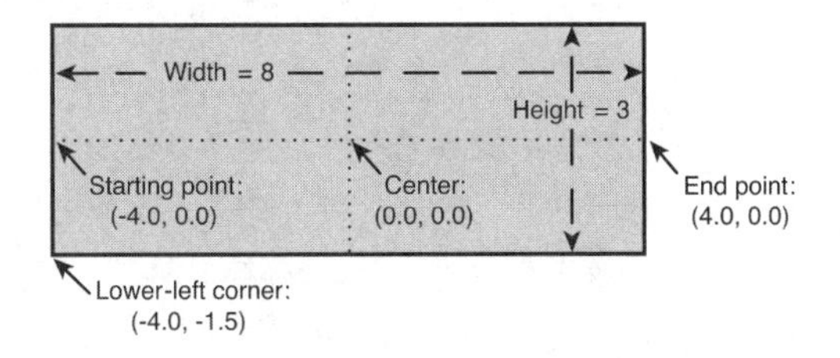

Figure B.6: Points on a Rectangle Macro

If this rectangle is solid (exposure = 1), its primitive can be defined in three ways:

- Determined by height and start/end points: `2,1,3,-4,0,4,0`
- Determined by width, height, and center: `21,1,8,3,0,0`
- Determined by width, height, and lower-left corner: `22,1,8,3,-4,-1.5`

It doesn't make any real difference which of the three methods is used. The choice depends solely on the tool and the designer's convenience.

B.5.3 Outline

An outline contains a series of connected lines that bound a closed shape. These lines have three requirements:

- The maximum number of lines is 4000.
- The last point must have the same coordinates as the first.
- Except at the final point, an outline can't intersect itself.

The full list of elements in an outline definition are given as follows:

- **Exposure**—Identifies whether the outline is added to (1) or subtracted from (0) the aperture.
- **Number of points**—The number of distinct points in the outline.
- **Starting coordinates**—The x-y coordinates of the first point.
- **Successive coordinates**—The x-y coordinates of the following points.
- **Ending coordinates**—The last point must have the same coordinates as the first.
- **Rotation angle**—An optional angle of rotation given in integer degrees.

An example will show how outlines are defined in practice. Figure B.7 depicts an outline with four points: (0, 4), (3, 0), (–1, –2), and (1, 2).

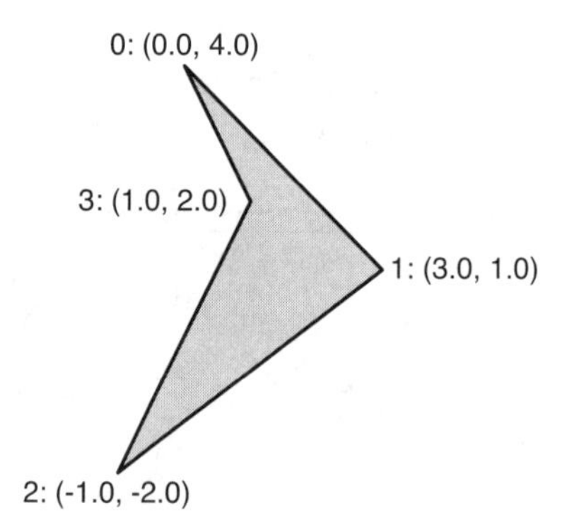

Figure B.7: An Outline Macro

Assuming the shape will subtract from the aperture (exposure = 0), the outline definition can be given as `4,0,4,0,4,3,0,-1,-2,1,2,0,4.`

Even though the number of points is given as 4, there are five pairs of coordinates in the definition. This is because the last point must have the same coordinates as the first.

B.5.4 Polygon

The polygon primitive is similar to the standard polygon discussed earlier. The number of sides must be between 3 and 12 and each side has the same length. The polygon's definition requires the X-Y coordinates of a center point and the outer diameter, which is twice the distance from the center to one of the polygon's points. The last parameter is an optional rotation angle around the origin.

Figure B.8 presents a six-sided polygon. It has an outer diameter of 6.0 and a counterclockwise rotation of 45 degrees.

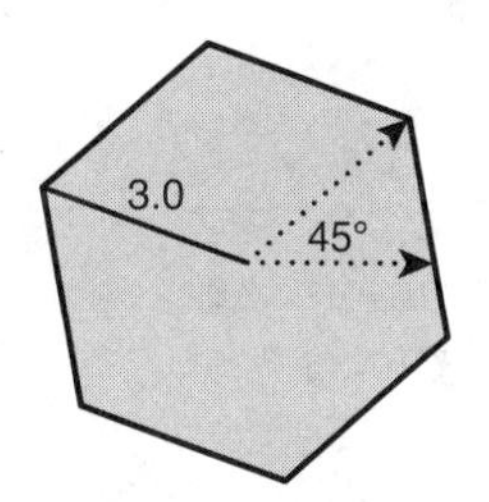

Figure B.8: A Polygon Macro

Assuming that the polygon's center is (2.5, 3.5) and that the shape will add to the aperture, the hexagon's definition can be given as `5,1,6,2.5,3.5,6.0,45`.

B.5.5 Moiré

A moiré is a shape made up of a crosshair overlapping a series of concentric rings. Old circuit board designs use them to ensure layer alignment. When the crosses and circles on two layers overlap, the layers are properly aligned. Modern board fabrication tools don't use moirés, but you may still encounter them in Gerber files.

A moiré definition doesn't explicitly state the number of rings. Instead, it requires four criteria:

- Diameter of the outermost ring
- Thickness of each ring
- Spacing between the rings
- Maximum number of rings

After providing this information, the definition sets the thickness and length of the moiré's crosshair. This may also provide a rotation angle.

Figure B.9 depicts a moiré with three rings. The outermost ring has an outer diameter of 3.0 and each ring has a thickness of 0.25, which is also the spacing between the rings. The lines in the crosshair have a thickness of 0.1 and a length of 2.25.

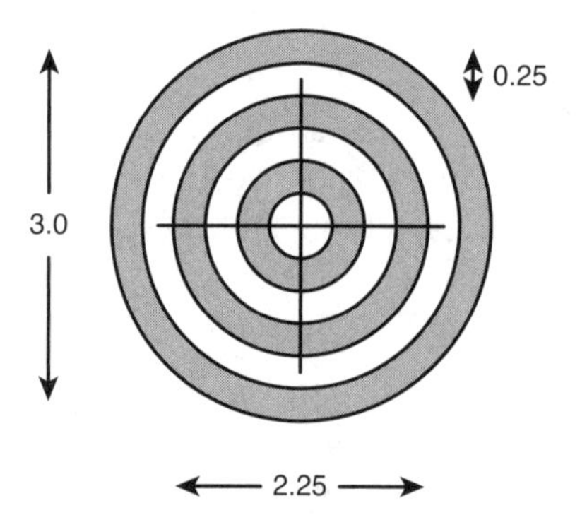

Figure B.9: A Moiré Macro

The moiré definition doesn't accept an exposure parameter, so the definition for the macro in the figure is `6,0.0,0.0,3.0,0.25,0.25,4,0.1,2.25`.

B.5.6 Thermal

Chapter 5, "Layout and Design Rules," explained how copper pour regions can be formed by drawing polygons in the board editor. When a component's lead needs to be soldered to a pad inside a copper pour, the pad is usually a thermal relief pad, commonly called a thermal. These pads increase solderability by connecting to the surrounding copper through narrow conductive paths.

The AM command can add thermals as primitives. A thermal's definition requires the coordinates of its center, the inner and outer diameters, and the thickness of the conductive paths. If an optional rotation parameter isn't provided, the conductive paths will be oriented vertically and horizontally.

Figure B.10 presents an origin-centered thermal with an inner diameter of 1.0 and an outer diameter of 1.5. The conductive paths have a thickness of 0.25.

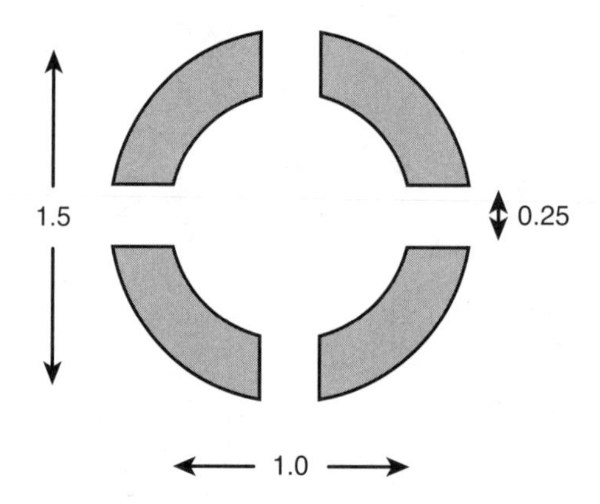

Figure B.10: A Thermal Macro

Like the moiré definition, the thermal definition doesn't accept a parameter for the exposure. Therefore, the definition of the thermal depicted in the figure can be given as `7,0,0,1.5,1.0,0.25`.

B.5.7 Apertures with Multiple Primitives

When you understand how to declare individual primitives, it's easy to combine them together into a custom aperture. The best way to understand this is through an example. Figure B.11 presents a custom aperture consisting of a diamond turned on its side. The left side has a circular hole and the right side has a pentagonal hole.

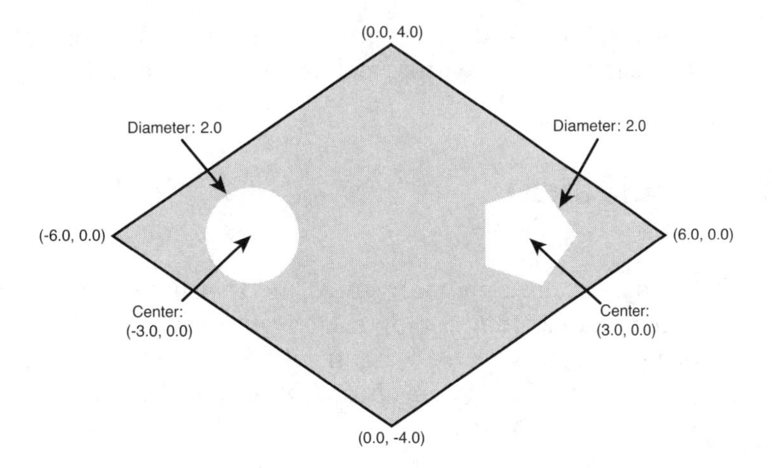

Figure B.11: A Custom Aperture with Three Primitives

The diamond's angles aren't equal, so it can't be drawn as a rectangle or polygon. Instead, a diamond-shaped outline is needed. The three primitives are defined in the following way:

- **Outline definition**—`4,1,4,-6,0,0,4,6,0,0,-4,-6,0`
- **Circle definition**—`1,0,2.0,-3.0,0.0`
- **Pentagon definition**—`5,0,5,3.0,0.0,2.0`

The following AM statement combines these definitions into a single custom aperture.

```
%AMCUSTOM*4,1,4,-6,0,0,4,6,0,0,-4,-6,0*
*1,0,2.0,-3.0,0.0*
*5,0,5,3.0,0.0,2.0*%
```

Note that each primitive definition is surrounded by asterisks. These definitions can be listed on the same line or on different lines.

After a custom aperture is defined, it must be associated with a D-code with the AD statement. The following statements present an entire file that displays the custom aperture in Figure B.11.

```
%FSLAX24Y24*%
%MOIN*%
%AMCUSTOM*4,1,4,-6,0,0,4,6,0,0,-4,-6,0*
*1,0,2.0,-3.0,0.0*
*5,0,5,3.0,0.0,2.0*%
%ADD10CUSTOM*%
D10*
X000000Y000000D03*
M02*
```

Here, the AD statement assigns the custom aperture (CUSTOM) to D10. Then it sets D10 as the current aperture and draws the shape at the origin.

B.6 Conclusion

PCB designs are commonly stored and transmitted in Gerber files, which are difficult for both humans and computers to read. Alternatives have been proposed, such as Mentor Graphics' ODB++, but the RS-274X format remains the only format everyone agrees on. For this reason, anyone who reads PCB designs directly should know how Gerber files are formatted.

In essence, a Gerber file consists of a series of statements, each of which provides information about a single layer of a circuit board. These statements can be divided into three categories: those that provide global information about the layer, those that define apertures, and those that draw shapes defined by the apertures.

Before a shape can be drawn, its aperture must be associated with a D-code and the D-code must be made active. Apertures come in two types: standard and custom. Standard apertures include circles, rectangles, obrounds, and polygons. Custom apertures are made up of building blocks called primitives, which include circles, rectangles, outlines, polygons, moirés, and thermal relief pads.

Index

B

C

D

F

G

H

I

J-K

L

M

N

O

P

Q-R

T

U

V

W

X-Y-Z

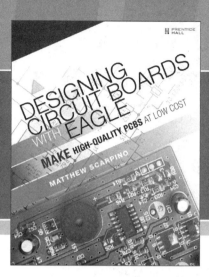